In Commemoration of the 60th Birthday of M Veltman

GAUGE THEORIES

Past and Future

World Scientific Series in 20th Century Physics

For information on Vols. 1–28, please visit http://www.worldscientific.com/series/wsscp

World Scientific Series in 20th Century Physics – Vol. 1

In Commemoration of the
60th Birthday of M Veltman

GAUGE THEORIES

Past and Future

Ann Arbor, Michigan   May 16 – 18, 1991

Editors

R Akhoury *(Univ. Michigan)*
B de Wit *(Utrecht)*
P van Nieuwenhuizen *(Stony Brook)*
H Veltman *(DESY)*

World Scientific
Singapore • New Jersey • London • Hong Kong

Published by

World Scientific Publishing Co. Pte. Ltd.

5 Toh Tuck Link, Singapore 596224

USA office: 27 Warren Street, Suite 401-402, Hackensack, NJ 07601

UK office: 57 Shelton Street, Covent Garden, London WC2H 9HE

British Library Cataloguing-in-Publication Data
A catalogue record for this book is available from the British Library.

World Scientific Series in 20th Century Physics — Vol. 1
GAUGE THEORIES — PAST AND FUTURE
In Commemoration of the 60th Birthday of M Veltman

ISBN 978-981-02-1028-1

Preface

On May 16-18, 1991, a conference was held in Ann Arbor, Michigan, to celebrate the 60th birthday of Tini Veltman and to honor his many contributions to Quantum Field theory. A majority of the articles in these proceedings are based on talks given at this conference; a schedule of which is included. Dieter Maison was unable to attend but commissioned a romantic and allegorical poem in Tini's honor which was read at the Banquet. This, together with a translation by Leo Stodolsky, is also included. In addition we have decided to reprint an article entitled "Relation Between the Practical Results of Current Algebra and the Originating Quark Model" which is based on lectures that Tini gave at Copenhagen in July 1968. This paper outlines clearly Tini's viewpoint at that time and brings out the arguments which led him to believe in an underlying Yang-Mills structure of the weak interactions.

We would like to thank the various organizations that provided the funds that made it possible to host this conference. There are: The University of Michigan Physics Department, College of Literature, Science and Arts, Office of the Vice-President for Research and the Graduate School; in addition, the U.S. Department of Energy and the National Science foundation. Gary Krenz, Ada-Mae Newton, Barbara Nieman, and Susanne Marsh helped with the organizational details. Special thanks go to Kate Wills without whose efforts this conference would not have been possible.

R. Akhoury
B. de Wit
P. van Nieuwenhuizen
H. Veltman

1. M. Veltman
2. L. Lederman
3. M. Schwartz
4. P. van Nieuwenhuizen
5. D. Robinson
6. H. van Dam
7. J. Smit
8. B. Roe
9. R. Akhoury
10. H. Veltman
11. B. DeWit
12. B. Zumino
13. W. Repko
14. H. Neal
15. N. Cabibbo
16. D. Errede
17. S. Errede
18. R. Errede
19. W. Bardeen
20. C. Korthal Altes
21. R. Gastmans

22. K. Gaemers
23. G. Passarino
24. E. Remiddi
25. M. Duncan
26. B. Haeri
27. G.-Lin. Lin
28. K. Adel
29. P. van Baal
30. E. de Rafael
31. M. K. Gaillard
32. A. Kataev
33. S. Zhang
34. S. Hong
35. K. Ellis
36. A.C.T. Wu
37. V. Baluni
38. J. Liu
39. S. Rudaz
40. A. Compagner
41. F. Berends
42. E. Tomboulis

43. Y.-P. Yao
44. J. Weyers
45. A. van de Ven
46. K.T. Hecht
47. F. Halzen
48. T. Weiler
49. Y. Tomozawa
50. C. Kyriazidou
51. H. Contopanogos
52. H. Steger
53. R. Gustafson
54. R. Garisto
55. J.J. van der Bij
56. J. Vermaseren
57. R. Sands
58. L. Liu
59. L. de Wit
60. M. Dryburg
61. D. Nitz
62. F. Yndurain
63. L. Jones

TABLE OF CONTENTS

DER VULKAN (TINICATEPETL)

by

Dieter Maison

Einen Vulkan erkennt man leicht
an dem Rauch der ihm entweicht,
meist qualmt and stinkt er fuerchterlich,
man sieht's von fern und huetet sich.
Gar grausig ist sein Inneres,
wie Hoellenglut und Schlimmeres.
Schon mancher der voll Uebermut
herausgefordert seine Wut
versucht ihm auf den Leib zu ruecken
und gar in seinen Schlund zu spucken
musst jaemmerlich das Weitte suchen,
da half kein bitten, beten, fluchen.
Doch wer sich naehert mit Verstand
dem bietet er ein fruchtbar Land,
ringsum die Felder quantisiert
und schoen geeicht, wie sich's gehoert.
Hier spriessen Zet und Dubelweh,
schoen renormiert und - ohne Schmaeh -
sogar das Ro - ein Para-Meter -
gedeihet hier, das weiss ein jeder.
Doch auch exotisch Kraut, herrjeh,
vom Quark - ein Tee??
Wenn man den schluckt, das ist gewiss,
bekommt man Bauchweh und Duennschiss.
Es waechst auch Dorniges, oh Jammer,
am schlimmsten hier das fuenfte Gamma.
der leuchtende Kaktus-Veltmanis.
Man kommt von ferne ihn zu sehen
und deierlich herumzustehen.
Schon sechzig Jahr' man glaubt es kaum,
gibt es nun diesen Stachelbaum.
Man wuenscht ihm gern nochmal soviel,
auch wenn er es vielleicht nicht will.
     Und der Vulkan? Wenn auch schon alt,
er ist im Innern noch nicht kalt,
ihr seht ja wie er schmaucht und raucht
und schmaucht und raucht....

THE VOLCANO (TINICATEPETL)

A free translation of some choice bits by

Leo Stodolsky

The great volcano is easy to make out
See the fire and smoke he lets out
......
Inside glow the fires of hell and smokey terror
Which have burned many that approached in error

But those who have come ready to understand
Have found waiting a fertile and welcoming land

The beautiful vector fields all about
Gauged and quantized right, without a doubt

Z and W renormalized and growing tall
Rho parameter, grazing Higgs and all

Certainly, there have been some thorny plants
Like gamma fives stuck up your pants

But all in all it's hard in practice
To get one by on that old cactus

......
We wish him sixty and the double
With all that means for our trouble

So if you think the volcano's getting old
Look out friends, inside it's sure not cold

****

Comments on the Occasion of the Symposium Celebrating the Sixtieth Birthday of Martinus Veltman

HOMER A. NEAL

Department of Physics
University of Michigan
Ann Arbor, Michigan 48109

I wish to welcome each of you to today's symposium in honor of our colleague, Martinus Veltman. Knowing Tini as I do, I am sure that he would rather be somewhere else today, but, Tini, this is the price you must pay for being such an outstanding physicist. We will try, as best we can, to minimize our accolades, in view of your modesty. But the various discussions of the day will nevertheless be rather painful for you, I am sure. So, my best advice for you is to sit back and enjoy every moment of it, for the words come sincerely from the hearts of these who admire you and who feel honored to have had the opportunity to have known you over the years of your extraordinary professional career.

Colleagues, we are assembled here today to pay tribute to the enormous accomplishments of Tini Veltman, and to let him know how honored we are to count him as a member of the University of Michigan Department of Physics and as a major force in our field. I will divide my remarks into four categories: One topic deals with what excellence means in the context of our field. Next I want to comment on Tini as a scholar. This will be followed by remarks on Tini as a teacher. Finally, I will speak about Tini as a colleague. Inasmuch as many of you have known Tini many years longer than I, my remarks will certainly not be all-inclusive, and we look forward to learning more about the colorful pieces of his life's mosaic.

Before I speak about Tini's accomplishments let me first give you a description of the metric I am using. Excellence is a word that has become so commonplace that its meaning has become obscure. In a University setting one hears daily that "our department is striving for excellence", that funds need to be raised to provide "a margin of excellence", and so on. So what would you attach to the phrase "Tini Veltman is a physicist par excellence".

Our jobs as university professors can be simply described as that of advancing our field, teaching our students, and rendering service to the University, the State, and Nation. This age-old definition aptly describes what many of us do, and even do quite well. But this definition fails when applied to people like Tini.

Tini has not only advanced his field – he has literally redefined our field, and has cleared a path through the jungle with machete knives for others to follow.

Tini has not only taught students – but he has picked some of the brightest students and helped to make them world leaders in the field. Tini has not only rendered service to the field, but has helped create the theoretical and experimental tools needed by the field as it attacks the next frontier of problems.

So my use of the words excellent, outstanding, exceptional, and so on, can not do full justice to specifying the nature of Tini's impact. In terms of Tini's scholarly accomplishments I, as an experimental high energy physicist, could attempt, in my own words, to describe what Tini's impact on physics has been. But I would be on much safer ground if I were to read to you the following citation prepared by some of Tini's colleagues.

"Since the discovery of Yang-Mills theories in 1954, the problem of its renormal-izibility had been a major unsolved problem until its complete resolution in 1972. It had been pursued, without success, by some of the brightest minds of the times including Schwinger, Feynman, Lee, Yang, Weinberg, and Glashow. The various steps involved in this proof culminated in 1972 and, as is the case with all major scientific discoveries, was a result of the contributions of several people. However, the approach, or the particular point of view which eventually triumphed, was per-sistently developed by M. J. G. Veltman over a period of several years and carried to its logical conclusion for the case of a spontaneously broken Yang-Mills theory by G. 't Hooft. Further, a very important ingredient, namely the regularization procedure necessary for an all orders proof was discovered jointly by 't Hooft and Veltman. These discoveries paved the way for a calculable theory of the weak interactions and in particular experimental confirmation of these developments have come from the close agreement between the theoretical calculation of the vector boson mass shifts in the standard model, performed by Veltman, and the UA1, UA2 observations. The proof of the renormalizibility of Yang-Mills theories had a major impact not only in the theory of weak interactions but also in the whole field of elementary particle physics. Suddenly, Yang-Mills theories became credible and this led to the possibility of describing other interactions, by means of such a theory. The viability of the SU(3) color gauge theory of strong interactions (QCD) as a renormalizable theory was put forward and this in turn was used to explain many of the observed phenomena such as scaling and the R ratio in e+ e- scattering. In brief, it ushered in the gauge theory era which is the basis for all of present day elementary particle physics, i.e., this field would not be what it is now were it not for this important discovery, made possible by the vision and persistence of M. J. G. Veltman.

In addition to the above, 't Hooft and Veltman have made other important contributions to the field of elementary particle physics. Together, they were the first to apply the dimensional regularization method to the study of divergences in the theory of gravitation which in turn motivated important developments in quantum gravity. (Besides his investigations of the renormalizability of Yang-Mills fields) Veltman had earlier made important contributions through his discovery of the divergence equations which indicated an underlying Yang-Mills structure of the weak interactions. Subsequently, and on a point of reasonability, he has made major advances in the study of radiative corrections to the standard model and in particular he introduced the Rho parameter which has since turned out to be very useful in the analysis of the low energy predictions of the model. More recently, his investigations of the heavy Higgs limit of the standard model have introduced new and exciting possibilities concerning a second threshold in weak interactions. The discoveries of the standard model have been widely acknowledged but the model itself would not have any predictive power unless it was proved renormalizible and, further, a systematic procedure developed by which higher order radiative corrections in the model could be calculated."

Tini's impact on physics was recently recognized by the State University of New York at Stony Brook through the bestowal of an honorary degree. The citation for that degree also refers to the seminal role of Tini's work.

As these two testimonials illustrate, we are honoring here today someone who has had a profound stamp on his field.

Veltman As A Teacher

As is often the case, great scientists can be outstanding teachers, both in the classroom and in the painstaking task of preparing graduate students to become future leaders of the field. During his years at Michigan, Tini has taught courses in Advanced Quantum Mechanics and Quantum Field Theory. Students have often commented on the clarity of his presentations, the rigor of the coursework, and the excitement of learning from one who has achieved a high mastery of his field.

Another measure of one's effectiveness as a teacher can be found by looking at the achievements of former students. No one I know can produce a list of former students more stellar than:

G. 't Hooft (University of Utrecht), B. deWit (University of Utrecht), P. van Nieuwenhuizen (Stony Brook), J. van der Bij (University of Amsterdam), R. Phelps (University of Michigan), Y. Chao (SLAC), A. Hill (University of Bologna, deceased), Piet Hut (Institute for Advanced Study), and others. POST DOCS: G. Passarino (University of Torino), M. Consoli (University of Catania), and D. Ross (University of Southampton).

Veltman As A Colleague

In this section I will speak from a personal perspective – since, in the end, our assessment of our colleagues as friends remains a subjective matter. Indeed, I am sure we will hear many such personal assessments during the course of the Symposium.

As Chairman of our Department, I have found Tini to be a person who cares deeply about the future of the Department, and one who has labored over the years in helping us launch our Astrophysics Initiative, helped to recruit future faculty in particle theory, and who gave generously of his time in serving on the Departmental Executive Committee.

His penchant for examining issues from angles that do not always coincide with the beam line, and his ploy of responding with answers that are different from what one might expect, always makes discourse with him a new and exciting challenge.

His disdain of all that is showy and flashy and without true content serves as a useful reminder to everyone in the Department that, in the end, we only exist as occupants of our position as professors of physics for one reason – and, as noted in a recent interview with Tini in an article in the LS&A Magazine – that reason is to find the secrets of nature. That pure and simple goal must constantly guide us in everything we do.

In closing, Tini on behalf of your colleagues in the University of Michigan Department of Physics, I want to extend to you our hearty congratulations on this occasion of your 60th birthday. We are extremely pleased that several years ago you chose to come to Ann Arbor and we look forward to many more years as your colleagues.

UM – TH – 92 – 01
February 19, 1992

The Largest Time Equation and Long
Distance Behavior in Gauge Field Theories

R. AKHOURY

Randall Laboratory of Physics
University of Michigan
Ann Arbor, MI 48109-1120

ABSTRACT

Unitarity and causality as expressed through the largest time equation are used to study soft and mass (collinear) singularities in gauge field theories. The origin and cancellation of these singularities in cross-sections is discussed both for abelian and non-abelian gauge theories. Exponentiation of soft divergences is also discussed in abelian theories. At the basis of this work on long distance behavior of gauge theories are certain diagrammatic identities that follow rather simply from the largest time equation. The discussion provides a unified treatment of soft divergencies and mass singularities and is common for both abelian and non-abelian gauge theories.

INTRODUCTION

For about a decade, starting with the early sixties, M. Veltman developed the basic theoretical tools which would be the cornerstones, indeed the foundations, upon which later developments in Gauge theories would be based. The extent of these contributions has been truly broad, covering a wide range of applicability and usefulness from the ultraviolet to the infrared domains of field theories. To the cognoscenti of Gauge theories, it is known how the developments of cutting rules, largest time equation[1], Bell-Treiman transformations, and the rudiments of what is now known as the R-gauge formulation[2] have been an essential part of the study of the ultraviolet behavior of gauge field theories. Further, dimensional regularization[3], developed in collaboration with G. 't Hooft, has been used as a regulator method both in the infrared and ultraviolet. It is perhaps less well known how fundamental the largest time equation is in the study of the infrared behavior of gauge theories. In this paper an effort is made to bring out this correspondence more clearly. Most of the results obtained herein are probably not original and in this sense it may be considered a review. Its purpose is to convince the reader of the important role played by unitarity and causality as implemented through the largest time equation, in the analysis of long distance behavior in gauge field theories. Many of the previously known results and perhaps some new generalizations are seen to follow rather straightforwardly from applications of this equation. The instances of its applicability as well as the cases where it is not, are discussed in detail below. It should be mentioned that the possibility of using the largest time equation to study infrared behavior was known to the discoverer of the equation itself (and to P. van Nieuwenhuizen). Thus, this paper is an attempt to make precise a connection already known to M. Veltman, and is offered as a birthday present from someone who has found much to learn from him.

The organization of the paper is as follows: In the next section (section II) we consider both the origin of infrared divergences in general and also the content of the various forms of the largest time equation. The physical picture[7] at the pinch singular point for infrared divergence is discussed which will be useful in making the connection with the use of the largest time equation. It is also the basis of a power counting procedure for estimating the infrared degree of divergence of a Feynman graph which is given in Appendix A. In section III, we start with a discussion of purely soft divergences. Consistent with the physical picture, the largest time equation is used to obtain a formula for the sum of the cuts of a time ordered Feynman diagram. To be definite, in IIIa, we consider annihilation processes like $\gamma^* \to e^+e^-$ + soft photons in QED with massive fermions, though it is clear that the same method will work for any renormalizable field theory. A version of the largest time equation is applied to relevant cut vacuum polarization diagrams and the cancellation of infrared divergences in this process is demonstrated to all orders. The cancellation is true on a diagram by diagram basis for a given uncut vacuum polarization diagram. Crucial to this cancellation mechanism is a certain diagrammatic identity which is algebraic in nature and follows rather straightforwardly from the largest time equation. Next in section IIIb we discuss scattering processes like electron potential scattering, Compton scattering, electron-electron scattering and so on. Here it is convenient to consider once again the diagrams obtained by squaring the amplitudes, summing over final states, but now we also need to close the initial legs of the resulting diagram, thus forming the K-diagrams of Kinoshita[4] and Nakanishi[5]. To a well defined set of such K-diagrams the largest time equation is applied and cancellation of soft divergences in emission processes (Bloch-Nordsieck mechanism) is demonstrated for $e^-\gamma^* \to e^-\gamma^*+$ soft photons, $\gamma + e^- \to \gamma + e^-+$ soft photons, again on a diagram by diagram basis for a given uncut K-diagram (set). The differences of these with, say, electron scattering are discussed. Next in section IIIc we discuss the generalizations necessary for non-abelian gauge theories by considering soft divergences in detail for the moment only, and point out the origin of the failure of the Block-Nordsiek mechanism in certain situations like quark-quark scattering. When initial state absorptions are taken into account, also, then once again there is complete cancellation of soft divergences. In section IV we discuss the mass divergences or collinear singularities arising in QED with massless electrons and in QCD with massless quarks and gluons (and also with massive quarks and massless gluons). The cancellation of mass singularities in annihilation processes as well as in potential scattering are demonstrated using the largest time approach. The non-cancellation in QCD for a fixed initial state, is discussed in the scattering of massless particles. Finally, in section V we present another proof of the exponentiation of soft divergences in QED[6] which is motivated by our discussion of section III. All the discussions outlined above are carried out with a minimal of calculational effort and a uniform treatment of soft and collinear divergences in any field theory (including abelian and non-abelian gauge theories) is given. It should be pointed out that since the cancellations (whenever they take place) occur between different cuts of a single uncut K-diagram (or vacuum polarization diagram), the group and spin factors are the same as that of the original diagram and so for purposes of our discussion there is no essential distinction between abelian and non-abelian theories. The only difference arises as discussed below, when there is a non-cancellation within a K-diagram set. In the abelian theory in such cases, for soft divergences, there is cancellation between topologically different K diagrams which fails for QCD since the group factors are different. This demonstrates the simplicity and power of the largest time approach. In section VI we present our conclusions.

The main result of this paper concerning the cancellation of the infrared singularities which we restate in the conclusion is the following. For any finite energy line which may be a quark, electron or gluon (color and spin summed) which is between adjacent hard interactions with an ordered sequence of soft interactions inbetween, the largest time equation can be used to show the cancellation of both soft and collinear singularities in the sum over cuts as we go around each soft interaction. The identities which implement this are given in equations (3.9) and (4.1). This result is obtained in steps by considering annihilation and two body scattering processes, however, its generality will be obvious.

IIa: A physical picture for infrared divergences in field theories.

We begin with a review of what is known about the origin and nature of soft and collinear divergences. The particular approach or point of view that turns out to be closely suited to our eventual purpose in this paper is that of ref. [7] which we outline below. We will consider $1PI$ (one particle irreducible) diagrams only since any graph can be decomposed into $1PI$ and tree subdiagrams and our discussion will be seen to trivially apply to the latter.

Consider a $1PI$ diagram in momentum space and its associated expression which is an integral over independent loop momenta of products of Feynman propagators and vertices. We will assume throughout this work that the theory under consideration has been rendered ultraviolet finite by a suitable off-shell renormalization procedure. The above mentioned Feynman integral thus diverges only at those points where Feynman denominators vanish. Of these singular points, only those where the undeformed contours (defined in the integral through the $i\epsilon$ prescription) are pinched (henceforth called pinch singular points) can give rise to mass singularities, both soft and collinear. For each such $1PI$ diagram, a reduced diagram can be constructed by contracting all off-shell lines to a point. The internal lines of such a reduced diagram can be divided into sets of zero momentum lines (k_i) and finite momentum lines (q_i). Singularities at $k_i = 0$ always trap the contour, and for finite momentum lines, the contour is trapped whenever the Landau equations are satisfied for some choice of Feynman parameters α_j; i.e.,

$$q_j^2 = -m_j^2, \qquad \sum_{\substack{\text{finite} \\ \text{mom.loops}}} \alpha_j q_j = 0. \qquad (\alpha_j > 0)$$

$$q_j = q_j^* \qquad \text{(real external momenta)} \tag{2.1}$$

In references [7] and [8] a physical interpretation of a reduced diagram at a pinch singular point is given based on the above considerations. According to this, all such reduced diagrams at a pinch singular point describe physically realizable scattering processes in which particles propagate freely between vertices, energy flows forward in time and there are no situations where energy flows in a loop. Each vertex is associated with a space time point and the diagram represents an ordered sequence of successive interactions. Vertices are divided into two classes: soft and hard vertices. A vertex is soft if all the lines it connects are at threshold in incoming and outgoing channels; otherwise it is hard. Some of their properties are now listed: A soft vertex preserves the direction of spatial momentum flow whereas a hard vertex does not. At a soft vertex, the number of constituents of a "jet" of parallel moving lines may change, however the total energy momentum is the same as before the action of the vertex where the total number of jets is also conserved. At a hard vertex, the jets may change direction as well as their number. Finally, it follows from the definition that all finite energy lines at a soft vertex are either all massive or all massless. An example of a hard (H) and soft (S) vertex, in $\gamma^* \to e^+e^-\gamma$ is shown in Fig. 1, for massless electrons. In this and in all other diagrams, finite energy lines are denoted by solid lines and zero momentum ones by dashed lines. Particle types are not distinguished, so for example a solid line may represent quarks or gluons or electrons. Dashed lines, however, because of power counting (Appendix A) can only be vectors.

As an application, we will discuss some consequences of the above physical picture for reduced diagrams for $1 \to n$ (annihilation) processes which could be obtained by cutting a vacuum polarization diagram. We will discuss first a theory with totally massless particles and then massive particles will be included. For a reduced diagram at the pinch singular point of the above mentioned decay graph, if H_1 denotes the vertex where external time-like momentum enters and a number of on shell lines emerge, then H_1 is the one with the earliest time and all other vertices are soft. This is easily seen as follows. Since the reduced diagram has to represent a physically realizable process then clearly H_1 is the earliest time vertex since energy has to flow forward in time and H_1 is the only source of positive energy. To show that there are only soft vertices we consider two situations. Emission or absorption of zero momentum lines can take place anywhere in the reduced diagram but such vertices are by definition soft. Now consider two finite energy lines emerging from H_1; if they reinteract at some vertex V then they must have collinear spatial momenta. If V is a three point vertex, then the third line must also be collinear since a massless particle can only decay to two massless collinear particles. Similarly for 4 and higher point vertices and thus all subsequent vertices after H_1 are soft ones which preserve the direction of spatial momenta. The reduced diagram therefore describes the evolution of jets and zero momentum lines, the spatial momentum flow as well as the number of jets being determined at H_1. The jets and zero momentum lines interact at soft vertices and the number of jets remains the same though the constituents of each may change. A typical such reduced diagram is shown in Fig. 2.

Inclusion of massive particles changes the picture somewhat. If we are interrested in situations where massive particles appear in the final state relevant for example for massive electrons or quarks, then the QED case is trivial; there being only soft divergences. For non-abelian theories, below the four particle threshold H_1 is again the only hard vertex. This is because massless finite energy lines cannot annihilate into massive ones since it would require non-collinear massless lines to come together at a hard vertex. Thus a typical reduced diagram would look like Fig. 3. Here massive lines are denoted by a double line. In all subsequent diagrams no such distinction is made.

A power counting procedure may be developed to estimate the degree of divergence at the pinch singular point.[7] It is based on identifying a minimal set of variables (normal variables) necessary to put all lines on shell at a pinch singular point. This analysis is presented in an Appendix for annihilation and scattering processes. The results from this Appendix will be quoted in the various sections as needed. The common feature that emerges for annihilation[7] and two body scattering processes[9,10] for a purely massless theory are that the divergences are at most logarithmic and that scattering takes place at a single hard vertex. All other vertices in the reduced diagrams are soft ones. This will be important in our subsequent analysis.

In concluding this section we would like to make some comments on the question of cancellation of infrared divergences. The proposal of Kinoshita[4] and Lee and Nauenberg[11] is that quantities that are finite in the zero mass limit are obtained by summing incoherently over a complete set of degenerate states for both initial and final states. There is another proposal made some time ago by Bloch and Nordsieck[12] for soft divergences, where finite cross sections can be obtained by summing over final state degenerate states only. An example is the well known cancellation of real and virtual emissions in electron potential scattering. In sections III and IV we will investigate these various mechanisms to all orders in perturbation theory.

IIb: The largest time equation, unitarity, and causality.

In ref. [1] M. Veltman showed how unitarity and causality could be implemented in field theories on a graph by graph basis. The method is based on just simple inspection of Feynman graphs and the basic equation is the largest time equation; no complicated discussions involving Landau singularities or analyticity properties are necessary. We briefly review these developments following closely the discussions and notations of refs. [1], [2] and [14] (see also [19]).

Let $F(x_1, \ldots x_n)$ denote the expression for a Feynman diagram with n vertices and containing several propagators:

$$\Delta_{ij} = \Delta(x_i - x_j) = \theta(x_{i0} - x_{j0})\Delta_{ij}^+(x_i - x_j) + \theta(x_{j0} - x_{i0})\Delta_{ij}^-(x_i - x_j)$$

$$= \frac{1}{(2\pi)^4 i} \int d^4k \, \frac{e^{ik(x_i - x_j)}}{k^2 + m^2 - i\epsilon} \tag{2.2a}$$

$$\Delta_{ij}^* = \theta(x_{i0} - x_{j0})\Delta_{ij}^- + \theta(x_{j0} - x_{i0})\Delta_{ij}^+ \tag{2.2b}$$

$$\left(\Delta_{ij}^{\pm}\right)^* = \Delta_{ji}^{\mp} \tag{2.2c}$$

$F(x_1, \ldots x_n)$ contains apart from coupling constants, a factor i for each vertex. Define new functions with one or more x_i underlined, ex. $F\left(x_1, x_2, \ldots, \underline{x}_i, \ldots, \underline{x}_j, \ldots, x_n\right)$ which is obtained from $F(x_1, \ldots, x_n)$ by the following rules.

a. $\Delta_{ij} \to \Delta_{ij}$ if x_i, x_j not underlined

b. $\Delta_{ij} \to \Delta_{ij}^+$ if x_i underlined, x_j not

c. $\Delta_{ij} \to \Delta_{ij}^-$ if x_j underlined, x_i not

d. $\Delta_{ij} \to \Delta_{ij}^*$ if x_i, x_j both underlined

e. for each underlined x_i change the factor i in the corresponding vertex to $(-i)$.

These lead to the largest time equation:

$$F\left(x_1, \ldots x_i, \ldots \underline{x}_j, \ldots, x_n\right) = -F\left(x_1, \ldots \underline{x}_i, \ldots \underline{x}_j, \ldots x_n\right) \tag{2.3}$$

if x_{i0} is the largest time. Introduce a notation such that the vertices corresponding to underlined variables are circled. As an example consider $F(x_1, x_2, x_3)$ and if x_{30} is the largest time then one possible largest time equation is shown in Fig. 4.

An equation that follows from the largest time equation but is true for any particular time ordering is

$$\sum_{\text{underlining}} F\left(x_1, \ldots \underline{x}_i, \ldots \underline{x}_j, \ldots, x_n\right) = 0. \tag{2.4}$$

From the definitions:

$$F\left(\underline{x}_1, \underline{x}_2, \ldots \underline{x}_i, \ldots \underline{x}_n\right) = F\left(x_1, x_2, \ldots x_n\right)^*$$

Thus, we may write

$$F\left(x_1, \ldots x_n\right) + F\left(x_1, \ldots x_n\right)^*$$
$$= -\overline{\sum}_{\text{underlining}} F\left(x_1, \ldots \underline{x}_i, \ldots \underline{x}_j, \ldots x_n\right) \tag{2.5}$$

where the bar reminds us to exclude in the sum the terms with all x_i not underlined and the one with all underlined. Next we can multiply the F's by appropriate wave functions and integrate over all x to get the corresponding functions $\overline{F}$ in momentum space which depend on the external and internal loop momenta. These will be composed of the functions

$$\overline{\Delta}^{\pm}(k) = \frac{1}{(2\pi)^3}\theta(\pm k_0)\delta\left(k^2 + m^2\right) \tag{2.6}$$

$$\overline{\Delta}(k) = \frac{1}{(2\pi)^4 i}\frac{1}{k^2 + m^2 - i\epsilon} \tag{2.7}$$

Thus, for example, a line joining an uncircled vertex to a circled one and carrying momentum k towards circle will be represented by $\frac{1}{(2\pi)^3}\theta(k_0)\delta\left(k^2 + m^2\right)$. Therefore energy always flows towards the circled vertices. The relevant part of the Feynman rules are given in Fig. 5. (The fourth Feynman rule in this will be discussed later.) Many diagrams are zero on account of the conflict between energy conservation at each vertex and the fact that energy is always forced to flow from uncircled to circled vertices. An example which will be relevant later is given in Fig. 6. The rule that therefore emerges for non zero diagrams in that circled vertices form connected regions which contain at least one external vertex. Because of the above property, we may draw a boundary line which separates the "shadowed" region with the circled vertices from the "unshadowed" one. Then, in momentum space, eq. (2.5) corresponds to

$$\overline{F}\left(k_1, \ldots, k_n\right) + \overline{F}\left(k_1, \ldots, k_n\right)^* = -\sum_{\text{cutting}} \overline{F}_c\left(k_1, \ldots, k_n\right) \tag{2.8}$$

where $\overline{F}_c$ contains at least one circled and one uncircled vertex. This is the cutting rule. It is the unitarity relation for a single diagram. For our purposes, we will also need an expression of causality that arises from application of the largest time equation.

Consider again the quantity $F(x_1, \ldots, x_n)$ and suppose $x_{i0} < x_{j0}$, then the largest time equation applies separately to the graphs with x_i circled and with it not. Then:

$$\theta\left(x_{j0} - x_{i0}\right)\left[F(x_1, \ldots x_n) + \sum_{\substack{\text{underlining} \\ \text{except } x_i}} F_c\left(x_1, \ldots \underline{x_k}, \ldots x_n\right)\right] = 0 \qquad (2.9a)$$

where F_c contains at least one circled vertex. We can split it further thus:

$$\theta\left(x_{j0} - x_{i0}\right)\left[F(x_1, \ldots x_n) + \sum_{\substack{\text{underlining} \\ \text{except } x_i, x_j}} F_c\left(x_1, \ldots x_i, \ldots x_j, \underline{x_k}, \ldots x_n\right)\right.$$

$$\left. + \sum_{\substack{\text{underlining} \\ \text{except } x_i \\ \& x_j \text{underlined}}} F_c\left(x_1, \ldots x_i, \ldots \underline{x_j}, \underline{x_k}, \ldots x_n\right)\right] = 0 \quad (2.9b)$$

Suppose now that $x_{i0} > x_{j0}$, then we have a similar equation with i, j interchanged. Adding the two we get

$$F(x_1, \ldots x_n) = -\sum_{\substack{\text{underlining} \\ \text{except } x_i, x_j}} F_c\left(x_1, \ldots x_i, \ldots x_j, \underline{x_k}, \ldots x_n\right)$$

$$- \theta\left(x_{j0} - x_{i0}\right) \sum_{\substack{\text{underlinings except } x_i \\ \& x_j \text{ underlined}}} F_c\left(x_1, \ldots x_i, \ldots \underline{x_j}, \underline{x_k}, \ldots x_n\right)$$

$$- \theta\left(x_{i0} - x_{j0}\right) \sum_{\substack{\text{underlinings except } x_j \\ \& x_i \text{ underlined}}} F_c\left(x_1, \ldots \underline{x_i}, \ldots x_j, \underline{x_k}, \ldots x_n\right). \quad (2.10)$$

Let us look at a special case of the above for a graph with two underlined external vertices x_1, x_2, then we have, (anticipating an eventual fourier transform to mom. space):

$$F\left(x_1, x_2\right) = -\theta\left(x_{20} - x_{10}\right) F^+ - \theta\left(x_{10} - x_{20}\right) F^- \qquad (2.11)$$

F^+ is sum over all underlinings except x_1, but x_2 underlined and F^- the same except the roles of x_1, x_2 are interchanged. F^+ in momentum space corresponds to the sum over all cuttings in which energy flows from the left vertex (x_{10}) to the right one (x_{20}), and F^- to the opposite situation.

Similarly if $x_{10} < x_{20}$ we have

$$F(x_1, x_2) = -F^+, \qquad (2.12)$$

as follows also directly from (2.9). These will be useful for our purpose later.

By multiplying (2.11) by appropriate sources, in momentum space, we can write using the standard representation of the $\theta(x)$ function

$$F(p) = \frac{1}{2\pi i} \int \frac{dk_0}{k_0 - i\epsilon} \left[\theta(k_0 + p_0) F_C(p + k) + \theta(k_0 - p_0) F'_C(p - k) \right]_{\vec{k}=0} . \qquad (2.13)$$

The right hand side contains two terms $F_C(p + k)$ and $F'_C(p - k)$ which are cut graphs in which energy flows from vertex 1 to vertex 2 (left to right) in the first and in the reverse order in the second. These are shown below: For example for a single propagator between sources[14] we have

$$(2\pi)^4 i \frac{1}{p^2 + m^2 - i\epsilon} = \frac{(2\pi)^8 i^2}{(2\pi)^3 2\pi i} \times$$

$$\int_{-\infty}^{\infty} dk_0 \frac{1}{k_0 - i\epsilon} \left[\theta(k_0 + p_0)\delta\left((p + k)^2 + m^2\right) + \theta\left((k_0 - p_0)\delta(k - p)^2 + m^2\right) \right]_{\vec{k}=0} (2.14)$$

In concluding this subsection we would like to make some remarks relevant to the applications of the above methods in the next sections. We first note that all of the above results, in particular equations (2.3), (2.4), (2.5), (2.9), (2.10), (2.11), (2.12) are valid if the spatial components of the various momenta are held at fixed values in momentum space. This is because $(\Delta^{\pm}(x_0, \vec{k}))^* = \Delta^{\mp}(x_0, -\vec{k})$, $\Delta(x_0, \vec{k}) = \theta(x_0)\Delta^+(x_0, \vec{k}) + \theta(-x_0)\Delta^-(x_0, \vec{k})$ and all the above arguments go through. The above, together with the rule for vertices guarantee that a largest time equation will hold. Thus, we could for example consider the functions $F(x_{01}, \ldots x_{0n}, \vec{k}_1, \ldots \vec{k}_n)$ for which the various propagators would be $\Delta^{\pm}(x_0, \vec{k})$, $\Delta(x_0, \vec{k})$, $\Delta^*(x_0, \vec{k})$ and equations like (2.3 - 2.5), (2.9 - 2.12) would still be valid, at a fixed $\vec{k}$. Also for the Fourier transforms of these functions $\overline{F}(\omega_1, \ldots \omega_n, \vec{k}_1, \ldots \vec{k}_n)$, the time ordering arguments leading up to the equations (2.8), (2.13) go through so that it too will satisfy the analogous equations at fixed $\vec{k}$ values . All that is required is that when the corresponding energies are integrated over, the function satisfies the largest time equation. This fact will be relevant for us later. The second remark concerns the fact that in the next section we will apply the results of this subsection to the reduced diagrams of subsection IIa. Is this justified? Especially since some or all of the vertices of a reduced diagram may be effective vertices (as distinct from elementary ones) obtained by shrinking all the off-shell lines of the Feynman diagram at the pinch configuration to a point. The answer is in the affirmative, since the only property of vertices that is required is that the vertex in the shadowed region be of same magnitude but opposite in sign to the one in the unshadowed region. This is still valid for the effective vertices since by definition none of the propagators comprising them are on shell and hence there are no real intermediate states. Finally we should add that we will have occasion to use, the earliest time equation for which similar equations as (2.3) will be valid.

III. Soft divergences in gauge field theories

In this section we will investigate the nature and the conditions under which there is a cancellation of soft divergences in gauge theories. We will mostly consider the Bloch-Nordsieck approach however we will comment on the case including initial state absorptions later. First we will discuss the abelian theory with massive fermions beginning with processes like $\gamma^* \to e^+ e^-$ + soft photons, and then discuss scattering processes. A certain identity will be derived using the largest time approach which is crucial in implementing the cancellation of infrared divergences. We will next discuss the generalizations for QCD which are really very straightforward except for collinear divergences, a discussion of which we postpone until the next section. The failure of the Bloch-Nordsieck program for quark scattering is discussed and comparison with electron scattering is made.

IIIa. Cancellation of soft divergences in cut vacuum polarization diagrams

We are interested in annihilation processes like $\gamma^* \to e^+ e^-$ + soft photons, with radiative corrections. The arguments leading to eq. (3.2) apply to all mass singularities, soft and collinear. Thus for the moment we are not restricting ourselves only to soft divergences. At the pinch singular points, a typical reduced diagram is shown in Fig. 7. In obtaining this, we have used certain results of sections IIa and Appendix A. These results are the following: There is only one hard vertex H_1 and all other vertices are soft. The soft infrared divergences arise when only soft vector particles attach to the external jet lines at three point vertices. The cross section for such $1 \to n$ processes are given by cut vacuum polarization diagrams. For the contribution of a particular state n (see Fig. 8) we can write

$$\sigma_{(n)} = \int \Gamma_L^{(n)} \, d\tau_n \, \Gamma_R^{(n)*} \tag{3.1}$$

where $d\tau_n$ is the phase space and $\Gamma_L (\Gamma_R)$ are the vertex functions on the left (unshadowed) or right (shadowed) regions. The reduced diagrams of $\Gamma_{L,R}$ at the pinch singular point are of the form discussed above and in Fig. 7.

Recall that at the pinch singular point, the spatial momentum flow as well as the number, energies and directions of jets, are determined at the hard vertex H_1. This remains the same at all other subsequent interactions and so the above characteristics are the same for the $\Gamma_L^{(n)}, \Gamma_R^{(n)}$ of a particular cut n. Thus, we can speak of a pinch singular point of a cut vacuum polarization diagram. The power counting discussed in Appendix A tells us that the soft divergences are at worst logarithmic for such cut vacuum polarization diagrams and that when the momentum of all soft vector lines go to zero simultaneously. We now proceed to the derivation of the equation that demonstrates the cancellation of IR divergences. We will first derive a particular form and then more useful ones will be derived: (3.9) for soft divergence sand in section IV, (4.1) for collinear ones.

Recall that for the cut vacuum polarization diagrams introduced above at the pinch singular point energy always flows from H_1 to H_2 without any situations where it goes around in a loop. The spatial momentum flow being completely determined at H_1 and also the energies of each jet. Thus, in each line of the reduced idagram, energy flows in the same direction from left to right. Referring now to the discussion at the end of section

IIb, we will see that the largest time equation tells us that at the pinch singular point only can we write the following equation (the arrows indicate the direction of energy flow), for these reduced diagrams.

$$\sum_{\text{cuttings}} \quad = \text{free of IR divergences} \qquad (3.2)$$

In the diagrams on the left we have to include all with H_1 uncircled, H_2 circled and other vertices in various circle configurations, i.e., it is the sum of all cut diagrams in which energy flows from H_1 to H_2 only. This is enforced by means of appropriate sources at H_1 and H_2. These sources are denoted by crosses above and they vanish unless energy flows from H_1 to H_2 only. We emphasize that the above is valid for fixed values of all spatial momenta as well, as discussed at the end of section IIb.

Two ways to see the validity of this equation will be discussed. First we must introduce sources J at vertices H_1 and H_2, denoted by crosses above, which enforce the appropriate direction of energy flow. Consider now eq. (2.8) for a vacuum polarization like diagram in which energy is ordered such that it flows from H_1 to H_2 only. between sources $J(p)$ which vanish unless $p^0 > 0$ (p is incoming momentum at H_1). Then clearly we have

$$\qquad + \left(\qquad \right)^* \quad = \quad - \sum_{\text{cuttings}} \qquad (3.3)$$

For example, for a single propagator:

where

$$\underset{p}{\longrightarrow} \quad = (2\pi)^4\, iJ(p)\frac{1}{p^2 + m^2 - i\epsilon}J(p).$$

The complex conjugate applies to everything except the sources. Thus in terms of equation ($J(p)$ is non zero only for $p_o > 0$).

$$J\left[(2\pi)^4 i\frac{1}{p^2 + m^2 - i\epsilon} - (2\pi)^4 i\frac{1}{p^2 + m^2 + i\epsilon}\right]J$$
$$= J\left[\frac{i^2(2\pi)^8}{(2\pi)^3}\theta(p_0)\delta\left(p^2 + m^2\right)\right]J$$

which is clearly valid. The right hand side of course represents an on shell line in which energy flows forward ($p^0 > 0$), which is typical for all the lines in our reduced diagram. In this connection, one should remember that for each line energy flows in the same direction at the pinch singular point. The left hand side of (3.3) is free of infrared divergences. This can be seen in a variety of ways.

First, the diagrams on the left hand side are energy (or time) ordered vacuum polarization diagrams in which H_1 is the earliest time vertex and energy flows from H_1 to H_2 and its complex conjugate. Such a diagram cannot contribute to an infrared divergence (soft or collinear). This can be seen in many ways: For soft divergences the result follows since a finite energy line betwen two hard vertices cannot go on shell. For collinear divergences from the physical picture introduced in IIa, at a pinch singular point, which is relevant for infrared divergence, we want the two different jets j_1 and j_2 to originate at H_1 and come together again at H_2 while undergoing only soft interactions in between. This is clearly impossible. Another straightforward way to see its vanishing is to apply the largest or earliest time equation. Note that we are dealing with a situation discussed in (2.11) or (2.12). Consider a low order example which is easily seen to generalize. Refer to Fig. 6 and if H_1 is the earliest time vertex then we have (from the earliest time equation)

$$\text{(3.4)}$$

since the second term is zero by itself because of the conflict with energy conservation at the circled vertex, the first one is free of IR divergences. This will happen in general if there are no cut lines between H_1 and H_2 with H_1 being the earliest time vertex. Mathematically speaking, the left hand side of (3.3) can arise if the time component of the Landau equations is satisfied but not the spatial ones. Since each line of the diagram has energy flowing in the same direction, and if we divide the integration momenta into transverse, plus and minus components, then the poles in the complex plane for the minus components can all be assembled on one side of the axis. Thus there is no pinching in this case. For a more analytical proof of this see [4] and [18].

Actually, a stronger result can be obtained thus: Refer to equations (2.13) and (2.14). We have of course to put in the sources to enforce the required unidirectional energy flow from H_1 and H_2. Now we have noted that at the singular point, energy flows forward for all lines in the diagram which are on shell and their momenta real. Thus, in equation (2.13), the region relevant to a singular point is $k_0 \to 0$ and in this region alone we can write an uncut energy ordered diagram as a sum of cuts of the diagram. The uncut diagram vanishes again for the same reasons as above.

As we will see equation (3.2) gives the cancellation of IR singularities within each uncut vacuum polarization diagram, i.e., on a graph by graph basis for a given topology. Because of the two approaches to deriving this equation, the cancellation is seen to be a consequence of unitarity and causality. An identity that implements this cancellation will be derived. This identity is another statement of the largest time equation and its validity is easily seen to be true by induction, thus independently verifying (3.2). Henceforth we will call the right hand side of (3.2) "zero" meaning that there are no IR divergences. There are of course IR finite contributions which we are not interersted in at the moment.

Let us now specialize only to soft divergences and begin by considering some lower order examples of (3.2):

$$\left(\, \right) + \left(\, \right) +$$

$$\left(\, \right) = 0 \qquad (3.5)$$

$$\left(\, \right) = 0 \qquad (3.6)$$

Actually, as we will see the two terms grouped together in (3.5) vanish individually. In fact, the above are just the well known examples of the cancellation between real and virtual soft photon emission in QED to lowest non trivial order. In the next order we will similarly have:

$$\left(\, \right) +$$

$$\left(\, \right) +$$

$$\left(\, \right) = 0 \qquad (3.7)$$

$$\left(\quad + \quad + \quad + \quad + \quad \right) = 0 \qquad (3.8)$$

We will now see that what is at the core of the cancellation of the various terms grouped together are special cases of an identity, which also follows from the largest time equation. This identity is more useful than (3.2) and is really the more interesting one for soft divergences. Before we discuss this however, we would remark that the spin factors which came as a Dirac trace are the same for all terms arising from cutting an uncut vacuum polarization diagram of given topology. Further, we should remark that the logarithmically divergent part of the integral only comes from the $(-i\gamma \cdot p + m)$ factor and omitting the photon momenta in the numerator. Given this we will ignore the spin factors below since they are common to all diagrams between which we want to show the cancellations. Consider fig. 10, which represents any one of the jet lines between two hard vertices, between which momentum flows from H_1 to H_2 and to which are attached an ordered set of soft photons of momenta $k_1, \ldots, k_n$ all of which are taken to be flowing towards S. Energy flows in the same direction for every line of the diagram. S represents a Green's function of soft vectors which for QED is disconnected but can have connected pieces in QCD. We will not assume any particular property of S except that it is composed of soft vertices only. Our arguments below are independent of the specific nature of S, a fact which allows for the straightforwardness of the generalizations to non-abelian gauge theories.

From the largest time equation applied at the pinch singular point only follows:

$$(3.9)$$

The right hand side vanishes (*i.e.*, is free of soft divergences) because an uncut fermion line between two hard interactions (with only soft interactions in between) can never go on shell which is necessary for a soft divergence. This actually guarantees the vanishing of the various terms in eq. (3.5)-(3.8) separately for the various groups in brackets. Actually we can go a step further to derive an algebraic identity that will be very useful later when we cannot apply the largest time equation directly. This follows from the following observation: In S we will be cutting certain photon lines and we are combining various terms in some of which certain number of photon lines are cut and in some not. However, we can derive an equivalence between the cut and uncut soft photon lines as far as the infrared divergences go, based on the following relations:

$$\frac{1}{i}\frac{1}{k^2 - i\epsilon} = \pi\delta(k^2) + \frac{1}{2i}\left(\frac{1}{k^2 - i\epsilon k_0} + \frac{1}{k^2 + i\epsilon k_0}\right)$$

$$-\frac{1}{i}\frac{1}{k^2 + i\epsilon} = \pi\delta(k^2) - \frac{1}{2i}\left(\frac{1}{k^2 + i\epsilon k_0} + \frac{1}{k^2 - i\epsilon k_0}\right)$$

$$2\pi\theta(k_0)\delta(k^2) = \pi\delta(k^2) + \frac{1}{2i}\left(\frac{1}{k^2 - i\epsilon k_0} - \frac{1}{k^2 + i\epsilon k_0}\right)$$

$$2\pi\theta(-k_0)\delta(k^2) = \pi\delta(k^2) + \frac{1}{2i}\left(\frac{1}{k^2 + i\epsilon k_0} - \frac{1}{k^2 - i\epsilon k_0}\right). \tag{3.10}$$

For the case under discussion we can neglect the contributions of the terms in brackets since the various poles can all be assembled on one side of the real axis. This is really a consequence of the fact that at the pinch singular point the vertices are ordered and the largest time equation can be applied. We can thus get an identity only for the cut fermion lines and in terms of equations:

$$\sum_{m=0}^{n}\prod_{i=0}^{m-1}\frac{1}{(p - q_i)^2 + m^2 - i\epsilon}\delta\left((p - q_m)^2 + m^2\right)\prod_{j=m+1}^{n}\frac{1}{(p - q_j)^2 + m^2 + i\epsilon} = 0. \tag{3.11}$$

where $q_i \equiv k_1 + k_2 + \ldots + k_i$ and $q_{i=0} = 0$. It is easier to visualize this equation in the eikonal approximation ($q_i^2 = 0$) where it can be written as:

$$\sum_{m=0}^{n}\prod_{i=0}^{m-1}\frac{1}{p \cdot q_m - p \cdot q_i - i\epsilon}\prod_{j=m+1}^{n}\frac{1}{p \cdot q_m - p \cdot q_j + i\epsilon} = 0. \tag{3.12}$$

For example we have:

$$n = 1 \qquad \frac{1}{-p \cdot k_1 + i\epsilon} + \frac{1}{p \cdot k_1 - i\epsilon} = 0$$

$$n = 2 \qquad \frac{1}{-p \cdot k_1 + i\epsilon}\frac{1}{-p \cdot (k_1 + k_2) + i\epsilon}$$

$$+ \frac{1}{p \cdot k_1 - i\epsilon}\frac{1}{+p \cdot k_1 - p \cdot (k_1 + k_2) + i\epsilon}$$

$$+ \frac{1}{p \cdot (k_1 + k_2) - i\epsilon}\frac{1}{p \cdot (k_1 + k_2) - p \cdot k_1 - i\epsilon} = 0$$

These two actually provide the cancellations in (3.5) and (3.7) separately of the terms grouped together in ( ). The identity (3.12) can be independently proven by induction thus providing an alternate check of our equation (3.2).

Before concluding this section, we would like to discuss a technical point concerning the treatment of self energy graphs. This will happen if in (3.11) or (3.12) we consider say, n soft vectors with $k_n = -\Sigma_{i=1}^{n-1}k_i$. We will briefly indicate the trick of introducing a fake momentum, δ, to circumvent this problem.[15] For this we need to consider the equation (3.9) with now the last photon leg's momentum as negative of the sum of the other photon momenta, $i.e.$, a situation like

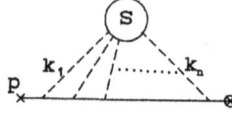

The problem now is that the extreme most fermion propagators above, next to the sources are formally on shell. We know that these poles are cancelled by the zeroes of, Σ, the self energy, on shell, after mass renormalization. Also, as discussed in, [17], only one self energy and not both are to be counted in the cross section. To get around these problems we introduce a prescription for going off-shell by adding a fake momentum, δ, at the highest numbered vector vertex above. As indicated $\delta \geq 0$, and is real. The physical result is obtained in the limit $\delta \to 0$, since

$$\frac{1}{i\gamma \cdot (p+\delta) + m}\Sigma(p)u(p) = 0 \text{ at } p^2 = -m^2.$$

This, in fact, ensures that only one self energy be included in the sum of graphs. The effect of the other is to cancel the wave-function renormalization factors in the definition of the cross-sections. The above tells us that there are internal cancellations in such situations. By introducing the fake momentum δ we need not worry about the internal structure of the graphs because we can now use (3.11) or (3.12) to show cancellation for a finite δ. This is the key point of this discussion. Thus, since the above graph is zero, we are not overcounting by including it in the sum. The equation (3.12) now reads the same in this case except for the replacement $k_1 + \ldots + k_n = -\delta$. Thus for $n = 2$, we have: (we can omit the $i\epsilon$ in the terms with δ alone since it is positive semi definite)

$$\frac{1}{-p \cdot k_1 + i\epsilon}\frac{1}{p \cdot \delta} + \frac{1}{p \cdot k_1 - i\epsilon}\frac{1}{p \cdot k_1 + p \cdot \delta + i\epsilon}+$$
$$\frac{1}{-p \cdot \delta}\frac{1}{-p \cdot \delta - p \cdot k_1 - i\epsilon} = 0 \tag{3.13}$$

and so on. This and the generalization for $n = 4$ are responsible for the cancellations in (3.6) and (3.8). We remind the reader that we have gone through this exercise of trying to exhibit the cancellation independent of the internal structure and properties of S because then the generalization to non-abelian theories will be straightforward.

We have thus demonstrated in this section how simply the cancellation of soft divergences in annihilation processes can be proved as a consequence of (3.9). For some more details of examples of explicit calculations see sec V. The cancellation, it should be noted takes place for different cuts of an uncut vacuum diagram, is independent of spin factors, and further takes place for the integrands themselves. We will next consider scattering process.

IIIb. Soft divergences in scattering processes

Let us begin by considering electron potential scattering. In particular, we wish to investigate the process $e^-\gamma^* \to e^-$ + soft photons. Notice first that this process can be obtained from the annihilation process by bringing the final state positron as an initial state electron. The reduced diagram for this process will also contain only one hard vertex which determines the initial jet and momentum distribution. Upon summing over final states we can write the reduced diagram relevant for the cross section as cuts over a certain forward scattering amplitude as shown in Fig. 11. Thus we can talk of the pinch singular points of the cut forward amplitude. The results of Appendix A tell us that the soft divergences are at most logarithmic once again, and that only when soft vectors attach to fermions at three point vertices. To show the cancellation of soft divergences in the sum over final state emissions only, we proceed thus: We first consider the diagram obtained by joining the final states of the diagram and its complex conjugate, i.e., the diagram like Fig. 11 without the cut. We next consider a diagram in which the "initial state" lines carrying momentum p are also connected together. This resulting diagram looks like a vacuum to vacuum Feynman diagram except that the initial legs connected together are really cut lines, i.e., we do not integrate over the momenta of the initial state lines but only average over their spins, as is usual. This is because in a cross section we usually consider initial particles of well defined momentum. We can now consider a set or collection of all such diagrams which give the same vacuum to vacuum Feynman diagram when the cut in the initial state lines is removed. An example is the set in Fig. 12a which gives the same vacuum to vacuum diagram in Fig. 12b. (In 12a the initial cuts are indicated.) Such a set will be called a K-diagram set. The K-diagrams were initially introduced by Kinoshita[4] Nakanishi.[5] We can now discuss the cancellation of soft divergences within a K-diagram set. As regards, soft photon emission where also the final state electron line is involved such as shown in Fig. 13, there is absolutely no problem. For a fixed initial cut, we can sum over all final cuts involving the line with momentum p' and this will be guaranteed to vanish on account of eq. (3.9). Equation (3.9) is valid for the final state line with momentum p' because energy always flows from the hard vertex H_1 to the hard vertex H_2 which lies at its two ends. Thus if line p' is never cut, it can never contribute to an infrared divergence since it cannot go on shell, guaranteeing thereby that the right hand side of eq. (3.9) is zero. For emissions involving only the initial state line, this is no longer true since a K-diagram is not a vacuum to vacuum diagram. (Recall that there is always an initial state cut in a K-diagram and no integration over initial momenta). Thus the largest time equation such as eq. (3.9) cannot be directly used for such cases. This is because we are not entitled to use the arguments after equation (3.3). The circled vertices can form connected regions now with the external initial line. If the K-diagram were a vacuum diagram then the LHS of (3.3) would again be zero. The point is that H_1 is no longer the earliest time vertex in the diagram. Only for those parts of the diagram (final state interactions) for which it is still possible to think in this way can one usefully apply the largest time equation to exhibit the cancellations. However, we may still use the identity eq. (3.11) which has been abstracted out of (3.9), but we must ensure that we can neglect the contributions of the extra terms in (3.10). This will enable cancellation of soft divergences in this case within a K-diagram set, if the above condition is met, which is the case of potential scattering.

An example is that the soft divergence in the sum of the diagrams shown in Fig. 12a vanishes due to the identity (3.12) or (3.13). Of course, in such a case one has to make sure that the replacement of cut and uncut soft propagators by (3.10) still is such that we may treat them both in an identical manner. For the potential scattering case, we are ensured of this from the crossing symmetry of the theory. This because we know that everything works correctly, as required by unitarity for $\gamma^* \to e^+e^-$ + soft photons and $\gamma^* e^- \to e^-$ + soft photons is related to it by changing an outgoing positron to incoming electron. This interchange does not disturb the cancellation mechanism since the fermion momentum in (3.12) is common to all terms. The only problems we can have with crossing symmetry is in regard to the photons.

Consider next some other situations where the above arguments remain valid. An example is Compton scattering. Here also, the results of Appendix A tell us, that scattering always takes place at a single hard vertex. As an example, consider Fig. 14 and we see that the reduced diagram at the pinch singular point corresponding to it is given in Fig. 15. This because the fermion line between the two hard photons can never go on shell. Thus the situation concerning the infrared divergent part become identical to the case of potential scattering considered above and need not be treated separately. The general case works exactly in the same way as can be verified using the results of Appendix A. In this case and in other situations a property of closed fermion loops needs to be discussed. This is that due to gauge invariance, the Green's function of massless vectors emanating from a closed fermion loop vanishes as any of their momenta goes to zero. Thus, for example, in the diagram of Fig. 16, none of the momenta q_i can be zero for there to be soft divergence and since no soft fermions are allowed by power counting, the reduced diagram becomes that of Fig. 17 with one hard vertex, H.

We will now discuss a case which exhibits a different behavior, vis a vis, the cancellation of soft infrared divergences. In such cases which includes, for QED, $e^+e^- \to \gamma^*$ + soft photons, electron-electron scattering, the cancellation mechanism outlined above breaks down. For the abelian theory, the soft divergences still cancel, but for it to be complete one has to include contributions from topologically different K-diagram sets. In QCD there is a break down[16] of the Bloch-Nordsiek mechanism. Let us discuss $e^+e^- \to \gamma^*$ + soft photons first. Return for the moment to $\gamma^* \to e^+e^-$ + soft photons where everything is guaranteed to work as a consequence of unitarity and causality. From time reversal invariance, the situation should be similar for, soft photons + $e^+e^- \to \gamma^*$. This would require absorption of soft photons in the initial state whereas the process we started to consider required the emission of photons in the final state. Thus, the two processes are not directly related by time reversal alone, but by it together with reversal of the soft photon momenta. In the eikonal approximation, in the fermion denominators, replacing

$k \to -k$ amounts effectively to changing the sign of $i\epsilon$

$$\left(\frac{1}{2p \cdot k + i\epsilon} \longrightarrow \frac{-1}{2p \cdot k - i\epsilon} \right).$$

Thus, now we will, in general, no longer be able to assemble all the poles on one side of the real axis for the extra terms, once we make the replacements (3.10) in the cut and uncut propagators, *i.e.*, the identity (3.12) or (3.13) does not imply the cancellation of soft divergences within a K-diagram set. The noncancellation of soft divergence in a given K-diagram set for $e^+ e^- \to \gamma^* +$ soft photons, will arise in the clash of the $i\epsilon$ prescriptions because of the fact that in the replacements (3.10) we are not allowed to omit the contributions of the extra terms in brackets. The non cancelling pieces are proportional to them alone. How then do the soft infrared divergences in QED cancel? This happens only because the noncancelled parts between two topologically distinct K-diagram sets are the same but of opposite sign. The abelian Ward Identity may be used to make this explicit. Thus, for the total process, the soft IR divergences again cancel but not within a topologically identical set of K-diagrams. Such an effect which comes about due to the clash of $i\epsilon$ signs can begin only at the two loop level. As discussed in section V, exponentiation involves adding together the contributions from diagrams with crossed photon lines. At two loops for example, we need to add together (a) and (b) and (a′), (b′) and (c′) of that section. For the $e^+ e^- \to \gamma^* +$ soft photons, the K diagram sets are similar to the ones just mentioned above; these are topologically distinct but the uncancelled pieces are identical in the two sets of diagrams and they cancel out in the sum separately. Thus, exponentiation goes through even though there are non cancelling pieces within a K-diagram set.

This example clearly brings out the fact that the peculiarities with the Bloch-Nordsieck mechanism are only in regard to initial state particles. Our analysis of electron potential scattering has shown us that cancellation of soft divergencies when the final state electron line is involved, is guaranteed by the largest time equation. For emission from initial state lines we need to consider K-diagram sets which involve a cut in the initial line and are not like the vacuum diagrams. Here the largest time equation, though valid, is not so useful since we cannot argue the vanishing of the right hand side of (3.3), (3.4) or (3.9). We may still use the identity (3.11) or (3.12) however, we must verify that when we make the substitutions (3.10) in the soft propagators then the terms in brackets do not contribute. In annihilation processes they do not and neither to the related processes of potential and Compton scattering. This is largely a result of unitarity, causality for the former, and symmetry and kinematics for the latter. For processes like $e^+ e^- \to \gamma^* +$ soft photons, electron-electron scattering, they do, and then the cancellation of soft divergences if it occurs, must be between (topologically) different K-diagram sets. To emphasize the fact that only the initial state interactions are a problem with the Bloch-Nordsieck mechanism we consider a process like electron-electron scattering. The results of Appendix A for massive fermion scattering in the abelian theory tells us that the reduced diagram at the pinch singular point is such that scattering takes place at a single hard vertex and the only logarithmic soft divergence is when soft vectors attach to fermion lines at three point vertices. A logarithmically divergent configuration contributing to the scattering process cross section is then depicted as a cut forward scattering amplitude shown in Fig. 18. Energy always flows from hard vertex H_1 to H_2 and therefore all soft interactions involving final state fermions will cancel due to the largest time equation. In fact, for interactions involving purely the final state fermion (primed ones in Fig. 18), the situation is identical to cut vacuum polarization diagrams and for those that involve a final state fermion and one or more initial state ones, we can again use the largest time equation by fixing a given initial cut and summing over the final state ones. Thus the only soft interactions that cause divergent behavior and which cannot be shown to cancel from largest time equation alone are the ones involving only initial state fermions. A typical K-diagram relevant for these is shown in Fig. 19. Here again, precisely for the same reasons as mentioned above for $e^+e^- \rightarrow \gamma^* +$ soft photons, there is non-cancellation of soft divergences in a given K-diagram set, however the complete scattering cross section averaged over initial spin (including all K-diagram sets) is free of IR divergences[6], for this QED case.

IIIc. Extension to non-abelian gauge theories

For non abelian gauge theories, even if there are only massive quarks, in addition to soft divergences, there will always be collinear ones due to the massless gluons. The cancellation of such collinear divergences will be considered in section IV. For the discussion of this section all quarks are treated as massive. As far as the purely soft divergences go, all the arguments of section IIIa, IIIb go through, just replace electron by "jet" which can be quark or gluon one, and soft photon by soft gluon. A typical reduced diagram for $\gamma^* \rightarrow q\bar{q}$ + soft gluons is given in Fig. 20. Ignore the collinear divergence for the moment. The soft logarithmic divergence again only arises when soft vectors couple to jet lines (quarks or gluons) only at three point vertices. S now contains connected parts due to gluon self interaction. The reason why the arguments of IIIa go through for this case is twofold. Firstly, (3.2) and (3.9) follow from the largest time equation and are true also in this case and the cancellation expressed by these equations takes place between the different cuts of an uncut vacuum polarization diagram. Thus not only the spin factors but all the color factors are the same for the different diagrams taking part in the cancellation since they factor out as a trace. In particular, also the identities (3.11), (3.12) and (3.13) remain valid. Secondly, no particular property of the soft particle green function S was used in showing the cancellation. For massive quark the scattering[15] processes, we must not only average over initial spins but also the initial color and form the K-diagram set. The spin and color factors are the same in each set since they can all be obtained from one vacuum to vacuum diagram, i.e., they are common factors involving traces over Dirac matrices or color matrices. For potential scattering, we could consider scattering from a color singlet or non-singlet potential. The reduced diagram analogous to Fig. 11 for the abelian theory is shown in Fig. 21. Soft divergences involving the final state quark or gluon jets between the hard vertices H_1 and H_2 will again cancel as a consequence of the largest time equation. For soft gluon emissions involving the initial state, we know that in a K diagram set all the color factors are the same for the various diagrams comprising the set. Thus the identity

(3.11), (3.12) or (3.13) will guarantee the cancellation of soft divergences in this case also.

Thus, we see that as long as there is cancellation between diagrams in a topologically equivalent K-diagram set, there is no distinction between abelian and non-abelian theories. One has merely to replace electron for quark or hard gluons and soft photons for soft gluons in the discussion of section IIIa. In quark-quark scattering, an analogous situation as in electron-electron scattering will occur. The only possible non cancelling soft divergence will involve gluon emissions from initial state quarks. Here the replacements (3.10) also fail to provide the equivalence of cut and uncut soft gluon propagators. The non-cancelling pieces will contain at least one term in the brackets of eq. (3.10). However, now comes the crucial difference with electron scattering. Whereas there the soft divergence still cancelled between topologically inequivalent K-diagrams, that is not possible here since the group factors are different for diagrams that are topologically different. The same is true for $q\bar{q} \rightarrow \gamma^* +$ soft gluons, however, for, soft gluons $+ q\bar{q} \rightarrow \gamma^*$ all is well again since it is related by time reversal to $\gamma^* \rightarrow q\bar{q} +$ soft gluons. Thus the Bloch-Nordsieck mechanism breaks down in non-abelian theories[16] for processes involving initial state emissions, in general, even if the initial color is averaged. Other, related discussions of the cancellation of soft divergences is given in [7], [9], and [15].

IV. Collinear divergences in massless gauge theories

Collinear divergences can arise when massless particles can decay to two or three massless particles that move parallel to each other. We will, in this section discuss the case of QED with massless electrons and QCD with massless quarks and gluons. Only the new kind of divergences, i.e., the collinear ones will be treated; the discussion of soft divergences and their cancellation parallels the one given in section III. The crucial point for the subsequent discussion of the annihilation and two body scattering cases is that in each case, the reduced diagram consists of only one hard vertex where the spatial momentum flow, the number and directions of the jets is determined. Then in each jet, there are only soft vertices. An important point about this is that if finite energy lines meet at a soft vertex, then they must necessarily be all massive or all massless. In all the cases to be discussed below, power counting at the pinch singular point given in Appendix A, determines the logarithmic divergent configuration to contain only three or four point vertices of finite energy lines at each soft vertex. At the pinch singular point for both annihilation and two body scattering processes, the divergent configuration is one in which the jets have a self energy like structure. Regarding the cancellation of these collinear divergences, we will discuss as before separately the case of annihilation and scattering processes. The main mechanism of the cancellation (when it does occur) is as follows: In all the cases to be discussed which give a logarithmically divergent contribution to the cross section, we need to discuss either cut vacuum polarization diagrams or K-diagram sets, between two hard vertices H_1 and H_2. This should be clear from our discussion of soft divergences. For scattering processes we will also need to consider the "forward jets" to be discussed below. Now take a jet with its ends fixed at two hard vertices. This jet undergoes many interactions between these, all of which are soft. They only change the number of finite energy lines making up the jet. We will focus our attention on the soft vertices made up only of finite energy lines and hence only massless ones. The cancellation identity which is analogous to (3.9) for soft divergences is that the mass singularities cancel when we sum up the contributions from all the cuts of this jet whose ends are fixed and where in making the cuts we go around each massless vertex. This identity which is the cornerstone of the method will be discussed, beginning with annihilation processes where, as before, everything goes as prescribed by unitarity and causality.

We consider cut vacuum polarization diagrams for which a typical reduced diagram is shown in Fig. 20. From power counting, the collinear divergences occur within each jet individually as indicated by the self energy-like structures. The largest time equation tells us the validity of eq. (3.2). This ensures cancellation of collinear divergences. However, because of the nature of the jet structure outlined above we see that the analog of eq. (3.9), true to all orders is

$$\sum_{\text{cuttings}} \quad = 0 \qquad (4.1)$$

At the basis of this equation is the fact that in a reduced diagram energy always flows in the same direction in all the lines comprising the diagram. In the above, J may denote a complicated jet subdiagram and H_1 and H_2 are hard vertices. (We have again factored out the common color and Dirac traces.) Also, in the above, we consider all diagrams with H_2 circled, H_1 uncircled and sum over circled configurations for all the vertices in between. This equation follows from applying the largest time equation between the hard vertices H_1 and H_2, where energy flows from H_1 to H_2 only. The uncut diagram is zero or gives no collinear divergence since, as before, it will not correspond to a physically observable process. Alternately, we can also see this by applying the earliest time equation to it. In eq. (4.1) we must separate out the self energy effect of the jet as a whole. That is, for external lines we must consider only one self energy correction in the cross section, since $\bar{u}(p)\Sigma(p) = \Sigma(p)u(p) = 0$, on shell. See the related discussion at the end of section IIIa. Below, we will discuss, some simple examples to illustrate the cancellation mechanism. These examples can be straightforwardly extended to give an alternate proof by induction of (4.1); thus providing independent proof of it and (3.2).

Consider the reduced jet diagram shown below in (a). S is a soft three point vertex which may be momentum dependent, however, these vertex factors are common to all terms discussed below. We need to look at

$$\qquad = 0 \qquad (4.2)$$

Ignoring factors common to both diagrams, we have: $(k_1 = k_2 + k_3)$

$$(4.2a) = \frac{1}{i}\frac{1}{k_1^2 - i\epsilon}(2\pi)\theta(k_{20})\,\delta\left(k_2^2\right)(2\pi)\theta(k_{30})\,\delta\left(k_3^2\right)$$

$$(4.2b) = (-1)(2\pi)\theta(k_{10})\,\delta\left(k_1^2\right)\left(-\frac{1}{i}\right)\frac{1}{k_2^2 + i\epsilon}\left(-\frac{1}{i}\right)\frac{1}{k_3^2 + i\epsilon}$$

The (-1) in (4.2b) comes from the fact that (b) has an additional circled vertex than (a). We want to show the cancellation of pinched poles in the k_{i0} integrations which are assumed though not indicated. Using the identity (see eq. 3.10)

$$\frac{1}{i}\frac{1}{k^2 + i\epsilon} = -2\pi\theta(k_0)\delta(k^2) + \frac{1}{i}\frac{1}{k^2 - i\epsilon k_0} \qquad (4.3)$$

in (4.2b) and keeping only the terms that cause mass singularities we have (again omitting common factors) for the sum of the two contributions.

$$\frac{1}{k_1^2}\theta(k_{20})\delta(k_2^2)\theta(k_{30})\delta(k_3^2)+$$

$$\frac{1}{k_2^2}\theta(k_{10})\delta(k_1^2)\theta(k_{30})\delta(k_3^2) + \frac{1}{k_3^2}\theta(k_{10})\delta(k_1^2)\theta(k_{20})\delta(k_2^2).$$

Here we have omitted the other terms where the poles can be assembled on one side of the axis. The above may now be written as:

$$\frac{1}{8k_2 \cdot k_3}\left[\frac{\delta\left(k_2^0 - K_2\right)\delta\left(k_3^0 - K_3\right)}{K_2 K_3} - \frac{\delta\left(k_1^0 - K_1\right)\delta\left(k_2^0 - K_2\right)}{K_1 K_2} - \frac{\delta\left(k_1^0 - K_1\right)\delta\left(k_3^0 - K_3\right)}{K_1 K_3}\right] \qquad (4.4)$$

where $K_i = |\vec{k}_i|$. The collinear divergence occurs when $\vec{k}_2$ and $\vec{k}_3$ are parallel; *i.e.*; when $K_1 = K_2 + K_3$. Then (4.4) becomes

$$\frac{\delta\left(k_3^0 - K_3\right)\delta\left(k_2^0 - K_2\right)}{8k_2 \cdot k_3}\left[\frac{1}{K_2 K_3} - \frac{1}{K_2\left(K_2 + K_3\right)} - \frac{1}{K_3\left(K_2 + K_3\right)}\right] = 0.$$

Note that also the soft infrared divergence cancels in this sum since power counting tells us that the vertices in the jet subdiagram can be either three point or four point, it is instructive to see the cancellation also for a single four point vertex. Consider

$$(4.5)$$

In this case, we obtain for these diagrams after the substitutions (4.3)

$$\frac{1}{k_1^2}\delta_+(k_2^2)\delta_+(k_3^2)\delta_+(k_4^2) + \frac{1}{k_2^2}\delta_+(k_1^2)\delta_+(k_3^2)\delta_+(k_4^2)$$

$$+ \frac{1}{k_3^2}\delta_+(k_1^2)\delta_+(k_2^2)\delta_+(k_4^2) + \frac{1}{k_4^2}\delta_+(k_1^2)\delta_+(k_2^2)\delta_+(k_3^2)$$

$$+ \text{finite terms}$$

where $\delta_+(k^2) = \theta(k_0)\delta(k^2)$ which can be rewritten using the fact that the collinear divergence occurs at $K_1 = K_2 + K_3 + K_4$

$$\frac{1}{16(k_2 \cdot k_3 + k_2 \cdot k_4 + k_3 \cdot k_4)}\delta\left(k_2^0 - K_2\right)\delta\left(k_3^0 - K_3\right)\delta\left(k_4^0 - K_4\right)$$
$$\left\{\frac{1}{K_2 K_3 K_4} - \frac{1}{K_1 K_3 K_4} - \frac{1}{K_1 K_2 K_4} - \frac{1}{K_1 K_3 K_4}\right\}$$
$$= 0.$$

Since the jet subdiagram J contains only three or four point soft vertices the above discussion, together with induction can be used to provide an independent proof of (4.1). However, it should be noted that it is crucial that the terms containing the $\frac{1}{k^2 - i\epsilon k_0}$ piece after the replacement (4.3) for the uncut propagators not contribute to divergent pieces. For the annihilation process under consideration because the vertices are ordered with energy flowing forward in time from H_1 to H_2, this is guaranteed.

We have thus demonstrated how the largest time equation implies the cancellation of collinear singularities within the cuts of vacuum polarization diagram for single photon annihilation processes. (See [7] for a related discussion.) Since the cancellation like the case of soft divergences, takes place between the cuts of a single uncut diagram by (3.2) or (4.1), the spin and color factors are common to all the cut diagrams taking part in the cancellation process. Thus the non-abelian nature of the theory does not pose any special problems over the abelian one. We will next discuss some scattering processes. Consider first, massive quark potential scattering, where the reduced diagram relevant to the cross section is shown in Fig. 21, where only one gluon jet J is indicated for simplicity. Clearly (4.1) is valid since the uncut diagram has no mass singularity, it being physically impossible for a massive quark jet and massless gluon jet to start at H_1 and come together simultaneously at H_2. Thus, mass divergence in final state jets with ends between H_1 and H_2 cancel. We could now form the K diagram sets as before. In this case, however, we know that since the quarks are massive no finite energy gluon lines can be exchanged between the two "initial state" lines of momentum p in Fig. 21. Consider next the case of two particle inelastic scattering with massless quarks or gluons (for QED electrons) eg. $qq \to qq+$ gluons and massless quark jets. From the power counting in Appendix A we may draw a reduced diagram relevant to the cross section as in Fig. 22. Only two final state, wide angle, jets J between H_1 and H_2 are shown though there can be more. In addition there may be two forward jets J_1 and J_2 connecting the "initial state" lines of momenta p_1 and p_2. From power counting, the logarithmically divergent configurations contain only three or four point soft vertices in each jet also for the inelastic process there is just one hard vertex, i.e. in the cross section H_1 and H_2 are the only hard vertices. Finally, all lines attached to a hard vertex must belong to different jets. In the figure soft lines can attach to the jet lines at three point vertices, however they are omitted since they have been dealt with earlier. The mass divergences in the wide angle jets between H_1 and H_2 all cancel since their situation is exactly that of the cut vacuum polarization diagrams. Thus we are left only with the uncancelled divergence in the forward jets J_1 and J_2 and we may construct the K-diagram set as before. An example is shown in Fig. 23 where we have ignored the jets between H_1 and H_2 since their mass divergences cancel. Now since we do not integrate over the initial states, we cannot ensure the vanishing of the right hand side of (3.3) or (4.1). Cancellation of such mass divergences will in general fail.

Up until now, both for soft and collinear divergences, we have only considered the case of emission of either soft vectors or collinear jets. This is because usually, for a cross-section we consider the collisions of initial particles of well defined momentum. However, the Kinoshita, Lee Nauenberg theorem tells us that if one also sums over degenerate initial states then transition probabilities are completely free of collinear or soft divergences. In our formalism, this would correspond to considering instead of the K-diagrams, the corresponding vacuum diagrams. Now there is no obstacle to applying the largest time equations also to the initial legs and therefore, there will be no distinction between initial and final jets. (The initial legs can also be considered as uncut lines between two hard vertices.) The soft and collinear singularities will cancel here also in the same way as discussed above for the final state jets.

V. Exponentiation of the soft divergences in QED

The exponentiation of the soft divergences in abelian theories is by now a well discussed problem.[6] Nevertheless, we analyze it here to show the overall consistency and the relative straightforwardness of our approach. In the abelian theory, exponentiation follows as a result of the factorization of the infrared divergences. In order to demonstrate these one has to add the contributions from topologically different cut diagrams (vacuum polarization or K-diagrams as the case may be). This rather than the presence of the three and four point couplings is the major obstacle to exponentiation in non-abelian gauge theories. Our starting point for proving the exponentiation are the largest time equation (3.2) or (3.9) for the reduced diagrams at a pinch singular point, and then the identities (3.11), (3.12) and (3.13). In addition we will need some eikonal identities which will be given when needed. These identities may be proven from the method of contour integral. The approach will be to show the exponentiation for the virtual corrections and that for the real emissions as well as the cancellations will follow since, (1) we will pick out the virtual corrections from the cuts of a vacuum polarization diagram as they occur in (3.2) and (2) we will add the contributions from similar cuts from the diagrams obtained by interchanging the photon legs attached to the fermion lines. We begin by discussing some lower order examples:

In the lowest nontrivial order, the two cuts relevant for virtual corrections are from (3.5),

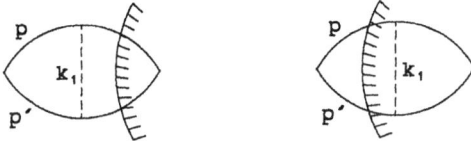

and the contributions are: (writing only the photon integrals)

$$C_0(ie)^2 \int \frac{d^4 k_1}{(2\pi)^4} \pi\delta(k_1^2) \left\{ \frac{1}{2p \cdot k_1 - i\epsilon} \frac{1}{-2p' \cdot k_1 - i\epsilon} + \frac{1}{-2p \cdot k_1 + i\epsilon} \frac{1}{2p' \cdot k_1 + i\epsilon} \right\}$$
$$(-i)2p_\alpha (-i)2p'_\alpha \tag{5.1}$$

The origin of the various terms needs explanation. C_0 includes the core diagram without any corrections, $i.e.$,

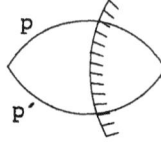

as well as possible momentum dependent effective vertex factors. The photon propagator has been replaced thus, in accordance with (3.10)

$$\frac{1}{i} \frac{1}{k^2 - i\epsilon} \to \pi\delta(k^2).$$

The terms in braces can be obtained from the identity (3.12). The factors $(-i)2p_\alpha(-i)2p'_\alpha$ originates from the soft photon fermion vertices thus: From Lorentz invariance, the soft photon-fermion vertex must be proportional to γ_μ. Next from the fermion propagators we keep only the pieces $(-i\gamma \cdot p + m)$, $(-i\gamma \cdot p' + m)$ in the numerator since by power counting the integral is at most logarithmically divergent and so soft momenta may be neglected in the numerator. Now in general we have terms like

$$(-i\gamma \cdot p + m)\, \gamma_\alpha \, (-i\gamma \cdot p + m) \ldots (-i\gamma \cdot p' + m)\, \gamma_\beta \, (-i\gamma \cdot p' + m) \ldots$$

which at $p^2 = -m^2$ will behave like

$$= (-i\gamma \cdot p + m)(-i)2p_\alpha \ldots (-i)2p'_\beta \, (-i\gamma \cdot p' + m).$$

thus, the complete numerator contributing to the divergent part will be independent of m and of the form shown in (5.1). We get for the divergent part:

$$e^2 D = e^2 \int \frac{d^4 k_1}{(2\pi)^4} \pi\delta(k_1^2) \left[-\frac{2p \cdot p'}{(p \cdot k_1)(p' \cdot k_1)} \right]. \tag{5.2}$$

Let us go to the next order. Henceforth, unless explicitly indicated, we will assume the following integral over all photon momenta $\int \frac{d^4 k}{(2\pi)^4} \pi\delta(k^2)$. We will first show the factorization, consider:

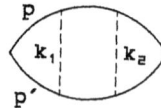

(a)

From the identity (3.12), the terms relevant for the virtual cuts are:

$$\frac{1}{p \cdot (k_1 + k_2) - i\epsilon} \frac{1}{p \cdot k_2 - i\epsilon} \frac{1}{-p' \cdot (k_1 + k_2) - i\epsilon} \frac{1}{-p' \cdot k_2 - i\epsilon} +$$

$$\frac{1}{p\cdot k_1 - i\epsilon}\ \frac{1}{-p\cdot k_2 + i\epsilon}\ \frac{1}{-p'\cdot k_1 - i\epsilon}\ \frac{1}{p'\cdot k_2 + i\epsilon} +$$

$$\frac{1}{-p'\cdot k_1 + i\epsilon}\ \frac{1}{-p\cdot(k_1+k_2) + i\epsilon}\ \frac{1}{p'\cdot k_1 + i\epsilon}\ \frac{1}{p'\cdot(k_1+k_2) + i\epsilon}.$$

Next we add to this the similar contributions from

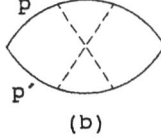

(b)

$$\frac{1}{p\cdot(k_1+k_2) - i\epsilon}\ \frac{1}{p\cdot k_2 - i\epsilon}\ \frac{1}{-p'\cdot(k_1+k_2) - i\epsilon}\ \frac{1}{-p'\cdot k_1 - i\epsilon} +$$

$$\frac{1}{-p\cdot k_1 + i\epsilon}\ \frac{1}{-p\cdot(k_1+k_2) + i\epsilon}\ \frac{1}{p'\cdot k_2 + i\epsilon}\ \frac{1}{p'\cdot(k_1+k_2) + i\epsilon}.$$

Note that here there are only two terms as compared to three in (a). This is because a diagram like

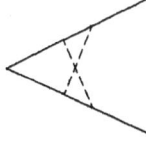

has only an overall divergence and no subdivergence like (a) does. This fact generalizes obviously for all such photon crossings. It is to be noted that this is automatic in our approach as distinct to [6]. Next, using the identity:

$$\frac{1}{\ell\cdot(k_1+k_2) + i\epsilon}\ \frac{1}{\ell\cdot k_2 + i\epsilon} + \frac{1}{\ell\cdot(k_1+k_2) + i\epsilon}\ \frac{1}{\ell\cdot k_1 + i\epsilon} = \frac{1}{\ell\cdot k_1 + i\epsilon}\ \frac{1}{\ell\cdot k_2 + i\epsilon}. \tag{5.3}$$

We get for the sum of terms from (a) and (b)

$$\frac{1}{2}\left(\frac{-2}{p\cdot k_1 p'\cdot k_1}\right)\left(\frac{-2}{p\cdot k_2 p'\cdot k_2}\right) \tag{5.4}$$

and for the complete contribution to the sum of diagrams (a) and (b)

$$C_0\frac{1}{2}\left(e^2\int\frac{d^4k_1}{(2\pi)^4}\pi\delta(k_1^2)\left[-\frac{2p\cdot p'}{p\cdot k_1 p'\cdot k_1}\right]\right)\left(e^2\int\frac{d^4k_2}{(2\pi)^4}\pi\delta(k_2^2)\left[-\frac{2p\cdot p'}{p\cdot k_2 p'\cdot k_2}\right]\right)$$

$$= C_0\frac{1}{2}\left(e^2D\right)^2. \tag{5.5}$$

Notice that what guarantees the factorization (5.4) and hence the exponential form (5.5) is the identity (5.3) which generalizes thus for n soft photons:

$$\sum_{\substack{\text{perms}\\1,\ldots,n}} \frac{1}{\ell\cdot k_1 - i\epsilon} \frac{1}{\ell\cdot(k_1+k_2)-i\epsilon} \cdots \frac{1}{\ell\cdot(k_1+\ldots k_n)-i\epsilon}$$

$$= \frac{1}{\ell\cdot k_1 - i\epsilon}\frac{1}{\ell\cdot k_2 - i\epsilon}\cdots\frac{1}{\ell\cdot k_n - i\epsilon}. \tag{5.6}$$

In the next order, we need to consider the various virtual cuts from the sum of the following diagrams:

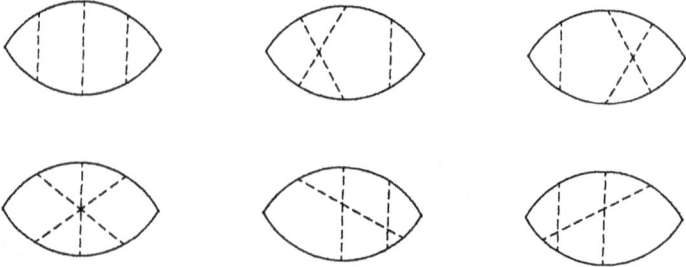

In this and in the higher order cases, one should symmetrize also in the attachments to the fermion line of momentum p. This is the origin of the $\frac{1}{3!}$ below. Using the identity (5.6) for $n = 3$, we now get:

$$C_0 \cdot \frac{1}{6}\prod_{i=1}^{3} e^2 \int \frac{d^4 k_i}{(2\pi)^4}\pi\delta(k_i^2)\left[-\frac{2p\cdot p'}{p\cdot k_i p'\cdot k_i}\right]$$

$$= C_0\frac{1}{3!}\left(e^2 D\right)^3.$$

It is instructive to consider also an example with self-energy correction.

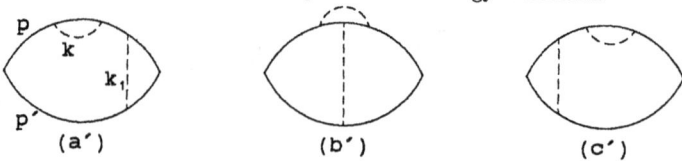

(a′)　　　　　　　(b′)　　　　　　　(c′)

The various terms coming from the cuts of these diagrams may be written down as before. For the diagrams $(a'),(c')$ it becomes necessary, as discussed in section 3, to introduce the fake momentum δ and to combine the $\frac{1}{p\cdot\delta}$ terms by means of the identity (3.13) which for this case reads:

$$\frac{1}{p\cdot\delta}\frac{1}{-p\cdot k + i\epsilon} + \frac{1}{-p\cdot\delta}\frac{1}{-p\cdot(k+\delta)+i\epsilon} = (-1)\frac{1}{-p\cdot k+i\epsilon}\frac{1}{-p\cdot(k+\delta)+i\epsilon}.$$

The remaining term each from (a') and (c') combine with the two virtual cuts of (b') to give a vanishing contribution. For this purpose the following identity is useful.

$$\frac{1}{p \cdot k + p \cdot k_1 - i\epsilon} \frac{1}{p \cdot k_1 - i\epsilon} + \frac{1}{p \cdot k + p \cdot k_1 - i\epsilon} \frac{1}{p \cdot k - i\epsilon}$$
$$= \frac{1}{p \cdot k - i\epsilon} \frac{1}{p \cdot k_1 - i\epsilon}$$

Again this falls out automatically in our approach and is a reflection of the zero momentum Ward Identity:

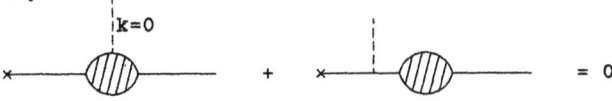

$$= 0$$

The generalization of this for one self energy and n other soft photons which is relevant for similar cancellations in higher order is:

$$\sum_{m=0}^{n} \prod_{j=0}^{m} \frac{1}{p \cdot k + p \cdot q_n - p \cdot q_j + i\epsilon} \prod_{j=m}^{n-1} \frac{1}{p \cdot q_n - p \cdot q_j + i\epsilon}$$
$$\frac{1}{p \cdot k + i\epsilon} \prod_{j=0}^{n=1} \frac{1}{p \cdot q_n - p \cdot q_j + i\epsilon}$$

where the q_i are defined as in section III.

We thus get for the various factors in the sum of diagrams $(a'), (b'), (c')$

$$\frac{1}{2!} 2 \frac{1}{p \cdot k \cdot p \cdot k} \left(\frac{-2}{p \cdot k_1 p' \cdot k_1} \right)$$

and for the total contribution

$$\frac{1}{2!} C_0 2 \left(e^2 \int \frac{d^4 k}{(2\pi)^4} \pi \delta(k^2) \frac{p^2}{p \cdot k \, p \cdot k} \right) \left(e^2 \int \frac{d^4 k_1}{(2\pi)^4} \pi \delta(k_1^2) \left[-\frac{2p \cdot p'}{p \cdot k_1 p' \cdot k_1} \right] \right).$$

Similar terms will come for self energy insertions on the p' line. It is now clear that the common infrared divergent factor taking into account self energy insertions as well is the straightforward generalization of (5.2), $i.e.$

$$e^2 B = e^2 \int \frac{d^4 k}{(2\pi)^4} \pi \delta(k^2) \left(\frac{p}{p \cdot k} - \frac{p'}{p' \cdot k} \right)^2.$$

It is also clearly indicated above that each step necessary for factorization to all orders has a well defined generalization. Thus, as can be verified by induction, we have proved for the annihilation cross section, the exponentiation rule:

$$C_0 \sum_{n=0}^{\infty} \frac{1}{n!} \left(e^2 B\right)^n = C_0 \, e^{\left(e^2 B\right)}.$$

This, of course, also shows exponentiation for form factors[20] and electron potential scattering, Compton scattering and so on. It is also clear from the above that a certain class of QCD soft divergences can be also shown to exponentiate. The advantage of our discussion above is that since our procedure works equally well for non-abelian theories, we are able to look at such classes rather simply. This, however, will not be discussed here.

In concluding this section we note again that it is the sum of diagrams like $(a), (b)$ or $(a'), (b'), (c')$ that factorizes hence exponentiate. In section (IIIb) we mentioned that for processes like $e^+ e^- \rightarrow \gamma^* +$ soft photons for which the K-diagram sets are identical to the case above, there is noncancellation of soft divergences within a topologically identical K-diagram set. This was basically because when the replacements (3.10) are made, the terms in brackets do not give zero. However, in QED, when we make a sum like $(a), (b)$ or $(a'), (b'), (c')$ then in this sum these extra terms cancel and exponentiation goes through even for such processes. This can be easily checked for the example mentioned. Finally, it should be mentioned, that even though we have shown factorization and exponentiation within the eikonal approximation, the above arguments are not at all limited to this. The exact identity (3.11) could serve just as well in showing exponentiation, though we do not discuss it here further.

VI. Conclusions

In this paper, we have studied the infrared divergence problem, in particular, the question of the cancellation of the soft and collinear singularities, from the viewpoint of the largest time equation. First, from the physical picture at the pinch singular point, the reduced diagram was determined, to which the largest time equation was applied. The particular form that ensured the cancellations (whenever it occurs) was crucially dependent on the fact that at the pinch singular point, energy flows in the same direction in every line of the diagram. Even though the reduced diagram structure together with the physical picture from which it follows, was an input independent of the use of the largest time equation, we feel that in fact the former follows, in some aspects, from the latter. This is because, in "deriving" the physical picture, Landau equations and the singularity structure of Feynman diagrams was used, which can also be shown to give rise to the cutting equations. This question deserves further study.

Next we summarize our results concerning the cancellation of the soft and collinear singularities. First the soft divergences. Here for all interactions involving the final state jet lines, the required form of the largest time equation may be directly applied to show cancellation. However, for soft emissions involving the initial state lines, in general for scattering processes, there will be a non-cancellation of infrared divergences within a K-diagram set. Concerning mass singularities, the situation is similar. Those mass singularities that arise from final state lines in the transition amplitude cancel in the transition probability. The only mass divergences present in the latter arise from self energy and vertex parts of the initial state lines as represented in the forward jets, or when zero mass particles are exchanged between initial and final states. If however, one takes into account also the initial state degeneracies then the situation is described by the Kinoshita and Lee-Nauenberg theorems when all infrared singularities cancel. Here our procedure, as in the other cases discussed in detail in this article, can be used to rather straightforwardly identify the Lee-Nauenberg cancelling sets.

The following general result which can be distilled from the discussions of this paper can be very simply put. Concerning the cancellation of infrared divergences (soft and collinear) whenever there is an uncut jet line (color and spin summed) between two adjacent hard vertices with an ordered sequence of soft vertices in between then the largest time approach can be used to show their cancellation in a sum over the cuts when we go around each soft vertex. The explicit identities that implement this are given in (3.9) for soft divergences and (4.1) for collinear ones. For annihilation and two body scattering processes discussed above the whole diagram or "half" of it can be represented thus.

In all of our discussions in this paper, we have treated the soft and collinear divergences separately. In as much as they arise from different regions in momentum space this is justified. The collinear divergences arise when the lines in a jet become parallel to one another. Consider for example the scattering processes. To separately handle the soft and collinear divergences we may restrict ourselves to the following momentum space regions: For the soft lines in the Green's function S, we do not want collinear divergences where the $\vec{k}_i$ can become parallel to the $\vec{p}_i$ or the momenta of the wide angle jets. This can be accomplished by restricting $\vec{k}_i$ such that

$$\left(1 - \frac{\vec{p}_i \cdot \vec{k}_j}{|\vec{p}_i| \cdot |\vec{k}_j|}\right)$$

is larger than $\mathcal{O}\left(|\vec{k}_j|\right)$. Similarly for $\vec{q}_i$, a typical finite energy momenta in the forward jets, we do not want soft divergences to arise. This can be accomplished by restricting them such that

$$\left(1 - \frac{\vec{p}_i \cdot \vec{q}_j}{|\vec{p}_i| \cdot |\vec{q}_j|}\right)$$

is less than $\mathcal{O}(|\vec{q}_j|)$. These restrictions on the spatial momenta does not invalidate our use of the largest time equation to ensure the cancellation of divergences. This is because as explained in section IIb, all the relevant equations are valid even when the spatial momenta are held at fixed values in momentum space.

Finally, we would like to note that the problem of exponentiation of the soft divergences in non-abelian gauge theories needs a thorough analysis to determine which class can be exponentiated and which not.

Acknowledgements

I am indebted to Prof. M. Veltman for his suggestions, and for making available his graphics program DIAGRAM. I am grateful to Dr. H. Veltman for an initial collaboration and for contributions to this work. I would like to thank Prof. G. Sterman for discussions and comments. This article relies heavily on his fundamental work on infrared divergences. This work was supported in part by the U.S. Department of Energy.

APPENDIX A

Power counting for infrared divergences

The power counting procedure developed here is based on the physical picture discussed in sec. IIa. The method is that from ref. [7], [9] and [10].

We will first discuss power counting for an energy ordered reduced diagram, *i.e.*, one in which energy flows in the same direction for every line in the diagram. Such a class of diagrams includes the reduced diagrams at a pinch singular point, but also others. This is done to treat annihilation and scattering processes all together as far as possible. Next by considering these processes separately, at the pinch singular point, we show that the infrared divergence (both collinear and soft ones) are at most logarithmic and obtain restructions on the vertex and jet structures of the respective reduced diagrams. Throughout we will be considering only a completely massless theory.

Let us consider a cut energy ordered vacuum polarization diagram or a forward scattering diagram at a singular point which is not necessarily pinched. If the momenta of on shell lines are real then the reduced diagram still consists of jets of parallel moving lines and soft lines, however the essential difference with the same diagram at the pinch singular point is that in the former case there are in general more hard vertices. This is because only at pinch singular points does a reduced diagram represent a physically realizable process. For such a diagram, we consider the soft and jet loops separately: All the lines making up a jet loop are finite energy whereas in soft loops at least one line should be a zero momentum one. Power counting is done in terms of the "normal" variables, which are the minimum set of variables necessary to put all lines on shell. For zero-momentum lines, these are just the four components of the loop momenta. For the finite energy lines, all such lines in a jet must be on shell and parallel to each other. For the normal variables associated with the internal jet loop momenta we may take to be two, however for external jet momenta, they may not be all independent [9], [10], so their number is not fixed yet. The infrared degree of divergence of such a reduced diagram with J finite energy lines, S zero-momentum lines, $L^{(s)}$ soft and $L^{(j)}$ jet loops is given by:

$$p = 4L^{(s)} + 2L^{(j)} + D + E - (2S + J) + N. \qquad (A.1)$$

(For infrared divergence $p \leq 0$.)

E is the number of external lines of the reduced graph.

N is the contribution from numerator factors and D is a measure of the contributions from the independent normal variables associated with the external jet momenta. Separating out the contributions from each jet (K is total number of jets) and using standard dimensional arguments for the soft lines we get:

$$p = \sum_{i=1}^{K} \left(2\ell_i^j - j_i + n_i^j \right) + D + E + b + \frac{3}{2}f \qquad (A.2)$$

where, b and f are respectively the number of soft boson and fermion lines.

We now consider the jet subdiagrams. Let us use the following notation for each jet i.

γ_i = total number of finite energy external lines
V_i = total number of vertices (including hard ones)
W_i = number of times only one finite energy external line is attached to a hard vertex
K_i = number of times more than one finite energy external line is attached to a hard vertex
$x_{i\alpha}$ = number of soft vertices with α jet lines
$y_{i\beta}$ = number of soft vertices with β jet lines and one or more soft lines.

Then,

$$j_i = \frac{1}{2}\left(\sum_{\alpha \geq 3} \alpha x_{i\alpha} + \sum_{\beta \geq 2} \beta y_{i\beta} + \gamma_i\right) \tag{A.3}$$

$$V_i = \sum_{\alpha \geq 3} x_{i\alpha} + \sum_{\beta \geq 2} y_{i\beta} + (W_i + K_i) \tag{A.4}$$

Further, $\ell_i = j_i - V_i + \Delta_i$, with $\Delta_i = 1$ if the jet is completely connected; otherwise $\Delta_i = $ number of disconnected pieces of the jet. Substituting in (A.2) we get:

$$p = \frac{1}{2}\sum_{i=1}^{K}\left(\sum_{\alpha \geq 3}(\alpha - 4)x_{i\alpha} + \sum_{\beta \geq 2}(\beta - 4)y_{i\beta} + 4\Delta_i - 3\gamma_i\right.$$
$$\left. + 4(\gamma_i - W_i - K_i) + 2n_i^j\right) + D + E + b + \frac{3}{2}f$$

We may show that in non-gauge theories:

$$n_i^j \geq \frac{1}{2}x_{i3} + \frac{1}{2}Z_i^{(0)}$$

where, $Z_i^{(0)}$ is the number of soft scalars emitted at three point vertices. In gauge theories, the above is still valid on a graph by graph basis only in certain noncovariant gauges where the longitudinal degrees of freedom are irrelevant. In any case, the unphysical degrees of freedom in the gauge particles are cancelled by some other mechanism, having nothing to do with infrared divergences. Crucial for this is the Ward Identity and so we have to ensure that a symmetry preserving regularization procedure is used. The axial gauge is an example where our power counting is valid on a graph by graph basis. We can further show the following inequalities.

$$(b + f) \geq \sum_{i=1}^{K}(y_{i3} + y_{i2})$$

$$f \geq \sum_{i=1}^{K} \left(y_{i2} - Z_i^{(0)} - Z_i^{(1)} \right)$$

where, $Z_i^{(1)}$ is the number of soft vectors emitted in the ith jet at three point vertices. Putting this together, we obtain

$$p \geq \frac{1}{2} \sum_{i=1}^{K} \left(\sum_{\alpha \geq 4} (\alpha - 4) x_{i\alpha} + \sum_{\beta \geq 4} (\beta - 4) y_{i\beta} \right.$$
$$\left. + 4\Delta_i - 3\gamma_i + 4 \left(\gamma_i - W_i - K_i \right) \right) + D + E$$
$$+ \frac{1}{2} \sum_{i=1}^{K} \left(b_i^{(0)} + b_i^{(1)} + Z_i^{(1)} \right) + \frac{1}{2} f \qquad (A.5)$$

Here, $b_i^{(0)}, b_i^{(1)}$ are respectively the total number of soft scalars and vectors.

From (A.5) we see that the most singular configuration corresponds to $\Delta_i = 1$, $\gamma_i = W_i$ and $K_i = 0$; $i.e.$, there are no disconnected jets and all lines that meet at a given hard vertex belong to different jets. In this case we have:

$$p \geq \frac{1}{2} \sum_{i=1}^{K} \left(\sum_{\alpha \geq 4} (\alpha - 4) x_{i\alpha} + \sum_{\beta \geq 4} (\beta - 4) y_{i\beta} - 3\gamma_i + 4 \right)$$
$$+ D + E + \frac{1}{2} \sum_{i=1}^{K} \left(b_i^{(0)} + b_i^{(1)} + Z_i^{(1)} \right) + \frac{1}{2} f \qquad (A.6)$$

Here, γ_i is the number of external finite momenta of the ith jet, each of which is carried by a single line ($\gamma_i \geq 2$). The number of independent finite momenta is $\gamma_i - 1$ and therefore the number of normal variables associated with it is $2(\gamma_i - 1) - 1$. (The -1 because of the overall freedom of choice of direction.) The problem is that in general they may not be all independent and only the latter are relevant for power counting.

Case 1: Cut vacuum polarization diagrams at the pinch singular point. Here $E = 0$, $i.e.$, there are no external lines of the reduced graph as a whole (we mean here the external lines of the vacuum polarization graph). Further, as discussed in sec. IIa, the physical picture at the pinch singular point tells us that $D = K$, the total number of jets, and there is only one hard vertex on either side of the cut $i.e.$, in (A.6) we can write

$$p \geq \frac{1}{2} \sum_{i=1}^{K} \left(\sum_{\alpha \geq 4} (\alpha - 4) x_{i\alpha} + \sum_{\beta \geq 4} (\beta - 4) y_{i\beta} \right)$$
$$+ \frac{1}{2} \sum_{i=1}^{K} \left(b_i^{(0)} + b_i^{(1)} - Z_i^{(1)} \right) + \frac{1}{2} f$$

$$+ \left\{ K - \sum_{i=1}^{K} \left[\frac{3}{2}(\gamma_i - 2) + 1 \right] \right\} \qquad (A.7)$$

and the terms in brackets vanishes since $\gamma_i = 2$. Thus, for cut vacuum polarization diagrams we may write down the following conditions for infrared divergence: The worst case of logarithmic divergence happens when the reduced diagram at the pinch singular point has only two hard vertices one on each side of the cut, there are no disconnected jets, each jet is attached to a hard vertex on either side by a single line and further:

$$b_i^{(0)} = 0$$
$$b_i^{(1)} = Z_i^{(1)}$$
$$f = 0$$
$$x_{i\alpha} = y_{i\beta} = 0, \qquad \alpha, \beta > 4.$$

Thus, soft vectors attach to jet lines only at three point vertices and is far as finite energy lines go, no soft vertex can be high order than four. Notice that this implies that each jet must constitute a self energy type diagram and momenta of all soft lines must simultaneously vanish. A typical reduced diagram is show in Fig. 21.

Case 2: Two particle inelastic processes or cut forward scattering diagrams at the pinch singular point. Here $E = 4$, which is the scattering diagram. Referring to equation (A.6), one can see that in this case the degree of divergence is bounded from below by

$$D + 4 - \sum_{i=1}^{K} \left[\frac{3}{2}(\gamma_i - 2) + 1 \right].$$

It has been shown in references [9] and [10] that $(D+4) = \sum_{i=1}^{K} \left[\frac{3}{2}(\gamma_i - 2) + 1 \right]$ and there is a logarithmic divergence only when $\gamma_i = 2$. Otherwise $(D+4) > \sum_{i=1}^{K} \left[\frac{3}{2}(\gamma_i - 2) + 1 \right]$ and there is no divergence. Thus once again there is only one hard vertex on each side of the cut in the forward scattering amplitude. In this case, however, the physical picture allows for two forward jets J_1 and J_2 attached to the initial state lines. A typical reduced diagram is as shown in Fig. (23) but together with the inclusion of soft vectors which couple to jets at three point vertices. There are any number of wide angle jets that can be produced at the initial hard vertex H_1. From (A.6) and the above discussion once again we see that there are only soft vectors which attach to jet lines at three point vertices and further, soft vertices consisting of finite energy lines can be at most of order 4. This together with the fact that in the logarithmically divergent configuration, jet lines are attached to hard vertices by a single line forces the jet structure in the reduced diagram to be self-energy like.

REFERENCES

1. M. Veltman, Physica 29, 186 (1963).

2. M. Veltman, Nucl. Phys. B7, 637 (1968).

3. G. 't Hooft, M. Veltman, Nucl. Phys. B44, 189 (1972).

4. T. Kinoshita, J. Math. Phys. 3, 680 (1962).

5. N. Nakanishi, Prog. Theo. Phys. 19, 159 (1958).

6. D. Yennie, S. Frautschi, H. Suura, Ann. Phys. 13, 379 (1961);
   G. Grammer, D. Yennie, Phys. Rev. D8, 4332 (1978).

7. G. Sterman, Phys. Rev. D17, 2773 (1978);
   ibid., D17, 2789 (1978).

8. S. Coleman, R. Norton, Nuov. Cim. 38, 438 (1965).

9. S. Libby, G. Sterman, Phys. Rev. D18, 4737 (1978);
   ibid., D18, 3252 (1978).

10. R. Akhoury, Phys. Rev. D19, 1250 (1979).

11. T.D. Lee, M. Nauenberg, Phys. Rev. 133, B1549 (1963).

12. C. Bloch, A. Nordsieck, Phys. Rev. 52, 54 (1937).

13. M. Veltman, Talk at Marseille Colloquium on Advanced Computing Methods in Theoretical Physics June 21-25, 1971.

14. G. 't Hooft, M. Veltman, Diagrammer, CERN Yellow report 73-9 (1973).

15. S. Libby, G. Sterman, Phys. Rev. D19, 2468 (1979);
    G. Sterman, V. Ganapathi, Phys. Rev. D23, 2408 (1981).

16. R. Doria, J. Frenkel, J.C. Taylor, Nucl. Phys. B168, 93 (1980).

17. I. Bialynicki-Birula, Phys. Rev. D2, 2877 (1970).

18. E. Poggio, H. Quinn, Phys. Rev. D14, 578 (1976);
    G. Sterman, ibid., D14, 2123 (1976).

19. E. Remiddi, Helvetica Acta 54, 364 (1981).

20. E. de Rafael, C. Korthal-Altes, Nucl. Phys. B106, 237 (1976).

46

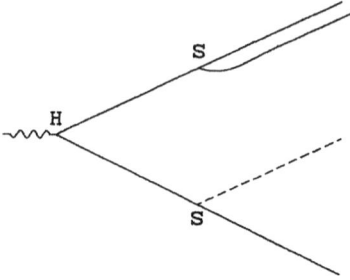

Fig. 1

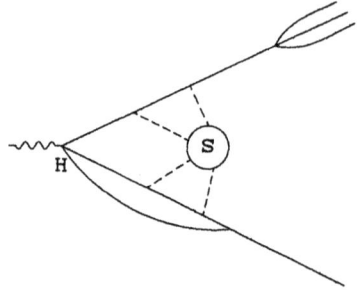

Fig. 2

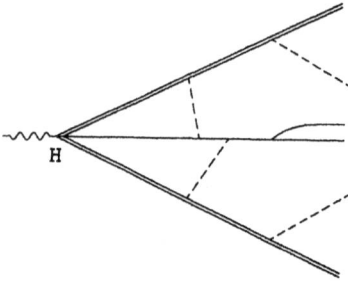

Fig. 3

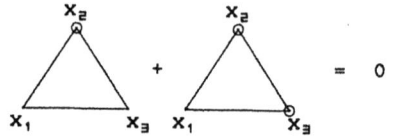

Fig. 4

$$\frac{1}{(2\pi)^4 i} \frac{1}{k^2 + m^2 - i\epsilon}$$

$$\frac{1}{(2\pi)^3} \theta(k_0) \delta(k^2 + m^2)$$

$$\frac{-1}{(2\pi)^4 i} \frac{1}{k^2 + m^2 + i\epsilon}$$

uncircled vertex: $(2\pi)^4 ig$

circled vertex: $-(2\pi)^4 ig$

Fig. 5

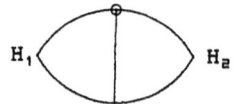

Fig. 6

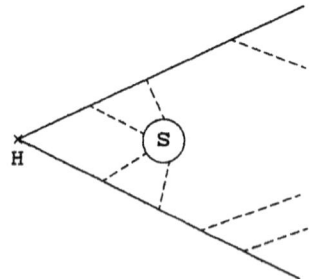

Fig. 7

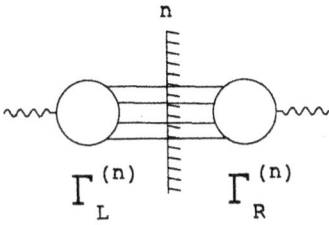

Fig. 8

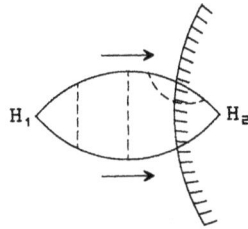

Fig. 9

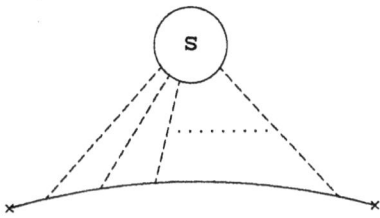

Fig. 10

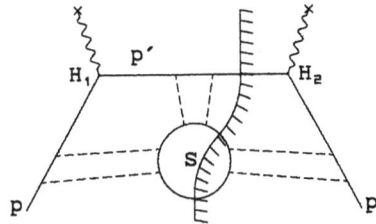

Fig. 11

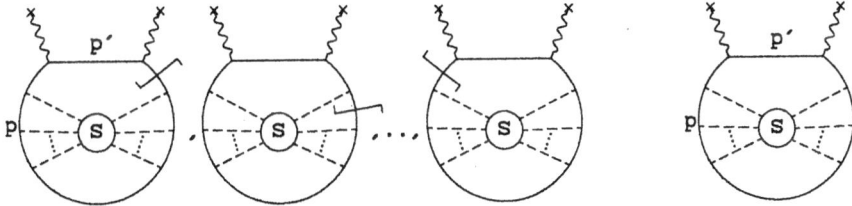

Fig. 12a

Fig. 12b

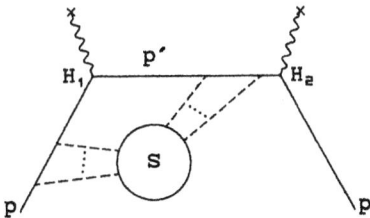

Fig. 13

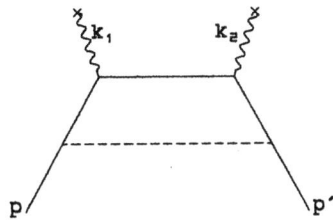

Fig. 14

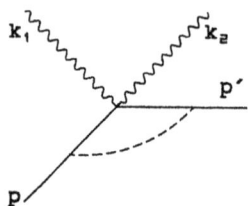

Fig. 15

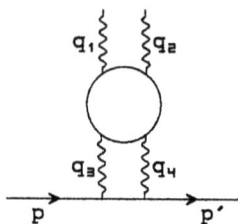

Fig. 16

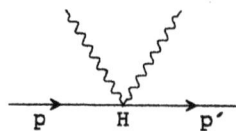

Fig. 17

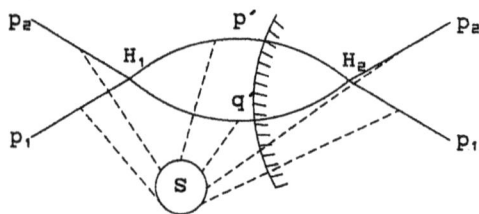

Fig. 18

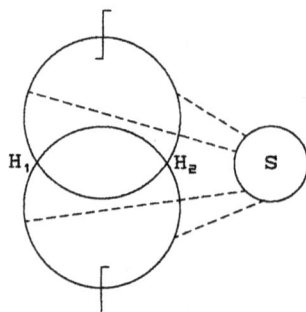

Fig. 19

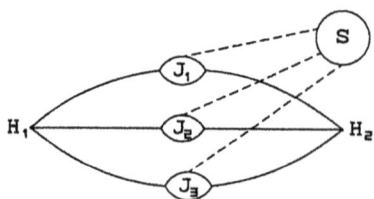

Fig. 20

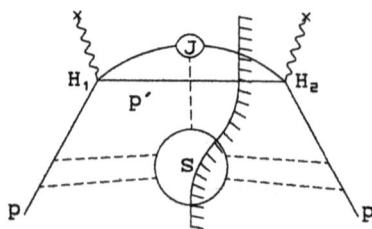

Fig. 21

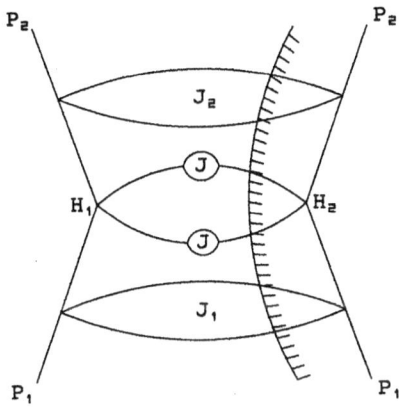

Fig. 22

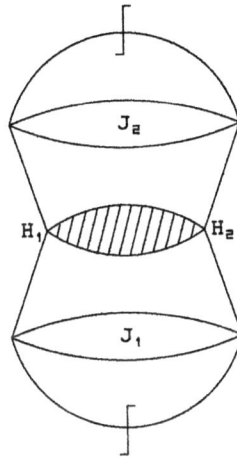

Fig. 23

Top Quark Condensate Models

WILLIAM A. BARDEEN
Fermi National Accelerator Laboratory
P.O. Box 500, Batavia, IL 60510

ABSTRACT

Gauge field theories with spontaneous symmetry breaking have been remarkably successful in providing a fundamental description of the electroweak interactions. Despite this success little is known about the mechanisms of electroweak symmetry breaking. This paper examines the possible role of heavy fermions, particularly the top quark, in generating the observed electroweak symmetry breaking, the masses of the W and Z bosons and the masses of all observed quarks and leptons.

1. Introduction.

Identifying the physical mechanism for the spontaneous breaking of the electroweak gauge symmetries has been a central challenge of elementary particle physics. During the present decade we should begin to establish the experimental evidence for the specific realization chosen by nature. Theoretical speculations have covered a wide range. The Standard Model presumes the existence of elementary scalar fields which condense to produce the observed symmetry breaking and result in a single physical Higgs particle whose mass is not determined. Alternative proposals have involved more complex Higgs sectors with multiple physical Higgs scalars, supersymmetric models with many additional physical states, or dynamical symmetry breaking models, such as technicolor, which rely on condensates of new technifermions and an entirely new sector of strong dynamics. The specific models are motivated by a variety of physical issues including renormalization, strong CP, natural gauge hierarchies, supersymmetry and superstrings. These models generally require the introduction of many new fundamental particles and their interactions.

This paper will focus on a different alternative which relies only on the presently observed particles and a heavy top quark to generate the electroweak symmetry breaking. Our work is motivated by Y. Nambu [1] who suggested that short range, attractive interactions between the fermions could generate the symmetry breaking, in analogy to the BCS theory of superconductivity. We will explore the physical consequences of these ideas. Fermion condensates of a heavy top quark will play a central role, and the expected masses for the top quark and physical Higgs particle, a top quark - antitop quark bound state, are principal predictions of the model. We will emphasize the reliability of these predictions for the minimal model and examine possible extensions to more complex situations.

The presentation of this paper follows the results of Bardeen, Hill and Lindner (BHL) [2]. The role of top quark condensates in electroweak symmetry breaking was also the central theme of the top mode model of Miransky, Tanabashi and Yamawaki (MTY) [3]. The more general theme of composite gauge vector bosons and composite Higgs bosons has been a recurrent theme in the theoretical literature [4].

After a brief discussion of the elements of the original BCS mechanism and its relativistic NJL generalization, the dynamical basis of the top condensate model is analyzed by comparing three different approaches to understanding the essential elements of the minimal version of the theory. In the following section, various critical elements of the theory are discussed along with possible generalizations of the minimal model.

2. The BCS and NJL Models.

The electroweak interactions are generated by gauge interactions where the electroweak symmetries are spontaneously broken by the vacuum structure. All of the masses of the observed elementary particles, from gauge bosons to quarks and leptons, are generated by this symmetry breaking. The Standard Model relies on a fundamental Higgs field to provide the symmetry breaking. Dynamical symmetry breaking replaces the elementary Higgs field by condensates of the more fundamental degrees of freedom. Dynamical symmetry breaking forms the basis of the BCS theory of superconductivity [5].

2.1. The BCS Theory.

In the BCS theory of superconductivity, the complex interactions between the electrons result in a residual local, attractive interaction between the electrons. This attractive interaction can cause the electrons to bind into Cooper pairs. Dynamical symmetry breaking occurs when the energy of a Cooper pair becomes negative, and the normal vacuum becomes unstable to spontaneous creation of electron pairs. The vacuum structure is modified, and a new BCS ground state forms with a condensate of electron pairs, $\langle e\uparrow e\downarrow \rangle \neq 0$. A gap forms at the fermi surface, the electron becomes "massive", and the Meisner effect excluding the magnetic field from a superconductor reflects the dynamical mass generated for the magnetic part of the photon interactions. It is clear that the basic elements of the BCS theory could provide the framework for dynamical symmetry breaking in the electroweak interactions as suggested by Nambu [1].

2.2. The NJL Model.

The attractive, local interaction which induced the dynamical symmetry breaking was given a relativistic generalization through the Nambu - Jona-Lasinio (NJL) model [6]. This model considers the effects of a local, chiral invariant interaction between the fermions of the theory. The model is described by the following Lagrangian,

$$L = \overline{\Psi} \, (i\partial) \, \Psi + G \, (\overline{\Psi}_L \Psi_R) \cdot (\overline{\Psi}_R \Psi_L) \tag{1}$$

where a cutoff, Λ, must be introduced to define the quantum theory. The interaction is attractive for $G > 0$.

If the interactions are not sufficiently attractive, $G < G_c$, the vacuum structure will choose the symmetric phase for the chiral symmetries. The fermions will remain massless, $m_f = 0$, and the fermions will not form chiral condensates, $\langle \overline{\Psi}_L \Psi_R \rangle = 0$.

For sufficiently attractive couplings, $G > G_c$, the four fermion interactions will induce a dynamical symmetry breaking. In the broken vacuum, the fermions will develop masses and chiral condensates which imply a gap equation the fermion masses, $m_f = - G*$

$\langle \overline{\Psi}_L \Psi_R \rangle \neq 0$. The symmetry breaking implies the existence of Nambu-Goldstone bosons. A scalar boundstate of the massive fermions is formed with the usual NJL relation, $m_S = 2 \, m_f$.

The NJL model is usually "solved" by using a bubble approximation for the dynamics. Fermions can develop dynamical masses, but all vertex corrections are surpressed. This approximation corresponds to the use of the BCS wavefunction in superconductivity. The gap equation follows from the mass relation, $m_f = - G^* \langle \overline{\Psi}_L \Psi_R \rangle \neq 0$, where the condensate is computed using an internal free fermion loop with the dynamically generated mass, m_f, and the cutoff, Λ. In bubble approximation, the fermion - anti-fermion scattering amplitude is computed as a sum of diagrams where the fermion bubbles are iterated in the direct channel. Because of the sensitivity to the cutoff, the bubble contributions to the condensate must be computed consistently with those in the scattering amplitudes, otherwise the cutoff will introduce an explicit breaking of chiral symmetries. A direct calculation using the bubble approximation confirms the existence of the appropriate Goldstone bosons and the NJL prediction of the scalar meson boundstate mass. The NJL model has been previously invoked for generating composite structures [4] and provides the fundamental basis for models involving condensates of the top quark.

3. Top Quark Condensate Models.

In the Standard Model, the electroweak symmetries are spontaneously broken by condensates of an elementary scalar Higgs field. As the allowed mass for the top quark has systematically increased, it is natural to speculate on a possible connection between a large top quark mass and source of electroweak symmetry breaking. Top quark condensate models carry this idea to the extreme. The Higgs sector of the Standard Model is totally eliminated in favor of local, attractive interactions between the fermions of the theory which will induce the electroweak symmetry breaking as in the NJL model. Because of its large mass, the top quark plays the central dynamical role. In the minimal model, electroweak symmetry breaking follows from top quark condensate alone.

The analysis of this section follows that of Bardeen, Hill and Lindner (BHL) [2] which was motivated directly by the suggestions of Nambu [1]. The idea of top quark condensates as the mechansim of electroweak symmetry breaking was also advocated by Miransky, Tanabashi and Yamawaki (MTY) [3]. MTY use a somewhat different dynamical basis for their analysis of the four-fermion interactions than that presented in the BHL paper and reach somewhat different results.

As mentioned above, the Higgs sector of the Standard Model is replaced by local, attractive interactions of the fundamental fermion fields. In the minimal model, the top quark plays an essential role in generating the electroweak symmetry breaking and the masses for the physical W and Z gauge bosons. The minimal model is described by the Lagrangian,

$$L_{fermion} = L_{kinetic} + G \, (\overline{\Psi}_{La}{}^A \, t_R{}^a) \cdot (\overline{t}_{Rb} \, \Psi_{LA}{}^b) \qquad (2)$$

where the composite operators are defined using a cutoff but preserving the electroweak gauge symmetries. $L_{kinetic}$ contains the kinetic terms for the fermions with the usual gauge couplings of the Standard Model. The four-fermion interactions in Eq.(2) represent the residual attractive interactions generated by a more fundamental dynamics existing above the cutoff scale. The four-fermion theory is not renormalizable, and physical quantities will be expected to depend on the cutoff scale even after renormalization of the independent coupling constants. Additional four-fermion interactions with weaker couplings could be added to generate masses for the lighter quarks and leptons but will have little effect on the dynmical symmetry breaking.

If these interactions are sufficiently attractive, $G > G_c$, then the electroweak symmetries will be spontaneously broken generating a mass for the top quark, $m_{top} > 0$, and a nontrivial top quark condensate, $\langle \bar{t} t \rangle \neq 0$. This symmetry breaking will also induce masses for the electroweak gauge bosons, $m_W \neq 0$ and $m_Z \neq 0$. We will also find that the physical Higgs particle of the Standard Model will be formed as a top quark - antitop quark bound state.

3.1. Bubble Approximation (NJL).

The standard method for analyzing the Nambu - Jona-Lasinio model (NJL) makes use of the bubble approximation. This method can be used as a first appoximation to the top quark condensate model as it contains the basic features of the composite structure produced by the dynamical structure of the theory. Phenomenological predictions will require the more complete analysis given in subsequent sections. The bubble approximation can be viewed as the large N_c limit of the theory where N_c is the number of colors, and $G*N_c$ is held fixed but all gauge couplings are neglected. The bubble theory has an exact solution in leading N_c approximation and $1/N_c$ corrections can be systematically computed.

The top quark mass is determined by the appropriate solution of the gap equation,

$$m_t = -(1/2) \ G \ \langle \bar{t} t \rangle$$

$$= G \ (N_c/8\pi^2) \ \{ \ \Lambda^2 - m_t^2 \ \log \ (\Lambda^2/m_t^2) \ \} \ m_t \tag{3}$$

with solutions,

$$m_t = 0$$

or

$$m_t^2 \ \log \ (\Lambda^2/m_t^2) = \Lambda^2 - 8\pi^2/(N_c*G) \tag{4}$$

with the massive solution being the preferred vacuum solution. If the top quark mass is to be much below the cutoff, $m_{top} \ll \Lambda$, then a fine tuning of the four-fermion coupling, G, is required to cancel the quadratic cutoff dependence. Dynamical symmetry breaking can only occur if $G > G_c = (8\pi^2/N_c) \ (1/\Lambda^2)$.

In the broken symmetry phase, the vector bosons become massive through the effects of the vacuum polarization diagrams including the contributions of the fermion bubbles which are needed to preserve the transversality consistent with the underlying electroweak gauge symmetry. The inverse W-boson propagator is given by

$$(1/g_2{}^2)D_W{}^{-1\mu\nu}(k) = (k^\mu k^\nu/k^2 - g^{\mu\nu}) \{ (1/g_2{}^2) k^2 - f_W{}^2(k) \} \qquad (5)$$

where the effective decay constant, f_W, is

$$f_W{}^2(k) = N_c (1/16\pi^2) \int_0^1 dx \, x \, m_t{}^2 \, \log(\Lambda^2/(xm_t{}^2 - x(1-x)k^2)) \qquad (6)$$

and $m_W{}^2 = g_2{}^2 f_W{}^2(m_W)$. The Z boson mass can also be computed in terms of its effective decay constant.

The computed values of the effective decay constants can be determined by the observed Fermi constant

$$W: \quad f_W{}^2 = (N_c/32\pi^2) \, m_t{}^2 \{ \log(\Lambda^2/m_t{}^2) + 1/2 \} = 1/4\sqrt{2} \, G_F$$

$$(7)$$

$$Z: \quad f_Z{}^2 = (N_c/32\pi^2) \, m_t{}^2 \{ \log(\Lambda^2/m_t{}^2) \} \approx f_W{}^2. \cdot$$

For a large cutoff, $\Lambda \approx 10^{15}$ GeV , the bubble theory predicts a value for the top quark mass, $m_{top} = 163$ GeV, while $m_{top} = 1$ TeV for a smaller cutoff, $\Lambda \approx 10$ TeV. The bubble theory also predicts the mass of the physical Higgs scalar boson. It is given by $m_{higgs} = 2 \, m_{top}$ which is the result usually quoted in the pure NJL theory.

We have seen that the elimination of the elementary Higgs sector has resulted in predictions for masses of both the top quark and the physical Higgs particle. Although qualitatively correct, the above predictions are strongly modified by the full electroweak dynamics.

3.2. Effective Field Theory.

From the bubble theory, we can infer that the effective low energy theory is the full Standard Model with composite Higgs fields. When viewed as the Standard Model, the coupling constants run with momentum scale at low energy, and the Higgs becomes static at high energy.

The effective field theory can be defined through the introduction of a static Higgs field, $H_A(x)$,

$$L_{fermion} = L_{kinetic} + G (\overline{\Psi}_{La}{}^A t_R{}^a) \cdot (\overline{t}_{Rb} \Psi_{LA}{}^b)$$

$$(8)$$

$$= L_{kinetic} - (\overline{\Psi}_{La}{}^A t_R{}^a)H_A - H^{+A}(\overline{t}_{Rb} \Psi_{LA}{}^b) - (1/G) H^+H.$$

Instead of integrating out the static Higgs field to produce the four-fermion interaction, we can instead integrate out the short distance physics, replacing the cutoff scale Λ with a lower normalization scale μ. The short distance physics will generate contributions to the effective action defined at scale μ,

$$L_{fermion} = L_{kinetic} - (\overline{\Psi}_{La}{}^A t_R{}^a)H_A - H^{+A}(\overline{t_{Rb}} \Psi_{LA}{}^b)$$

$$(9)$$

$$+ Z_H (D_\mu H)^2 - (1/2) \lambda_0 (H^+H)^2 - (1/G +\Delta M^2) (H^+H) + O(1/\Lambda^2).$$

In the bubble theory, the induced couplings are given by

$$Z_H = (N_C/16\pi^2) \log(\Lambda^2/\mu^2)$$

$$(10)$$

$$\lambda_0 = (2N_C/16\pi^2) \log(\Lambda^2/\mu^2)$$

and ΔM^2 has a quadratic dependence on the cutoff. From these results we can infer compositeness conditions on the running coupling constants,

$$Z_H = 1/gt^2 \to 0, \qquad \mu^2 \to \Lambda^2 \qquad (11a)$$
$$\lambda_0 = \lambda/gt^4 \to 0, \qquad \mu^2 \to \Lambda^2 \qquad (11b)$$

These conditions are exact in the bubble theory and are abstracted to the full theory where they should refect the approximate behavior of the effective running couplings. If Z_H becomes sufficiently large, then the effective top quark Yukawa coupling, gt, is small, and the effective field theory below scale μ is the weakly coupled Standard Model with a dynamical Higgs field.

3.3. Renormalization Group.

The long distance behavior of the four-fermion theory can be described by a weakly coupled gauge theory with a composite Higgs field. Renormalization group methods are an efficient way to sum infinite sets of diagrams. The leading terms are just the leading log contributions which are expected to dominate if there is a large hierarchy of scales, ie. m_{top} $<< \Lambda$. The renormalization group can be used to evolve the running couplings to high scales where they must be matched to the appropriate boundary conditions of the composite theory.

We can compare various treatments of the coupling constant evolution:

1] Bubble (NJL) theory includes only the fermion loop contributions. This theory generates a composite Higgs but suppresses gauge and Higgs loop contributions.

2] The usual large N_c limit of QCD requires $N_c \to \infty$, with $G*N_c$ and α_3*N_c fixed, neglecting all other gauge couplings in loops. This theory includes all planar QCD corrections, and the low energy behavior is affected by infrared fixed points and is ultimately a theory of hadrons, not quarks and gluons.

3] The full Standard Model includes the effects of virtual Higgs contributions as well as the full gauge boson corrections. The theory is dominated by infrared fixed points of the renormalization group.

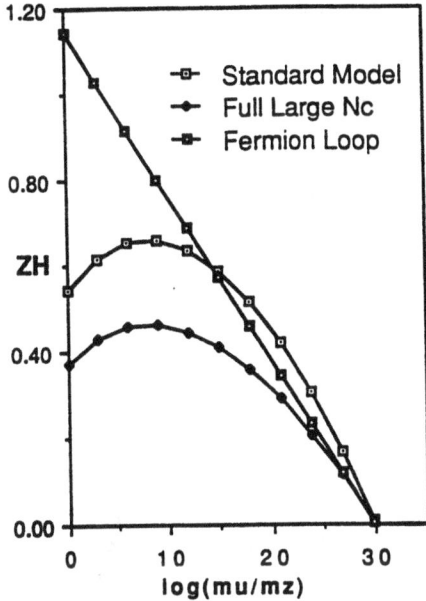

Figure 1. Higgs wavefunction/ top Yukawa coupling evolution.

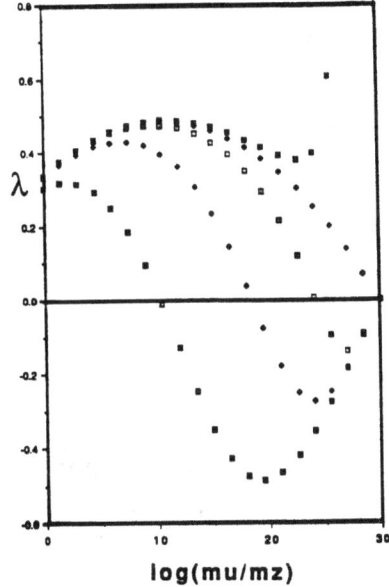

Figure 2. Higgs self-coupling evolution.

Figure 1 compares the three treatments for the running of $Z_H = 1/g_t^2$, from the Z boson mass scale to the cutoff scale, $\Lambda = 10^{15}$ GeV. This coupling directly determines the top quark mass as $m_{top} = g_t(m_{top})*v$. The turnover observed at low scales reflects the infrared fixed point behavior. The naive composite boundary condition was used at high scales although the perturbative renormalization group methods will break down as g_t becomes large. For a cutoff of 10^{15} GeV, the top quark mass predictons are

$$m_t(N_c \to \infty) \approx 270 GeV > m_t(SM) \approx 230 GeV > m_t(NJL) \approx 165 GeV.$$

Figure 2 compares various renormalization group trajectories for the coupling constant of the Higgs self-interaction which determines the physical Higgs particle mass. The infrared fixed point structure sharply focusses the running at low energy and provides a precise prediction of the mass. The trajectories which flow to negative couplings at high energy are ruled out by the expected vacuum instability of these solutions at short distance.
Using the naive compositeness conditions but the full Standard Model evolution, BHL

[2] obtained the following predictions for the top and physical Higgs particle masses for various composite scales, Λ. From a study of the compositeness conditions and the Standard Model evolution, it is expected that the theoretical ambiguities in the top quark mass predictions are only a few GeV [2].

Λ(GeV)	10^{19}	10^{15}	10^{13}	10^9	10^4
m_{top}	218	229	237	264	455
m_{higgs}	239	259	268	310	605

Table 1. Top quark and Higgs mass predictions in minimal model.

3.4. Conclusions for Minimal Model.

In the minimal model, the Higgs sector of the Standard Model is replaced by short range interactions of the top quark. If these interactions are sufficiently attractive, the electroweak symmetries are dynamically broken and the top quark becomes massive. The effective low energy field theory is the full Standard Model. The renormalization group methods give precise predictions for the top quark and the physical Higgs particle. The top quark must be quite heavy, $m_{top} > 220$ GeV. A high composite scale is favored, 10^{15} GeV $\rightarrow 10^{19}$ GeV, which might be identified with GUT or string model physics. The physical Higgs particle is expected to be only slightly heavier than the top quark, $m_{higgs} \approx 1.1 \, m_{top}$, in contrast to the NJL (bubble) prediction of $m_{higgs} \approx 2 \, m_{top}$. The infrared fixed point structure of the renormalization group equations stabilizes the predictions of the minimal model.

The minimal model predictions can be compared with constraints on the top quark mass coming from the various precision tests of the Standard Model. CDF provides a direct lower limit for the top quark mass, $m_{top} > 91$ GeV [7]. The strongest upper limits on the top quark mass come from deep inelastic neutrino scattering experiments and the collider measurements of the W boson mass. Langacker has reported the results of global fits to the present electroweak data [8]. He obtains the following upper limits,

$$m_{top} < 180 \text{ GeV (90\%CL)}, < 190 \text{ GeV (95\%CL)}, 210 \text{ GeV (99\%CL)}$$

where a Higgs particle mass of 250 GeV is assumed. This analysis depends on the careful understanding of the systematic errors for both theory and experiment for a wide range of processes. It is clear that the present analysis favors a lighter top quark than the expectations of the minimal top quark condensate model. However, we must await the discovery of the top quark as the determination of its mass will have crucial implications for the minimal top quark condensate model and perhaps the structure of Standard Model radiative corrections.

4. Comments and Extensions.

The minimal model of electroweak symmetry breaking discussed in the previous section can be related to a number of other approaches. It is important to examine the theoretical structure that makes it possible to have rather precise predictions of the top quark and Higgs particle masses. There are also many alternatives to the minimal model which still rely on the basic idea of short range, attractive interactions to generate the composite Higgs structure. In this section we will make a number of comments on the theoretical foundations of the model and discuss some of the most obvious extensions of the theory including a fourth generation and supersymmetry.

4.1. Infrared Fixed Points and Triviality.

The low energy behavior of the minimal model is governed by the full dynamics of the usual Standard Model. The low energy predictions are stabilized by the infrared fixed points (or more precisely pseudo-fixed points) of Standard Model renormalization group equations.

Fixed points were originally analyzed by Pendleton and Ross [9] and shown to provide a relation between the top quark Yukawa coupling constant and the running of the gauge coupling constants. If the evolution is to match the compositeness condition at high energy, then a different, but similar, relation between the couplings is achieved and the pseudo-fixed point discovered by Hill [10] dominates the low energy behavior. Figure 3 show the running top quark mass, or Yukawa coupling constant, as a function of normalization scale using a variety of couplings at a high energy cutoff scale, (A): $\Lambda = 10^{15}$ GeV or (B): $\Lambda = 10^{19}$ GeV. The renormalization flow of the top quark and Higgs coupling constants to the pseudo-fixed point was analyzed by Hill, Leung and Rao [11] and is shown in Figure 4 for a variety of intial conditions. The pseudo-fixed point structure makes the low energy predictions very insensitive to high energy boundary conditions and the precise value of the cutoff.

The Standard Model is said to be a trivial quantum field theory [12] as the running coupling constants grow at high energy and that the low energy couplings would vanish in a theory without cutoff. In this sense the Standard Model is not fully renormalizable as the cutoff can not be removed. Triviality diagrams, as in Figure 5, show the limitations on the low energy parameters, m_{top} and m_{higgs}, which follow from requiring that the effective theory remain perturbative up to the cutoff scale. Since the minimal model requires the naive compositeness condition, $Z_H \to 0$ or $g_t^2 \to \infty$ as $\mu \to \Lambda$, be satisfied, the top quark condensate model will lie on the boundary of the triviality diagram. In fact, it will only be consistent with the tip of the diagram with the largest allowed values of the top quark and Higgs masses for a given cutoff scale because the vacuum instability noted in Figure 2 eliminates possible solutions for lighter Higgs particle masses. For a wider class of models, the boundaries of the triviality domains may be interesting to analyze for the possible interpretations of composite structure.

4.2. Compositeness Conditions.

In the bubble (NJL) approximation, an exact connection was made between the fundamental four-fermion dynamics at short distance and the effective Standard Model theory relevant to the long distance dynamics. We have abstracted an ultraviolet boundary

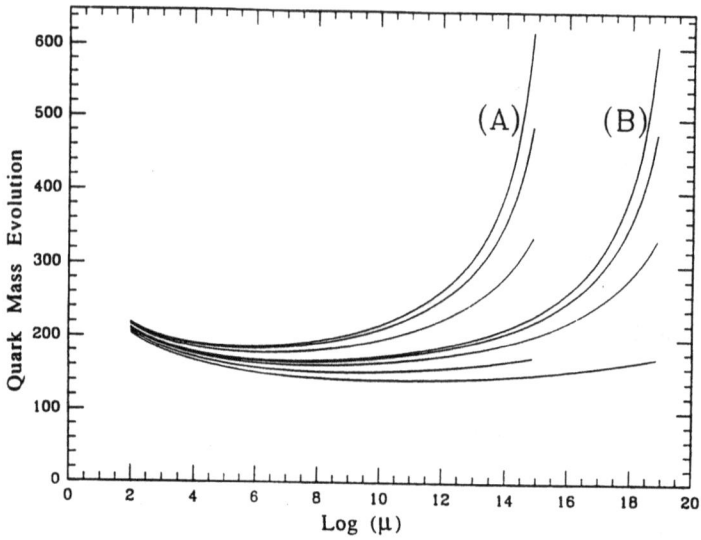

Figure 3. The quark mass evolution showing the pseudo-fixed point behavior for composite scales: 10^{15} (A) and 10^{19} (B) GeV.

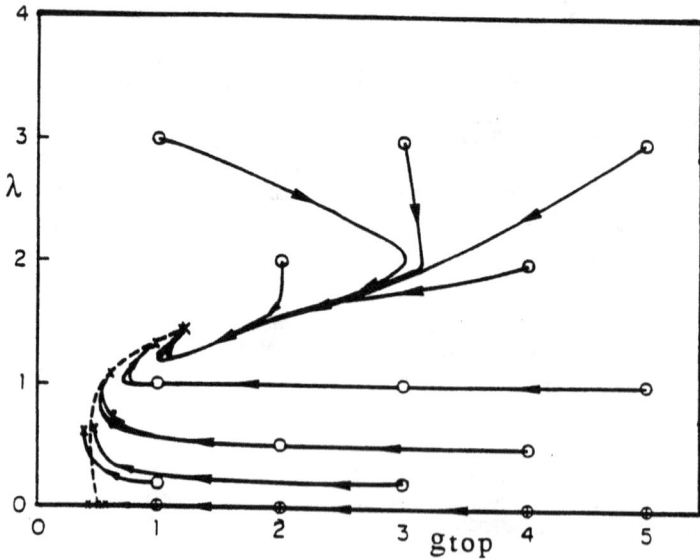

Figure 4. Renormalization flow to the pseudo-fixed point [11].

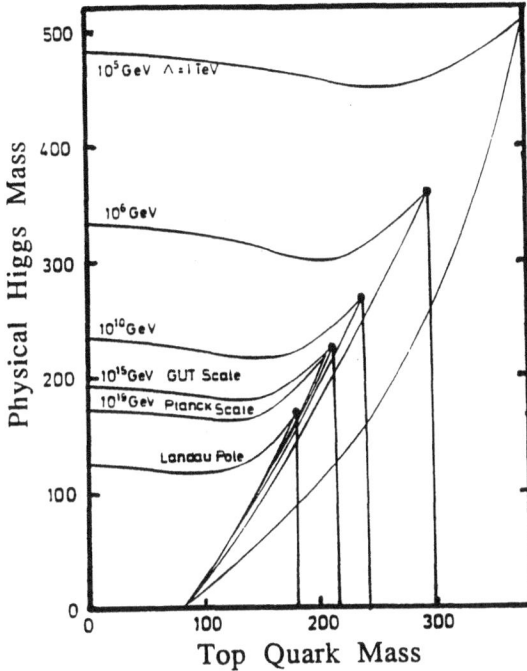

Figure 5. The triviality diagram for the Standard Model [12]. For each cutoff, the physical values of the top quark and Higgs particle masses must lie within the triviality domain. The compositeness condition is shown for each cutoff by the vertical line .

condition on the running of the Standard Model coupling constants to reflect the composite structure. This connection is made in a domain where both the effective Standard Model couplings, $g_t \to \infty$, and the four-fermion couplings, $G \to G_c$, are becoming nonperturbative. Physics near the composite scale, Λ, is expected be very sensitive to renormalization effects, strong operator mixing, etc.

The basic physical structure of the theory will be preserved so long as the critical coupling, $G \to G_c$, remains a second order phase transition. The fine tuning required to produce an electroweak scale much below the composite scale can always be achieved. The second order transition implies the existence of a dynamical Higgs field and the effective field theory to describe the low energy physics. The precise bound state structure (one Higgs doublet, two Higgs doublets, etc) may depend on the nonperturbative aspects of the fundamental theory. However, the effective field theory which includes the bound states as independent degrees of freedom, should provide a good description of the dynamics at scales sufficiently below the composite scale, $\mu < \Lambda/10$, $\Lambda/100$. The physics near the composite scale, $\Lambda/10 < \mu < \Lambda$, is nonperturbative and must be properly integrated out. Corrections to the bubble theory can be expected to be large, $O(1)$. However, these effects are expected to be small compared to the large logs generated by integrating out the physics below the the scale where the effective field theory becomes perturbative, eg. $\Lambda/10$. This expectation should be valid for the running couplings but not for the effective Higgs mass parameter which is subject to fine tuning and remains quite sensitive to even small modifications of the full theory. This sensitivity is irrelevant so long as fine tuning is possible and so long as a dynamical explanation of the fine tuning mechanism is not demanded.

To test the sensitivity to the specific choice of compositeness conditions, a model with higher derivative four-fermion interactions suggested by Suzuki [13] can be analyzed. The Lagrangian of Eq.(2) is replaced by

$$L = L_0 + G_0 \cdot ((\overline{\Psi}_L t_R + (\chi/\Lambda^2) \cdot D\overline{\Psi}_L D t_R)(\overline{t}_R \Psi_L + (\chi/\Lambda^2) \cdot D\overline{t}_R D \Psi_L)$$

$$(12)$$

$$= L_0 - (1/G_0) \cdot (H^+ H) - (\overline{\Psi}_L t_R + (\chi/\Lambda^2) \cdot D\overline{\Psi}_L D t_R)H$$

$$- H^+ (\overline{t}_R \Psi_L + (\chi/\Lambda^2) \cdot D\overline{t}_R D \Psi_L)$$

By integrating out the high momentum components of the fermion loops, the effective action becomes

$$L \to L_0 - (\overline{\Psi}_L t_R + (\chi/\Lambda^2) \cdot D\overline{\Psi}_L D t_R)H - H^+ (\overline{t}_R \Psi_L + (\chi/\Lambda^2) \cdot D\overline{t}_R D \Psi_L)$$

$$(13)$$

$$+ Z_H \cdot (DH^+ DH) - m^2 (H^+ H) - (1/2) \cdot \lambda_0 \cdot ((H^+ H)^2)$$

where the running couplings are given by

$$Z_H = (N_c/(4\pi)^2) \cdot \{ \ln(\Lambda^2/\mu^2) - 2 \cdot \chi + \chi^2/4 \}$$

$$(14)$$

$$\lambda_0 = 2 \cdot (N_c/(4\pi)^2) \cdot \{ \ln(\Lambda^2/\mu^2) - 4 \cdot \chi + 3 \cdot \chi^2 - (4/3) \cdot \chi^3 + \chi^4/4 \}$$

$$m^2 = 1/G_0 + \cdots \quad \text{(fine tuned)}$$

using bubble approximation for the explicit calculations. For scales sufficiently below the composite scale, the higher derivative Yukawa couplings may be neglected, $D\mu/\Lambda \ll 1$, and the theory evolves as the normal Standard Model as in the case of minimal model. However, the presence of the higher derivative interactions has modified the compositeness boundary conditions.

Using Eq.(14) for the evolution between scales Λ and $\Lambda/5$, the low energy effective theory can be computed using various approximations for the evolution (NJL(bubble), large Nc, Standard Model) below the scale, $\Lambda/5$. Figures 6 and 7 show the effects of the

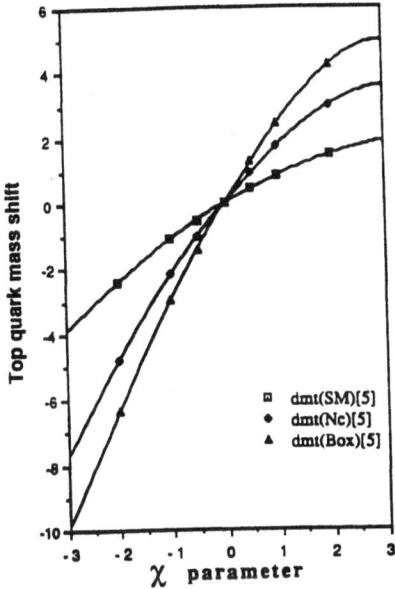

Figure 6. Top quark mass shift.   Figure 7. Higgs mass shift.

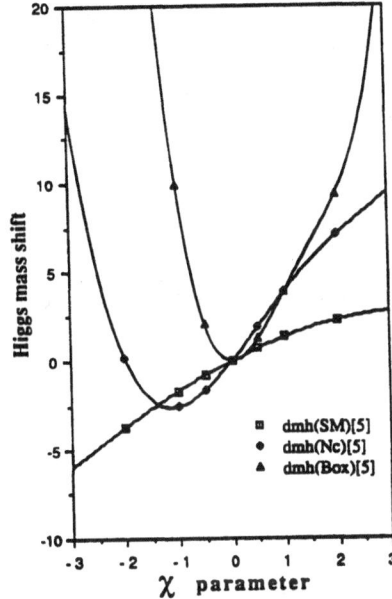

Shift due to modified UV boundary conditions for various treatments of the coupling constant evolution (SM, large N_c, box (NJL)) where χ is the coupling constant of the higher derivative interactions.

higher derivative interactions on the predictions of the top quark mass and the Higgs mass. For reasonable variations of the higher derivative coupling strength, $0 < \chi < O(1)$, the predictions of the Standard Model evolution are very stable with at most a few GeV shift in the masses.   It is the fixed point structure of the full Standard Model evolution that provides this stability. This example is used only to indicate the possible effects of the physical evolution of the effective theory near the composite scale.   Suzuki's analysis has been generalized by Hasenfratz et al [13] to a complete set of higher derivative interactions which they claim can reproduce, in bubble approximation, the complete range of couplings for an effective Standard Model at low energies. However, if dimensionless couplings of the generalized higher derivative interactions are $O(1)$ then there is little effect on the low energy dynamics as in the case of the Suzuki analysis. As mentioned earlier, a complete understanding of the initial evolution near the composite scale could require nonperturbative

analysis or a more complete description of the true short distance dynamics. This initial evolution modifies the boundary conditions for the subsequent Standard Model evoluton but has a limited effect on the ultimate predictions.

4.3. Fourth Generation Models.

If the top quark is found to be light, $m_{top} < 200$ GeV, then the top quark dynamics can not produce all the observed electroweak symmetry breaking, for $\Lambda < m_{planck}$. Additional symmetry breaking could come from a number of sources.   An obvious extension would be to consider condensates involving a fourth generation of quarks and leptons assuming the masses satisfy the ρ parameter bounds.   If the fourth generation is very heavy, then the composite scale could be much lower than the GUT scale considered for the minimal top quark condensate model. For low composite scales, the fine tuning problem is reduced, and the composite scale physics could be observable through the study of rare decays, FCNC, etc.

A degenerate fourth generation was considered by BHL [2]. The top contribution to electroweak symmetry breaking was neglected, and the degenerate mass for the fourth generation quark and the Higgs particle mass were computed for different composite scales, Λ, and are shown in Table 2.

Λ (GeV)	10^{19}	10^{15}	10^{13}	10^6	10^4
m_{quark}	199	206	212	277	388
m_{higgs}	235	248	258	365	553

Table 2. Mass predictions for the fourth generation model.

A fourth generation model with maximal mixing of the fourth generation quarks with the top quark was considered by Marciano [14]. He found that the top quark and the fourth generation up quark were nearly degenerate with a mass of 140 GeV.   The fourth generation bottom quark was somewhat heavier at 160 GeV.   Clearly the precise nature of the weak mixing will have an important impact on the predictions for a fourth generation model, and the compositeness conditions will only partly constrain these mixings.

The recent bounds from LEP on the number of light neutrinos implies that the fourth generation neutrino, if it exists, must be rather heavy, $m_{\nu 4} > 45$ GeV.   Heavy neutrinos could result from mixing structure in the neutrino mass matrix [15]. With the addition of right-handed neutrinos, it might be natural to expect that the dirac masses of the neutrinos are comparable to the charged lepton masses. A large Majorana mass for the neutrinos which does not break the usual electroweak symmetries could then be invoked which would suppress the masses of the observed neutrinos of the first three generations leaving the fourth generation neutrino heavy.   Hill et al [16] have suggested a class of models where the Majorana mass results from neutrino condensates produced by attractive interactions of the right-handed neutrinos.   Their predictions for the fourth generation masses are given as a function of the composite scale, Λ, in Figure 8 while the spectrum of neutrino masses are given in Figure 9. These results are for the case where the Majorana and normal Higgs VEV's are taken equal, $\beta = V_m/V_h = 1$; note that a low composite scale, $\Lambda < 10^6$ GeV, is required in this scheme. Right-handed neutrino condensates were also considered by Achiman and Davidson [17] as a mechanism for neutrino mixing with implications for composite DFS axions.

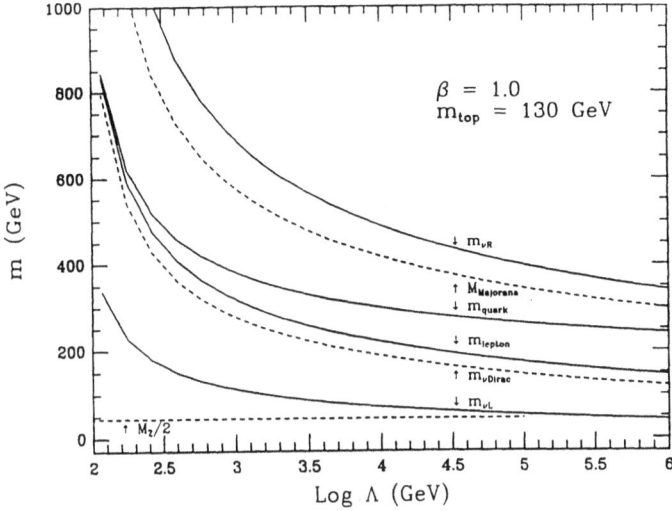

Figure 8. Quark and lepton masses of the four generation model
with neutrino and quark condensates as functions of the cutoff.

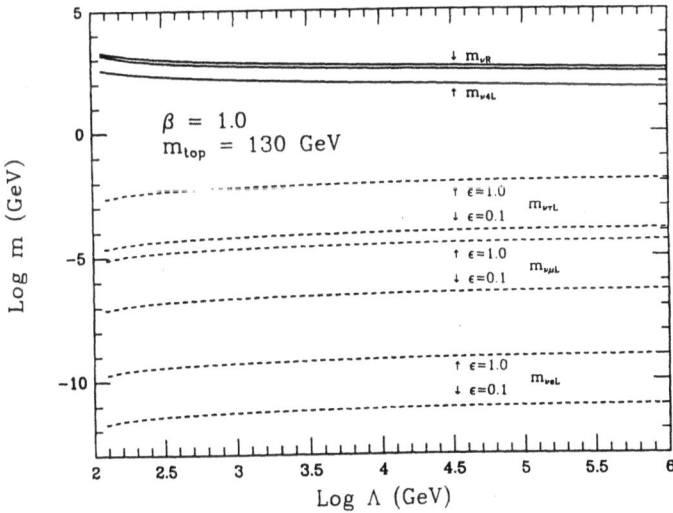

Figure 9. Neutrino mass spectrum of the four generation model
assuming comparable Dirac masses for charged and neutral leptons.

4.4. SUSY Extensions.

It is natural to consider the possible extension of the fermion condensate model to theories with supersymmetry. At the fundamental level, a supersymmetric extension of the local four-fermion interaction replaces the fundamental Higgs fields. This minimal supersymmetric extension of the BHL model would generate two composite Higgs supermultiplets. The naive compositeness boundary conditions require that only one of the the wavefunction normalization factors need vanish, $Z_H = 0$ and $Z_{H'} \neq 0$, to have both Higgs supermultiplets, H and H', be composite.

The results of even the minimal model are sensitive to the nature of supersymmetry breaking mechanisms. The renormalization group methods can be applied in two stages. From the composite scale, Λ, to the SUSY breaking scale, Δ, the couplings evolve supersymmetrically. At low energies the theory evolves as a normal gauge theory with additional Higgs representations.

A minimal version of a supersymmetric top quark condensate model was considered by Clark, Love and Bardeen (CLB) [18]. Only the top quark supermultiplet was involved in the dynamics with minimal SUSY breaking. The top quark becomes massive with the development of the corresponding condensate. CLB found that the top quark remained rather heavy with $m_{top} \approx 200$ GeV. The fine tuning problem is reduced as the SUSY breaking scale, Δ, determines the fine tuning rather than the composite scale, Λ, which could be much larger.

The minimal model of CLB was somewhat deficient as only the effective Higgs coupled to the top quark received a vacuum expectation value. The bottom quark would

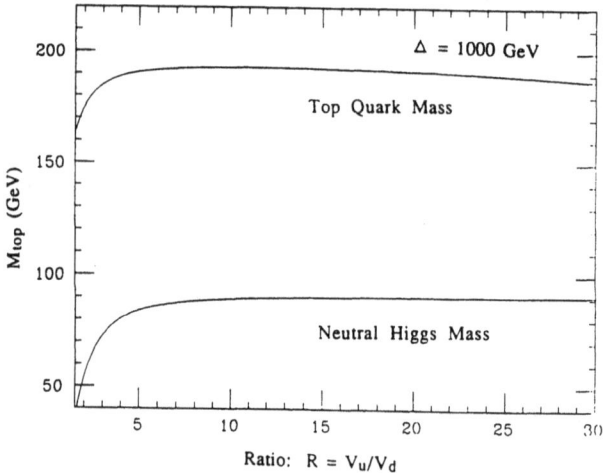

Figure 10. Top quark and physical Higgs particle masses for the minimal SUSY condensate model as a function of the ratio of the vacuum expectation values of the effective Higgs fields.

remain massless even if the effective Yukawa coupling were nonzero. Additional SUSY breaking must be added, as in the usual SUSY models, to the original interactions to produce vacuum expectation values for both effective Higgs multiplets. The prediction of the top quark mass is reduced by the ratio of the VEV seen by the top quark, V_u, to the VEV seen by the gauge bosons, $V = \sqrt{V_u^2 + V_d^2}$. The top quark mass depends on both the composite scale and SUSY breaking scale. The predicted values are given in Table 3 where the mass value must still be multiplied by the VEV ratio, $m_{top} = [\text{Table 3}] * (V_u/V)$.

With this standard SUSY breaking scheme, the top quark mass predictions [19] are shown in Figure 10 as a function of the Higgs VEV ratio, V_u/V_d. Also plotted are the mass predictions for the lightest neutral Higgs particle. These results can be related to the boundaries of standard renormalization group studies of supersymmetric models [20]. In the minimal models considered above, the Higgs particle mass must be less than the Z boson mass, 91 GeV, and more than the recent LEP lower bounds, 40 GeV [21]. These results imply that the top quark should be heavier than about 170 GeV in the minimal composite SUSY model with the standard SUSY breaking scheme.

$\Lambda \backslash \Delta$	10^2	10^4	10^6	10^8	10^{10}
10^6	259	290			
10^9	222	241	254	263	
10^{13}	203	215	224	231	236
10^{15}	198	208	216	222	227
10^{19}	191	200	206	211	214

Table 3. SUSY model predictions of the top quark mass (GeV) as functions of the composite scale Λ and SUSY breaking scale, Δ.

Supersymmetry is usually invoked to help explain the gauge hierarchy problem. The composite Higgs models considered in this section do not directly address this problem although the fine tuning scale is set by the SUSY breaking scale instead of the higher compositeness scale. However this scale dependence also implies that the effective couplings of the fundamental, higher dimension operators may have greatly enhanced couplings if they are to generate the composite Higgs structure; this may cause problems with the perturbative unitarity of the theory. Although these models focus on composite Higgs structure, other supermultiplets could be considered for compositeness on the same basis.

4.5. Multiple Composite Higgs.

Another natural extension of the minimal top quark condensate model is the possible formation of additional composite Higgs bound states. This extension can be achieved by considering additional local, four-fermion interactions between the fundamental fermion fields. These additional interactions can produce attractive interactions in more than one channel and result in new bound states. Normally one would expect that additional fine tuning would be required for the new states to generate physics at a sufficiently low scale.

The simplest model involves interactions of both the top and bottom quarks as given in the following Lagrangian,

$$L_{fermion} = G_u \, (\overline{\Psi}_{La}{}^A \, t_R{}^a) \cdot (\overline{t_{Rb}} \, \Psi_{LA}{}^b) + G_d \, (\overline{\Psi}_{La}{}^A \, b_R{}^a) \cdot (\overline{b_{Rb}} \, \Psi_{LA}{}^b)$$

$$+ G_{ud} \, (\overline{\Psi}_{La}{}^A \, t_R{}^a) \cdot (\overline{\Psi}_{Lb}{}^B \, b_R{}^b) \varepsilon_{AB} + h.c. \tag{15}$$

where the last term is needed to provide mixing and break the chiral symmetries which would result in electroweak axions that are presently ruled out by experiment. With sufficient fine tuning, this theory generates two Higgs doublets, $H_{uA} = (\overline{t_{Rb}} \, \Psi_{LA}{}^b)$ and $H_{bA} = (\overline{\Psi}_{Lb}{}^B \, b_R{}^b) \varepsilon_{AB}$. Whether the degree of fine tuning required to keep both Higgs multiplets light is possible to achieve may require a nonperturbative study of the phase transition structure of the fundamental four-fermion theories. Studies using bubble approximation and modified renormalization group methods seem to indicate a consistent picture with additional light composite Higgs states. Two doublet models were considered by a number of authors [22] with consequences for the mass predictions of the top and bottom quarks as well as the spectrum of observable Higgs particles. The compositeness conditions place interesting constraints on the effective potential of the composite Higgs fields with implications for their low energy dynamics.

As mentioned before, four generation models require a mechanism for the producing a massive fourth neutrino. This may be accomplished [15,16] by introducing right-handed neutrinos. Possible condensates of these right handed neutrinos would produce electroweak singlet composite Higgs fields [16,17]. Hill, Luty and Paschos [16] have studied a unified picture of singlet and nonsinglet condensates which could play a role in realistic four generation models. Achiman and Davidson [17] have emphasized the possible role of right-handed neutrino condensates in producing composite DFS axions and their role in model building.

4.6. UV Fixed Points and Reduction.

We have emphasized the role of infrared (pseudo-) fixed points in generating stable predictions for the top quark condensate models. The pseudo-fixed points [10] control the renormalization group evolution at low energy where the gauge coupling constants control the running of the top quark Yukawa coupling constant. If the composite scale is large, then the top quark mass predictions are insensitive to the composite boundary conditions and are dictated by the infrared pseudo-fixed points.

A more ambitious analysis of the relation between the various electroweak coupling constants is made in the reduction approach advocated by Zimmerman, et al. [23] and Marciano[24]. In this approach, it is assumed that the Standard Model couplings are not all independent but have a functional interdependence. For example, the top quark Yukawa coupling constant is determined as a function of the gauge coupling constants. The renormalization group is used to determine the possible relationships between the various Standard Model couplings. For the top quark, the results are similar to those obtained in the orginial fixed point analysis of Pendleton and Ross [9].

Focussing on the top quark mass, the renormalization group provides a relation between the running top quark Yukawa coupling, K_t, and the color coupling constant, α_3. The renormalization group equations can be integrated to give

$$K_t = 2 \ \alpha_3^{8/7} \ / \ (C + 9 \ \alpha_3^{1/7}) + \text{electroweak corrections} \tag{16}$$

where C is an integration constant with special values, 0 and ∞. $C = \infty$ might correspond to the situation of the light quarks. For $C = 0$, the coupling relation becomes analytic, and $C > 0$ is required if the Eq.(15) is to be nonsingular. If we identify $C = 0$ with the top quark situation, then the top quark mass is predicted to be 90-95 GeV [23, 24] which is just at the present lower limit of the direct search by CDF [7]. The same analysis would predict the Higgs particle mass of about 64 GeV which is still somewhat above the recent LEP lower bounds [21]. Refinements are not expected to change these predictions by large amounts.

The reduction approach chooses the special value, $C = 0$, to preserve the analytic structure of Eq.(15). From the renormalization group point of view, this constraint comes from requiring a nonsingular behavior up to infinite energies where α_3 vanishes. This contrasts with the top quark condensate models where the coupling constant becomes large at the composite scale which is taken to be less than the Planck scale. Hence, the solutions with $C < 0$ are required for top condensate models. The reduction method requires knowledge of the gauge coupling constant, α_3, for physical values corresponding to energies far beyond the Planck scale. Hence we can only view the constraints imposed by the reduction method as mathematical conditions imposed on the theory (perhaps as a result of hidden symmetries) and not as conditions on the physical running couplings. We will soon be able to determine the validity of the top quark mass predicition as the limits from CDF are improved or the top quark is discovered.

4.7. Schwinger-Dyson Equation Approach.

We have emphasized the use of renormalization group methods to give reliable predictions for the low energy parameters of the top condensate theory. The renormalization group is used to sum the leading contributions from an infinite set of Feynman diagrams describing the short distance physics. An alternate approach involves the direct solution of the Schwinger-Dyson equations for the behavior of the top quark self-energy function and use it to predict the top quark mass and the related electroweak symmetry breaking.

The Schwinger-Dyson equations have been studied by Barrios and Mahanta [25] and by King and Mannan [26]. Both groups study the effects of local four-fermion interactions on the solutions to the Schwinger-Dyson equations with the gauge interactions included in ladder approximation. The four-fermion coupling must be fine-tuned to generate a light top quark, $m_{top} \ll \Lambda$. The color gauge interactions increase the predicted value of the top quark mass from that obtained from the pure four-fermion theory in bubble approximation. The results are given in Table 4 for various values of the cutoff scale, Λ.

Λ (GeV)	$m_t(\alpha_3=0)$	$m_t(\alpha_3)$	$m_t(RG)$
10^5	379.5	442.9	438.2
10^9	229.2	306.2	310.6
10^{15}	165.2	257.3	261.6
10^{19}	143.9	242.0	246.1

Table 4. Top mass predictions from the Schwinger-Dyson approach compared with the renormalization group results.

Barrios and Mahanta have compared their Schwinger-Dyson results with the standard renormalization group procedure, as used by BHL [2], and achieve good agreement between the two methods so long as the same physics is considered.

The Schwinger-Dyson equation method as implemented by both groups [25,26] includes only the effects of the local four-fermion interactions and the color gauge interactions, both in ladder approximation. These results must be compared to the equivalent renormalization group calculations which are similar to the "large Nc" results of BHL. As emphasized by BHL, the full low energy dynamics must be included to make meaningful physical predictions. The virtual Higgs contributions and the full electroweak gauge interactions were essential to obtain reliable predictions of the top quark and Higgs particle masses. These physical contributions must be included in the Schwinger-Dyson approach before its results can be compared directly to data. The Schwinger-Dyson approach must also confront possible nonperturbative aspects of the short distance physics associated with the four-fermion interactions which may also affect the intial evolution of the renormalization group method. The low energy predictions are expected to be somewhat insensitive to the short distance structure because of the infrared (pseudo-) fixed point behavior at long distance, but it may be difficult to separate these effects in the more direct analysis of Schwinger-Dyson equations.

4.8. Long Distance Contributions.

The renormalization group method can be used to systematically study the contributions of physics below the composite scale. The BHL analysis uses the one loop anomalous dimensions to compute the effective action to use at low energies, $\mu \approx m_Z$. The low energy radiative corrections must be used for accurate comparison of the theory with experiment. For example, top quark self-energy will receive a low energy contribution from its QCD interactions which result in a mass shift,

$$\delta m_{top} = m_0 \, (\alpha_3/3\pi) \, (\, 4 + 3 \, \ln(\mu^2/m_0^2) \,) \tag{17}$$

where μ is the low energy normalization scale, α_3 is the QCD coupling constant and m, is the lowest order top quark mass. In the BHL calculation, the ln contributions were absorbed by evolving the top quark Yukawa coupling constant to the top quark mass scale instead of the m_Z normalization scale. A remaining $O(\alpha_3)$ contribution could affect the top quark mass predictions as emphasized by Kugo [27]. However, these QCD corrections may be largely cancelled by similar electroweak corrections as they are for the log corrections which are largely cancelled because of the infrared fixed point structure. A consistent calculation must incorporate the two loop contributions to the renormalization group evolution as well as the finite part of the one loop effects at the low energy scale. Since the two loop anomalous dimensions are known, the top quark mass predictions can be systematically improved.

5. Conclusions.

We have shown that the BCS mechanism can be used to connect the electroweak scale (m_W, m_Z) with the masses of the top quark and the physical Higgs particle. Electroweak symmetry breaking is produced by condensates of the top quark which are triggered by attractive, local interactions of the top quark. Stable predictions for the top quark and Higgs masses are achieved using the full Standard Model evolution which reflects the presence of infrared (pseudo-) fixed points in the renormalization group equations.

The minimal model predicts a heavy top quark, $m_{top} > 200$ GeV. For large composite scales, the physical Higgs particle is predicted to be only slightly heavier than the top quark, $m_{higgs} \approx 1.1 * m_{top}$ which contrasts with the NJL prediction of $m_{higgs}/m_{top} = 2$. The renormalization group analysis results in a rather precise prediction of the top quark mass, $m_{top} = 229 \pm 5$ GeV ($\Lambda \approx 10^{15}$ GeV) where the error reflects an estimate of the theoretical error coming from the composite boundary conditions. To achieve a relatively light top quark in the minimal model, the composite scale must be taken to be quite large and could be associated with the GUT scale, 10^{15} GeV, or the Planck scale, 10^{19} GeV. Even with this choice of composite scale, the minimal top quark model predicts masses which are somewhat larger than the range estimated by recent fits to the Standard Model radiative corrections, $m_{top} = 137 \pm 40$ GeV [8]. While the minimal top condensate model seems to be somewhat disfavored by the present data, we should wait until the top quark is discovered before reaching final conclusions about the validity of the minimal model or the normal radiative corrections.

There are many extensions to the minimal model which preserve the basic idea of electroweak symmetry breaking being generated by new short distance dynamics rather than additional fundamental degrees of freedom. The four generation version of the theory relaxes the constraint on a heavy top quark but requires a mechanism for producing a heavy neutrino for the fourth generation. Right-handed neutrinos could have an important dynamical role in generating mixing and producing a spectrum of neutrino masses. Condensates of right-handed neutrinos could have important phenomenological consequences and could be responsible for a new mechanism for axions.

Fermion condensates may also generate a more complex Higgs structure. Models with two Higgs doublets have been studied in some detail. The compositeness conditions place interesting constraints on the effective potentials of the dynamical Higgs fields. As stated above, neutrino condensates may also play a role in the low energy dynamics.

We have briefly discussed the relation of the condensate models with alternative approaches including coupling constant reduction which has much different predictions for the top quark and Higgs masses and the direct Schwinger-Dyson equation method which yields the same results as the renormalization group methods so long as the same physical input is achieved. We have not discussed early work [4] which focussed on the possible role of four-fermion interactions in generating composite vector mesons using methods associated with the NJL models. The role of electroweak gauge symmetry plays a crucial role in these models, and the dynamical structure of these models remains to be understood.

There are many remaining questions for the implementation of condensate models. The models require fine-tuning to produce an electroweak scale that is much below the composite scale as was the case in the normal Standard Model. It is interesting to speculate about possible mechanisms which could generate this fine-tuning dynamically. The physics at the composite scale is not renormalizable and is presumably generated by physics beyond the composite scale such as the fragments of a GUT model or superstring theory. These theories should generate many higher dimension interactions which are mostly irrelevant to the low energy physics. If the coupling constants of these interactions are dynamically determined, it is possible to imagine feed-back mechanisms could produce the apparent fine-tuning needed to produce the desired infrared structure of the full theory. The top quark condensate model makes definite predictions for the top quark and Higgs particle masses the flavor mixing structure observed in the quark sector can only be accommodated. The understanding of flavor mixing whether it is in the explored domain of the quarks or the unexplored domain of the neutrinos remains an outstanding problem for any fundamental theory.

6.   Acknowledgements.

I would like to thank the organizers of this Symposium for their hospitality and for giving me the opportunity to recognize Tini Veltman's many contributions to physics. A preliminary version of this talk was presented at the 1990 Nishinomiya Yukawa Memorial Symposium. This work would not have been possible without the continuing collaboration with Chris Hill and my other colleagues on the many different aspects of this research.

7.   References.

1.   Y. Nambu, Proceedings of the 1988 Kazimierz Workshop, Z. Ajduk et al., eds. (World Scientific, 1989); Proceedings of the 1988 Nagoya Workshop, M. Bando et al., eds. (World Scientific, 1989); EFI Preprint 89-08 (1989) (unpublished).

2.   W. Bardeen, C. Hill and M. Lindner, Phys. Rev. D41, 1647(1990).

3.   V. Miransky, M. Tanabashi and K. Yamawaki, Mod. Phys. Lett. A4, 1043(1989); Phys. Lett. B221, 177(1989).

4.   H. Terazawa, Phys. Rev. D22, 2921(1980); H. Terazawa, Y. Chikashige and K. Akama, Phys. Rev. D15, 480(1977), F. Cooper, G. Guralnik, and N. Snyderman, Phys. Rev. Lett. 40, 1620(1978); T. Eguchi, Phys. Rev. D14, 2755(1976); K. Kikkawa, Prog. Theor. Phys. 56, 947(1976); T. Kugo, Prog. Theor. Phys. 55, 2032(1976); J.D. Bjorken, Ann. Phys. 24, 174(1963).

5.   J. Bardeen, L. Cooper and J. Schrieffer, Phys. Rev. 108, 1175(1957).

6.   Y. Nambu and G. Jona-Lasinio, Phys. Rev. 122, 345(1961); W. Bardeen, C. Leung,

and S. Love, Phys. Rev. Lett. 56, 1230 (1986).

7. CDF Collaboration (F. Abe, et al.), Phys. Rev. Lett. 64, 142(1990); Phys. Rev. D43, 664 (1991); FERMILAB-PUB-91-280-E.

8. D. Kennedy and P. Langacker, Phys.Rev.Lett. 65, 2967 (1990); ERRATUM-ibid.66:395,1991.; P. Langacker, "Precision Tests of the Standard Model", U. Pennsylvania Preprint UPR-0435T (1990).

9. B. Pendleton and G. Ross, Phys. Lett. 98B, 291(1981).

10. C. Hill, Phys. Rev. D24, 691(1981).

11. C. Hill, C. Leung and S. Rao, Nucl. Phys. B262, 517(1985).

12. L. Maiani, G. Parisi and R. Petronzio, Nucl. Phys. B136, 115(1978); N. Cabbibo et al, Nucl. Phys. B158,295(1979); M. Lindner, Z. Phys. C31, 295(1986).

13. M. Suzuki, Mod. Phys. Lett. A5,1205(1990); 1990 Nagoya Workshop (1990); Anna Hasenfratz, Peter Hasenfratz, Karl Jansen, Julius Kuti, Yue Shen, Nucl. Phys. B36, 79(1991).

14. W. Marciano, "Dynamical Symmetry Breaking and the Top Quark Mass", Brookhaven Preprint, September, 1990.

15. C. Hill and E. Paschos, Phys. Lett. B241, 96(1990).

16. C. Hill, M. Luty and E. Paschos, "Electroweak Symmetry Breaking by Fourth Generation Condensates and the Neutrino Spectrum", Phys.Rev.D43, 3011 (1991).

17. Y. Achiman and A. Davidson, "Dynamical Axion of Dynamical Electro-Weak Symmetry Breaking", Phys.Lett. B261, 431 (1991).

18. T. Clark, S. Love and W. Bardeen, Phys. Lett. B237, 235(1990).

19. K. Sasaki, M. Carena, C. Wagner, T. Clark and W. Bardeen, "Dynamical Symmetry Breaking and the Top Quark Mass in the Minimal Supersymmetric Standard Model", FERMILAB-PUB-91/96-T.

20 H.P. Nilles, Phys. Rep. 110, 1(1984); J. Bagger and S. Dimopoulos, Phys. Rev. Lett. 55 920(1985); H. Okada and K. Sasaki, Phys. Rev. D40, 3743(1989).

21. OPAL Collaboration (M. Akrawy, et al.), Phys.Lett. B253, 511 (1991).

22. M. Luty, Phys. Rev. D41, 2893(1990); M. Suzuki, Phys. Rev. D41, 3457 (1990); C. Froggatt, I. Knowles and R. Moorhouse, Phys.Lett. B249, 273 (1990) ; J. Fröhlich and L. Lavoura, Phys. Lett. B253, 218 (1991); C. Hill, M. Luty and E. Paschos, Phys. Rev. D43, 3011 (1991).

23. J. Kubo, K. Sibold and W. Zimmerman, Phys. Lett. B220, 85(1989).

24. W. Marciano, Phys. Rev. Lett. $\underline{62}$ (1989).

25. F. Barrios and U. Mahanta, Phys. Rev. $\underline{D43}$, 284 (1991).

26. S. King and S. Mannan, Phys. Lett. $\underline{B241}$, 249(1990).

27. T. Kugo, contribution to Nishinomiya/Kyoto Symposium (1990).

FAST AND RELIABLE RANDOM NUMBERS
FOR EXTENSIVE MONTE CARLO CALCULATIONS

AALDERT COMPAGNER

Laboratory of Applied Physics
Lorentzweg 1, 2628 CJ Delft, the Netherlands

ABSTRACT

The total amount of correlation in an arbitrary sequence depends only on its length. For well-tempered pseudorandom sequences, the remaining correlations are of high order or cover a large distance. Efficient recipes to generate these sequences are described and their reliability as sources of random numbers at bit rates > 1 GHz is discussed.

1. Introduction

For reasons of efficiency and general control, the sequences of random numbers needed in Monte Carlo calculations have to be generated by deterministic algorithms such as the linear-congruence, the lagged-Fibonacci and the Tausworthe-shiftregister methods described in many reviews[1-4]. It is sufficient to discuss periodic sequences of N bits, with $n = {}^2\log N$ degrees of freedom. Hitherto most recipes involve less than 100 degrees of freedom, which is not enough for large-scale Monte Carlo calculations (a simple reason being that only $n/32$ subsequent random numbers of 32 bits can be truly independent). To extend these recipes to considerably larger values of n is not easy, and the necessary tests would be very time-consuming. Moreover, the meaning of these tests is not clear when an operational definition of randomness is absent.

These problems can be overcome by means of ensemble theory and by studying the complete hierarchy of correlation coefficients. Conservation laws for the total amount of correlation result, together with an interpretation of randomness in terms of entropy. As described in detail elsewhere[5-7], this leads to efficient recipes with $n > 10^4$ for well-tempered pseudorandom sequences, which a priori have good randomness properties.

2. The hierarchy of correlations

Consider a sequence of bits $\{y_i\}$ with $y_{i+N} = y_i$ or the corresponding parity sequence $\{b_i\}$ with $b_i = 1-2y_i$. All N translated versions of the sequence are assumed to be different. A correlation coefficient for a set I of order q and size s is defined by

$$C_{I(q, s)} = \frac{1}{N} \sum_{j=0}^{N-1} \prod_{i \in I} b_{i+j} , \tag{1}$$

where the parity product is averaged over the ensemble of all translated versions of the original sequence (other ensembles can be envisaged also). The set $I(q, s)$ is an ordered selection of indices $\{i_1, i_2, ... , i_q\}$, where q is the order of the set and $s = i_q - i_1 + 1 \geq q$ its size. The correlation coefficients obey a strong conservation law[6]:

$$\langle C_{I(q, s)}^2 \rangle \equiv \frac{1}{2^N} \sum_I C_{I(q, s)}^2 = \frac{1}{N} , \tag{2}$$

where the summation is over all 2^N different sets I of any order and size. When N is large, the mean-square value (2) is small, but this cannot hold for all individual contributions. Tests for randomness, though usually not stated in terms of correlation coefficients of any size and order, tend to select sequences for which the contributions to (2) that the tests happen to be sensitive for are small; other contributions, which remain hidden, are necessarily close to 1.

A sequence of N bits has $n = {}^2\log N$ degrees of freedom; at best, all 2^n different strings of n subsequent bits can be accommodated once along the sequence. When this is the case, all true ($q \neq 0$) correlation coefficients for $s \leq n$ vanish, and the sequence is called pseudorandom in a strict sense. In practice, most recipes for random-number generation give rise to binary sequences that fall into this class. When a fair coin is tossed N times, the resulting sequence is not strictly pseudorandom, but large correlation coefficients (close to ± 1) are expected to appear only when strings of more than n bits are taken into account.

When a string of n bits of a pseudorandom sequence is given, the next bit of the sequence must depend on the first bit and one or more of the other bits of that string, otherwise the number of degrees of freedom differs from n. In the simplest case, a q-bit linear-feedback production rule holds,

$$y_{i_1+n} = y_{i_1} \oplus y_{i_2} \oplus ... \oplus y_{i_q} , \tag{3}$$

with $i_q < i_1 + n$. The $n + 1$ bits that are involved form a completely correlated set $I(q+1, n+1)$, with $C_I = 1$. This is the case of shiftregister sequences of maximum length $N = 2^n-1$, which are almost pseudorandom (only the string of n zeros is missing). The set $I(q+1, n+1)$ is the first correlated set (since $C_{I(q,s)}$ almost vanishes for $s \leq n$), which is equivalent to the production rule. For instance, the first-correlated set $I(3,5) = \{1,2,5\}$ corresponds with the maximum-length sequence 111100010011010 with period $2^4-1 = 15$.

Many 2-bit feedback production rules, up to $n = 9689$, are known[8,9] that generate a maximum-length sequence. Since the first correlated set, of size $n + 1$, then is only of third order, the resulting pseudorandom sequence is not as random as can be. Maximum-length sequences produced by a linear-feedback rule with $q \approx \frac{1}{2} n$ feedback positions, scattered rather irregularly over the string of n bits, are the simplest example of well-tempered pseudorandom sequences, for which the smallest correlated sets (of size $s \geq n+1$) that have a non-vanishing correlation coefficient are of high order. Well-tempered pseudorandom sequences exhaust the main possibilities that are available to imitate randomness by a deterministic algorithm.

3. Discussion

Other production rules than the linear-feedback rule (3) exist that also generate pseudorandom sequences which possibly may be called well-tempered. Then, the first correlated set may have a non-vanishing correlation coefficient that is not equal to 1, although the strong conservation law must be obeyed. For linear-feedback rules, all contributions to the conservation law are due to completely correlated sets, which facilitates the analysis.

Iteration of the production rule (3) is a stochastic process which leads to other correlated sets of increasing size $s > n+1$ and still of high order if the initial q is large. It is very likely that non-vanishing correlation coefficients of low order can be avoided until sets of size s much larger than n come into play. The remaining problem is that maximum-length rules with $q \approx \frac{1}{2} n$ are not easy to find for $n \approx 10^4$.

Consider however the sequence that is obtained by a bitwise mod-2 addition of the maximum-length sequences generated by m different 2-bit rules, with periods that are relative primes. The resulting sequence is an approximate maximum-length sequence (an AM-sequence, as it was called[7]) and has an effective q-bit production rule with $q \approx 3^m$ feedback positions. The correlation properties are more complicated than but very similar to those of a true maximum-length sequence; in fact, still only the relatively small number of completely correlated sets need to be considered, all other sets having negligible correlation coefficients. When $m = 8$ of the 2-bit rules from Zierler's list[9] are combined, with n varying from 127 to 9689, a well-tempered pseudorandom monstrosity appears, with a period well above 2^{10000} and with a first-correlated set of order $q \approx 6000$. Nonetheless, an efficient implementation in software or hardware (or even in a VLSI chip) is not difficult, since only exor-operations are involved. When parallel techniques are used, the bit rates > 1 GHz necessary in large-scale Monte Carlo simulations can be realized.

The randomness properties of AM-sequences were compared with those of sequences generated by so-called optimal production rules that give rise to good 'figures of merit', as considered by André, Mullen and Niederreiter[10], for N below $\approx$ 100, where the necessary computations are not prohibitive. The results[7] of this comparison are in agreement with the expected behavior of AM-sequences. Recently, extensive tests were carried out by Berdnikov[11] for AM-sequences resulting when 4 or 8 rules of Zierler's list[9] are used, including one with n = 9689. The preliminary results of these tests are very acceptable; in fact, they appear to be similar to those obtained for one version (with n = 568 degrees of freedom) of the 'add-with-carry' rules that recently were proposed by Marsaglia and Zaman[12] (see also the discussion by James[4]).

Further tests and an analysis in terms of higher-order correlations are in progress.

4. Acknowledgements

It is a great pleasure to present this article, on high entropy rather than high energy, to Tini Veltman at the occasion of his 60th birthday. I am indebted to Dr. D. Wang of the Northwest University at Xian for many helpful discussions, and to Dr. A.S. Berdnikov of the Institute for Analytical Instrumentation at Leningrad for sending me his test results prior to publication.

5. References

1. D.E. Knuth, *The Art of Computer Programming*, Vol.2 (Addison-Wesley, 1981), Ch.3.
2. G. Marsaglia, *A Current View of Random Number Generators*, in: *Computer Science and Statistics*, ed. by L. Billard (North-Holland, Amsterdam, 1985).
3. B.D. Ripley, *Thoughts on Pseudorandom Number Generators*, J.Comp.Appl. Math. **31** (1990) 153.
4. F. James, *A Review of Pseudorandom Number Generators*, Comp.Phys.Comm. **60** (1990) 329.
5. A. Compagner, *Definitions of Randomness*, Am.J.Phys. **59** (1991), to be published.
6. A. Compagner, *The Hierarchy of Correlations in Random Binary Sequences*, J. Stat.Phys. **63** (1991) 883.

7. D. Wang and A. Compagner, *On the Use of Factorable Polynomials as Random-Number Generators*, Math. of Computation, submitted.

8. N. Zierler and J. Brillhart, *On Primitive Trinomials (Mod 2)*, Inform. and Control **13** (1968) 541, **14** (1968) 566.

9. N. Zierler, *On Primitive Trinomials whose Degree is a Mersenne Prime*, Inform. and Control **15** (1969) 67.

10. D.A. André, G.L. Mullen and H. Niederreiter, *Figures of Merit for Digital Multistep Pseudorandom Numbers*, Math. of Computation, **54** (1990) 737.

11. A.S. Berdnikov, private communication.

12. G. Marsaglia and A. Zaman, *A New Class of Random Number Generators*, Siam J. Sci. Stat. Computing (1991), to be published.

LARGE HIGGS MASS , TRIVIALITY AND
ASYMPTOTIC FREEDOM

M. CONSOLI

Istituto Nazionale di Fisica Nucleare, Sezione di Catania, Catania, Italy

ABSTRACT
A detailed analysis of the effective potential leads to the conclusion that massless $\lambda\phi^4$ theories undergo spontaneous symmetry breaking and are asymptotically free, not in contrast with known rigorous results. As a consequence, the shifted theory is trivial and the Higgs boson turns out to be naturally heavy ($\sim 2\ TeV$) in agreement with Veltman's hypothesis of a second threshold in weak interactions

1.Introduction

The role of the Higgs particle in connection with a second threshold in weak interactions was enphasized by Veltman[1]. By following his argument, the Higgs mass m_H represents, in a natural way, the cutoff of the Glashow model[2] just like the W mass in the case of the four fermion V-A theory. Indeed, the one-loop renormalizability of massive gauge theories suggests that their intrinsic cutoff might be very large as compared to the W and Z masses and the large Higgs mass one loop correction to the ρ parameter[3] in the standard model[4,5] has the same form as in the gauged non linear σ-model[6]. Very large values of the Higgs mass,however, are usually excluded on the basis of lattice computations[7] ($m_H < 640\ GeV$) providing evidences for the triviality of $\lambda\phi^4$ theories. These theories,in fact, are believed to be either not interacting or else inconsistent in the continuum limit, despite of the fact that no complete proof, covering all possible phases , exists. As discussed in the following, we shall present an answer to this problem ,which takes into account in an essential way, the occurrence of spontaneous symmetry breaking and such that the triviality of the shifted theory does not imply any upper limit on the Higgs mass.

In sect.2 , we shall show, by using effective potential techniques, that massless $\lambda\phi^4$ theories, undergoing spontaneous symmetry breaking, in four space-time dimensions, are asymptotically free in contrast with the perturbative calculations but not with the known rigorous results. Indeed , from ref. 8, one learns that asymptotic freedom provides the only possible non trivial realization of the theory, in four dimensions. In ref. 9, on the other hand, triviality is only proved rigorously in the symmetric phase, as the extension of the proof in the case of spontaneous symmetry breaking requires a delicate interpolation along the critical line of the bare parameters based on the validity of a weak coupling perturbative expansion (see sect. 4 of ref. 5.b).

In sect.3 ,we shall discuss how, _as a consequence_ of the asymptotic freedom of the massless theory,the shifted field is free in the continuum limit regardless of the magnitude of its mass.

Finally,in sect.4 we shall argue about the value of the Higgs mass in the standard model.

2. Asymptotic freedom of massless $\lambda\phi^4$ theories.

The starting point for our general analysis is the introduction of an effective potential (we shall consider theories described by a single coupling constant and without fundamental dimensionful parameters)

$$V = V(F) \tag{1}$$

which depends on some,space-time independent, "order parameter" F associated with the expectation value of some, local, operator $\hat{\Theta}$, i.e.

$$F = < \hat{\Theta} > \tag{2}$$

and, without loss of generality, the perturbative ground state corresponds to the case $F=0$.

Depending on the approximation used (loop expansion, variational method,...) $V(F)$, in general, does contain ultraviolet divergent terms. The usual procedure is to introduce a suitable regulator (lattice spacing a , ultraviolet cut-off Λ ,....) which gives a meaning to the "bare" parameters in $V(F)$; they represent, in an effective way, the result of integrating out the high frequency modes above the ultraviolet cut-off.

Renormalization Group (RG) invariance implies ($r = \Lambda = 1/a = ...$) the partial differential equation (our definition of $V(F)$ incorporates the subtraction of the perturbative ground state energy, i.e. $V(0) = 0$)

$$\left\{ r\frac{\partial}{\partial r} + \beta\frac{\partial}{\partial g} - \gamma F\frac{\partial}{\partial F} \right\} V(r, g, F) = 0 \tag{3}$$

where we have introduced the Callan-Symanzik β-function

$$\beta(g) = r \frac{\partial g}{\partial r} \qquad (4)$$

and the anomalous dimension $\gamma(g)$ defined by

$$r \frac{\partial F}{\partial r} = -\gamma(g) \, F \qquad (5)$$

Eqs. 3,4,5 define the renormalization procedure beyond perturbation theory.

The possibility of having non trivial minima is associated with the solution of

$$\left. \frac{\partial V}{\partial F} \right|_{r,g} = 0 \qquad (6)$$

With respect to g, depending on the theory under investigation, Eq. 6 may exhibit two distinct, characteristic features: it may require a critical value g_c for the coupling constant, as in the case of chiral symmetry breaking in massless QED[10], or it may possess non trivial extrema for arbitrarily small values of g ("essential instability").

The implementation of the absolute minimum condition Eq. 6, $F = \overline{F}(r, g)$, into the effective potential, provides the ground state energy density $W_0 = V(r, g, \overline{F}(r, g))$, which is a RG invariant quantity, satisfying

$$\left\{ r \frac{\partial}{\partial r} + \beta \frac{\partial}{\partial g} \right\} \, W_0 \, (r, g) = 0 \qquad (7)$$

In this approach, Eq. 7 defines the β-function of the theory, independently on any perturbative calculation, and $\gamma(g)$ can be deduced from Eq. 3 after β is known from Eq. 7.

Eq. 6 amounts to establish a simple proportionality relation between the two basic RG invariants

$$P = F \, exp \int^g dx \frac{\gamma(x)}{\beta(x)} \qquad (8)$$

$$\overline{r} = r \, exp - \int^g \frac{dx}{\beta(x)} \qquad (9)$$

associated with the various integral curves, Eqs. 4,5 in the (r, g, F) three-dimensional space.

Asymptotic freedom is, in this approach, a formal statement concerning the integral curves of Eq. 7. They are all driven toward the ultraviolet fixed point at $g = 0$ in the limit $r \to \infty$.

In the case of $\lambda\phi^4$ theory, the perturbative result for the β-function is

$$\beta_p(\lambda) = \frac{3\lambda^2}{16\pi^2} + O(\lambda^3) \qquad (10)$$

By using a variational method within the space of normalized gaussian state functionals [11] and regulating the ultraviolet divergences with an ultraviolet cutoff, or by employing the effective potential for composite operators of ref.12,and by using dimensional regularization[13], one finds indications for the occurrence of spontaneous symmetry breaking and a variational estimate of the β function $(0 < c \leq 1)$

$$\beta_{var}(\lambda) = -\frac{\lambda^2}{8\pi^2 c} + O(\lambda^3) \tag{11}$$

Clearly, in this situation, perturbation theory and stability analysis give opposite indications.Rather than entering the details of refs.11,13 we shall present,here,the discussion in the case of the one loop potential[14] ,which leads to the same results as the non perturbative methods of refs.11,13.

Let us define a massless $\lambda\phi^4$ theory as in ref.15 ,i.e. from the condition

$$\frac{d^2 V}{d^2 \phi}\bigg|_{\phi=0} = 0$$

in this case we obtain ,at the one loop level,[15] $(\lambda > 0)$,

$$V_{1-loop}(\phi^2) = \frac{\lambda}{4!}\phi^4 + \frac{\lambda^2\phi^4}{256\pi^2}\left(ln\frac{\lambda\phi^2}{2\Lambda^2} - \frac{1}{2}\right) \tag{12}$$

Motivated by the perturbative calculation in the symmetric phase where $\gamma = O(\lambda^2)$, we can find a first solution of eq. 3,

$$\gamma = O(\lambda^2) \tag{13}$$

$$\beta = +\frac{3\lambda^2}{16\pi^2} + O(\lambda^3) \tag{14}$$

Note that the one loop potential, in eq.12 has a minimum but the resummation of the leading logarithmic terms, coming from higher loops, associated with the above value of β makes the one loop minimum to disappear. The positive sign of β and the rigorous argument of ref.8 lead to the conclusion that, in this case, the theory is trivial.

Another solution of the RG eq. 3 is the following. Notice that, at the extrema of V

$$\frac{\partial V}{\partial \phi}\bigg|_{\phi=\bar{\phi}} = 0 \tag{15}$$

the anomalous dimension plays no role.Indeed $W_0(\lambda, \Lambda) = V(\lambda, \Lambda, \bar{\phi}(\lambda, \Lambda))$ satisfies eq.7. At one loop we obtain

$$\frac{\lambda\bar{\phi}^2}{2} = \Lambda^2 \, exp - \frac{32\pi^2}{3\lambda} \tag{16}$$

(which is equivalent to $P^2 = \bar{r}^2$ with P and $\bar{r}$ defined in eqs.8,9) and

$$W_0 = -\frac{M^4}{128\pi^2} \tag{17}$$

with $M^2 = \lambda\bar{\phi}^2/2$. Independently on any assumption about γ, we obtain from eq. 7

$$\beta_{ep} = -\frac{3\lambda^2}{16\pi^2} + O(\lambda^3) \tag{18}$$

which ,for weak coupling, implies asymptotic freedom. Going back to eq. (3) we obtain, in this case,

$$\gamma = -\frac{3\lambda}{32\pi^2} + O(\lambda^2) \tag{19}$$

or, to this order, $\gamma = \beta/2\lambda$.

Therefore, the RG invariance of the one-loop effective potential requires $\lambda\bar{\phi}^2$ to be a RG invariant quantity differently from the perturbative indications $Z_\phi = 1 + O(\lambda^2 ln\Lambda^2)$ or $\gamma(\lambda) = O(\lambda^2)$. At the same time the β-function in eq.18 yields a value of $c = 2/3$ in Eq. 11 in agreement with the variational arguments of refs. 11,13, $0 < c \leq 1$.

Notice that in this case, when β is negative there are no additional leading logarithmic corrections from higher loops in the effective potential (as in the case of Yang-Mills theories[16]) and, therefore, the effective potential of ref.15 does contain all the relevant effects. One should be careful when comparing our results with the existing two loop calculations of the effective potential. They make essential use of the one loop perturbative renormalization in the symmetric vacuum where β is positive and $\gamma = O(\lambda^2)$. At the same time,beyond one loop and beyond the gaussian approximation, a reliable estimate of the effective potential requires the solution of a non linear, integral gap equation[12] for the shifted field propagator. By following any other procedure the deep connection with the energy density is lost and one is producing a perturbative calculation not related, in any sense, to the stability analysis of the underlying quantum field theory and, as such, potentially wrong.

The crucial difference with respect to perturbation theory,

$$P^2 = \lambda\bar{\phi}^2 = RG \quad invariant \tag{20}$$

first noticed in ref.17 to renormalize the effective potential of ref.18, is a simple consequence of the fact that the renormalization properties of the background field and of the shifted field are unrelated (see sect.3) as in Yang-Mills theories[16]. By solving the renormalization group Eq. (3) with $\gamma = O(\lambda^2)$,which leads to the perturbative solution in eqs.13,14, one confuses the anomalous dimension of the shifted field with that of the constant, background field which is related to the β-function of the theory by the simple proportionality relation $\gamma = \beta/2\lambda$ (analogously to the relation $\gamma = \beta/g$ of Yang-Mills theories[16]).

Eq. 19 can be easily checked by explicit calculation in the shifted theory as follows.

By inserting Eq. 16 into Eq. 12 we obtain the "renormalized" one loop potential

$$V_{1-loop}(\phi) = \frac{\lambda^2 \phi^4}{256\pi^2} \left(ln\frac{\phi^2}{\bar{\phi}^2} - \frac{1}{2} \right) \tag{21}$$

which is the generating functional for 1-PI one loop diagrams at zero external momenta as deduced from the general expression

$$V(\phi) = -\sum_{n=0}^{\infty} \frac{1}{n!} (\phi - \bar{\phi})^n \, \Gamma^{(n)}(0, ..., 0) \tag{22}$$

Therefore, from the relation (remember that ϕ carries an anomalous dimension)

$$\frac{d^2V}{d^2\phi}\Big|_{\phi=\bar{\phi}} = \frac{m_R^2}{Z_\phi} \tag{23}$$

we obtain

$$Z_\phi = \frac{16\pi^2}{\lambda} \tag{24}$$

$$\gamma = \frac{\beta}{2\lambda} \tag{25}$$

due to the well known relation

$$\gamma = -\frac{1}{2}\Lambda \, \frac{\partial}{\partial\Lambda} \, ln Z_\phi \tag{26}$$

Analogous results for β and γ hold in the gaussian approximation [11,13]

$$V_{Gauss}(\phi) = \frac{\lambda^2 \phi^4}{576\pi^2} \left(ln\frac{\phi^2}{\bar{\phi}^2} - \frac{1}{2} \right) \tag{27}$$

up to a finite renormalization ($Z_\phi^{Gauss} = 24\pi^2/\lambda$).

The special role of the "classical" field should not surprise. Indeed, the non perturbative nature of Z_ϕ has a well known counterpart in many particle Bose systems due to the singularity in the quantum measure associated with the mode $k=0$. Finally, numerical evidences that Z_ϕ is not correctly predicted by perturbation theory can be found in ref. 19. In conclusion, both in the one loop effective potential and in the gaussian approximation, we obtain the following results :

1) There is a non trivial minimum at $\phi = \bar{\phi}$ in the effective potential for massless $\lambda\phi^4$ theories.

2) This minimum can be trusted as the relevant β-function is negative in the weak coupling limit, thus implying asymptotic freedom.

3) There is a non trivial, infinite rescaling of the background field.

The above results, remarkably, do not contraddict the rigorous results of refs. 8,9 and thus classically scale invariant, but quantum mechanically spontaneously broken $\lambda\phi^4$ theories are non trivial in the sense that they generate a scale which remains fixed in the continuum limit as required for a consistent interacting theory.

Finally,it is worth to recall here the results of ref.20 ,in support of our conclusions.By using a conformal mapping technique to eliminate the infrared singularities the interaction lagrangian for massless ϕ_4^4 is shown to exist as a hermitian operator in a dense domain of the free field Hilbert space and the S-matrix is shown to be non trivial.The conclusion of ref.20 is that " by virtue of these features there is no apparent non linear relativistic quantum field theory for which there is any greater indication of comprehensive non trivial existence than massless ϕ_4^4 theory".

3. The triviality of the shifted theory.

The results of sect.2, do not imply, by themselves , the existence of a non trivial S-matrix for the shifted theory. One can easily understand this point by noticing that, in the continuum limit, the bare coupling constant vanishes and one must introduce an infinite rescaling of the shifted field, as well, to obtain a non zero scattering amplitude. A non perturbative answer requires to explore the renormalization group equations in the one and two particle sectors of the Fock space of the shifted field $h(x)$. In this case, from the results of ref. 21, one obtains that the energy of the one particle states, $(p^2 + m_R^2)^{1/2}$, is automatically finite due to asymptotic freedom. Analogously, the two-particle state energy, in the center of mass system, acquires its minimum value at $M_2(0) = 2m_R$ up to terms which vanish in the infinite cutoff limit. Therefore ,no wave function renormalization is introduced for the shifted field. Similar results can be obtained by employing the usual definition $Z_h = 1 + \partial\Pi/\partial Q^2|_{Q^2=m_R^2}$, which yields $Z_h = 1 + O(\lambda) + O(\lambda^2 ln(\Lambda^2/\lambda\phi^2))$, and which goes to 1 in the infinite cutoff limit where $\Lambda \to \infty$, $\lambda \to 0$ and $\lambda ln(\Lambda/m_R) = fixed$. Notice that Z_h is the relevant quantity to renormalize the term $(\partial\phi)^2$ in the full effective action associated with non translationally invariant field configurations which cannot be described in terms of a bose condensate in the mode k=0.

Therefore the shifted field is free in the continuum limit, all interaction effects being reabsorbed into its mass and no non trivial S-matrix exists in this case.

Still,spontaneous symmetry breaking indicates a non trivial dynamics of the underlying massless theory which may be relevant for the standard model.Even in the absence of any self-interaction of the shifted field a non vanishing vacuum expectation value is a convenient device to generate the vector boson masses without destroying the crucial property of renormalizability.This conclusion was already obtained in ref.19.

4. The Higgs mass in the standard model.

In the case of the standard model our conclusions in sects.2 and 3 are still valid provided the bare self-coupling constant of the massless scalar theory $\lambda(\Lambda)$ is _defined_ to include all possible contributions from gauge fields and fermion loops as well. The only condition for the consistency of the analysis is that at some value of the ultraviolet cutoff Λ this global coupling has to be positive and small $(0 < \lambda(\Lambda)/16/\pi^2 << 1)$. Whatever definition of this bare parameter one employs, the associated "one loop" corrections will produce an effective potential of the form in eq.12 thus inducing spontaneous symmetry breaking and driving the quantity $\lambda(\Lambda) \to 0$ in the limit $\Lambda \to \infty$. As a consequence, starting from a massless theory, the scalar sector has no intrinsic ultraviolet cutoff and the Higgs mass is a free parameter which cannot be constrained by theoretical arguments, just like the gluon condensate in pure Yang-Mills theories.

At the same time the triviality of the shifted theory implies that the Higgs particle would be stable in the absence of the gauge couplings since its coupling to the massless Goldstone bosons vanish identically . Notice that ,in the absence of gauge bosons, and according to the Goldstone theorem, a property of the continuum theory, the massless Goldstone bosons are *exact* eigenstates of the hamiltonian, regardless of the value of the Higgs mass. Our result, implying that these states are free in the infinite cutoff limit, is completely consistent with the exact result differently from the usual triviality argument which predicts a strongly interacting Goldstone sector for large values of m_H.

Finally, by comparing with the experiments ,it is not difficult to obtain a semiquantitative estimate of the Higgs mass. By introducing the "renormalized" field

$$v = \phi(0) = \frac{\phi}{(Z_\phi)^{1/2}}$$

and keeping track of the different values of Z_ϕ in the two cases, both the one loop potential in eq.21 and the gaussian effective potential in eq.27 can be expressed as

$$V = \pi^2 v^4 \left(ln\frac{v^2}{\bar{v}^2} - \frac{1}{2} \right) \tag{28}$$

This particular form allows for a simple comparison with the Fermi constant which is related to the renormalized vacuum expectation value at low energy.In this case,from the relation

$$\frac{d^2V}{dv^2}\Big|_{v=\bar{v}} = m_H^2 \tag{29}$$

we obtain

$$m_H^2 = 8\pi^2\bar{v}^2 \tag{30}$$

or

$$m_H \sim 2 \ TeV \tag{31}$$

given the experimental result

$$\bar{v} = (\sqrt{2}\,G_F)^{-1/2} \sim 246\ GeV \tag{32}$$

The prediction in eq.(31) is not easy to test at the present. However, it is very well compatible with the results for the ρ parameter from the Z line shape. To this end we can introduce the parametrization[22] for the leptonic widths

$$\Gamma(Z \to l^+ l^-) = \Gamma_l = \rho \frac{G_F M_Z^3}{24\pi\sqrt{2}}(1 + (1 - 4sin^2\theta_{eff})^2)(1 + \frac{3}{4}\frac{\alpha}{\pi}) \tag{33}$$

$$\Gamma(Z \to \nu\bar{\nu}) = \rho \frac{G_F M_Z^3}{12\pi\sqrt{2}} \tag{34}$$

where $sin^2\theta_{eff}$ represents the value of $sin^2\theta_w$ relevant for the neutral current couplings of the Z to quarks and leptons and can be expressed in terms of ρ and the Z mass through the relation[23] $(\mu^2 = \pi\alpha/G_F/\sqrt{2} = (37.2802..GeV)^2)$

$$sin^2\theta_{eff} = \frac{1}{2}(1 - \sqrt{1 - \frac{4\mu^2}{\rho M_z^2}\frac{\alpha(M_z)}{\alpha}}) \tag{35}$$

Leaving out the Z hadronic width as a free parameter we obtain from our simultaneous fit to the hadronic and leptonic line shapes from LEP[24-27]

$$M_z = 91.176 \pm \underline{0.020}\ GeV$$

$$R_{had} = \frac{\Gamma_h}{\Gamma_l} = 20.96 \pm 0.11$$

$$\rho = 0.996 \pm 0.004$$

$$\chi^2/dof = 142/(180 - 3)$$

equivalent to

$$\Gamma_l = 83.1 \pm 0.4\ MeV$$

$$\Gamma_{inv} = 495 \pm 8\ MeV$$

$$\Gamma_h = 1743 \pm 9\ MeV$$

$$\Gamma_z = 2487 \pm 9\ MeV$$

By retaining in the ρ parameter the two potentially large effects due to a heavy Higgs[3] and to a heavy top[28] one finds

$$\rho = 1 + 3\frac{G_F m_t^2}{8\pi^2\sqrt{2}} - 3\frac{G_F M_z^2 sin^2\theta_{eff}}{4\pi^2\sqrt{2}} ln(\frac{m_H}{M_z}) \tag{36}$$

From our fit to ρ and the present experimental limit from CDF $m_t > 89\ GeV$ at the 95% C.L. it is clear that a negative contribution in eq.36 (i.e. a large Higgs mass) is welcome to obtain a good agreement with the experimental data.

Independently of this last remark on radiative corrections, it is apparent that our results provide the natural framework to accomodate a heavy Higgs particle ,as suggested in ref.1, and can be extremely important for the phenomenology at the future hadron colliders.

AKNOWLEDGEMENTS

I thank J.M.Cornwall,K.Huang,R.Jackiw and J.Polonyi for many useful discussions.

References

1. M.Veltman,*Acta Phys.Pol.* **B12** (1981) 437.

2. S.L.Glashow,*Nucl.Phys.* **22** (1961) 579.

3. M.Veltman,*Acta Phys.Pol.* **B8** (1977) 475.

4. S. Weinberg, *Phys. Rev. Lett.* **19** (1967) 1264;
   A.Salam, Proc. 8th Nobel Symp., N. Svartholm ed. (Almqvist and Wicksell, Stockolm, 1968), p. 367.

5 S.L.Glashow,J.Iliopoulos and L.Maiani, *Phys.Rev.* **D2** (1970) 1285.

6. J.van der Bij and M.Veltman, *Nucl.Phys.* **B231** (1984) 205.

7. J.Kuti,L.Lin and Y.Shen, *Phys.Rev.Lett* **61** (1988) 678

8. J.Frohlich, *Nucl. Phys.* **B200** (1982) 281.

9. a) M.Luscher and P.Weisz, *Nucl. Phys.* **B290** (FS20) (1987) 25; b) M.Luscher and P.Weisz,*Nucl. Phys.* **B295** (FS21) (1988) 65.

10. P.I.Fomin,V.P.Gusynin,V.A.Miransky and Yu.A.Sitenko, *Riv.Nuovo Cim.* **6** n.5 (1983) 1.

11. P.Castorina and M.Consoli, *Phys. Lett.* **B235** (1990) 302.

12. J.M.Cornwall,R.Jackiw and E.Tomboulis, *Phys. Rev.* **D10** (1974) 2428.

13. V.Branchina, P.Castorina, M.Consoli and D.Zappalà, *Phys. Rev.* **D42** (1990) 3587.

14. M.Consoli and D.Zappala','"Renormalization group and stability analysis of $\lambda\phi^4$ theories",Advanced Study Institute on Vacuum structure in Intense Fields" ,July 31-August 10,1990, Cargese,H.M.Fried and B.Muller eds.,Plenum Press 1991,Vol.225,pag.333.

15. S.Coleman and E.Weinberg, *Phys. Rev.* **D7** (1973) 1888.

16. L.F.Abbott, *Nucl. Phys.* **B185** (1983) 189.

17. P.M.Stevenson and R.Tarrach, *Phys. Lett.* **B176** (1986) 436.

18. M.Consoli and A.Ciancitto, *Nucl. Phys.* **B254** (1985) 653.

19. K.Huang,E.Manousakis and J.Polonyi, *Phys.Rev.* **D35** (1987) 3187.

20. J.Pedersen, I.E.Segal and Z.Zhou, "Massless ϕ_d^q quantum field theories and the nontriviality of ϕ_4^4", MIT preprint, April 1991.

21. P.M. Stevenson, *Phys. Rev.* **D 32** (1985) 1389.

22. M.Consoli,S.Lo Presti and L.Maiani,*Nucl.Phys.* **B223** (1983) 357.

23. M.Consoli,W.Hollik and F.Jegerlehner,contribution to Z Physics at LEP1, G.Altarelli,R.Kleiss and C.Verzegnassi eds, CERN/89-08,Vol.1,pag.1.

24. ALEPH Collab.,D.Decamp et al.,preprint CERN-PPE/91-105,July 1991.

25. DELPHI Collab.,P.Abreu et al.,preprint CERN-PPE/91-95,July 1991.

26. L3 Collab.,B.Adeva et al.,L3 preprint 28,February 1991.

27. OPAL Collab.,M.Z.Akrawy et al.,preprint CERN-PPE/91-67,April 1991.

28. M.Veltman, *Nucl.Phys.* **B123** (1977) 89

STRONG CP-VIOLATION
AND EFFECTIVE CHIRAL LAGRANGIANS

Eduardo de RAFAEL

Centre de Physique Théorique, Section 2
CNRS Luminy, Case 907, F-13288 Marseille Cedex 9

This talk is a review of recent work which Toni Pich and I have been doing in connection with the question of strong CP-violation in the Standard Model [1].

The source of strong CP-violation is the extra term which has to be added to the usual QCD Lagrangian in order to take into account non-perturbative properties of the QCD vacuum, the so called θ_0-term

$$\theta_0 \frac{g_s}{32\pi^2} \sum_a \frac{1}{2} \epsilon_{\mu\nu\rho\sigma} G_{(a)}^{\mu\nu}(x) G_{(a)}^{\rho\sigma}(x) \ , \tag{1}$$

which involves the product of the gluon field strength tensor $G_{(a)}^{\mu\nu}(x)$ with its dual; i.e., the same quantity which appears in the axial $U(1)_A$ anomaly and gives the divergence of the flavour-singlet axial current.

The new term (1) violates P, T and CP and may lead to observable effects in flavour conserving transitions. It may generate, in particular, a sizeable neutron electric dipole moment (nEDM), which very refined experiments have constrained down to a very high precision [2,3]. With the standard definition ($F_{\mu\nu}(x) \equiv \partial_\mu \mathcal{A}_\nu(x) - \partial_\nu \mathcal{A}_\mu(x)$)

$$\mathcal{L}_{nEDM} = \frac{d_n^\gamma}{2} \bar\psi_n(x) i\gamma_5 \sigma^{\mu\nu} \psi_n(x) F_{\mu\nu}(x) \tag{2}$$

for an electric dipole moment coupling, the most accurate measurements give [3]

$$d_n^\gamma = (-3 \pm 5) \times 10^{-26} e \ cm \ ; \tag{3}$$

and a further improvement in sensitivity by a factor of 10 is still expected from the same experiment within the next two years [4].

In fact, the observable parameter is not quite θ_0 in (1) but rather the combination

$$\theta \equiv \theta_0 + \arg(\det\mathcal{M}) \ , \tag{4}$$

where $\mathcal{M}$ denotes the full mass matrix emerging from the Yukawa couplings of the light quarks in the electroweak sector of the Standard Model. In the absence of the θ_0-vacuum term in the QCD-Lagrangian, the phase $\arg(\det \mathcal{M})$ could be reabsorbed by a simple $U(1)_A$ rotation of the quark fields. However, because of the $U(1)_A$ anomaly, the θ_0-vacuum is not invariant under $U(1)_A$ transformations. The $U(1)_A$ rotation which eliminates arg(det $\mathcal{M}$) from the mass term generates a new θ_0-vacuum value, which is the combination appearing in eq.(4) above.

There has been some recent confusion in the literature as to whether or not one can reliably compute low energy effects which may originate in the QCD θ_0-term. In view of this, we have taken a fresh new look at this subject from the point of view of the chiral effective Lagrangian formulation of the strong CP-violation sector of QCD. The question of how to calculate long-distance effects like $\eta \to \pi\pi$ and d_n^γ, to some approximation at least, translates then into the question of to what level in a systematic chiral expansion there appear new constants in the effective Lagrangian not fixed by symmetry arguments alone, and unknown from phenomenology. The details of our analysis can be found in ref.[1]. Here, I shall limit myself to stress some basic points.

For three light flavours u, d, s, and to lowest non-trivial order in the $1/N_c$-expansion (N_c = number of colours) the chiral symmetry breaking effect induced by the $U(1)_A$ anomaly can be taken into account in the effective Lagrangian, through the term (see Di Vecchia and Veneziano [5] and Witten [6])

$$\mathcal{L}_{U(1)A} = -\frac{f_\pi^2}{4}\frac{a}{N_c}\left\{\frac{i}{2}[\log(\det\tilde{U}) - \log(\det\tilde{U}^\dagger)]\right\}^2,\qquad(5)$$

which breaks $U(3)_L \otimes U(3)_R$ but preserves $SU(3)_L \otimes SU(3)_R \otimes U(1)_V$. Here, $\tilde{U}$ denotes a unitary 3×3 matrix which parametrizes the Goldstone excitations over the vacuum. Under the chiral group $U(3)_L \otimes U(3)_R$, the field matrix $\tilde{U}(\phi)$ transforms as $\tilde{U} \to g_R \tilde{U} g_L^\dagger$ ($g_{R,L} \in U(3)_{R,L}$). It is convenient to factor out from $\tilde{U}(\phi)$ its vacuum expectation value, i.e.

$$\tilde{U}(\phi) = <\tilde{U}> U(\phi),\qquad(6)$$

with $<U>= 1$. A useful parametrization for $U(\phi)$, which we shall adopt, is then

$$U(\phi) \equiv \exp(-i\sqrt{2}\Phi(x)/f_\pi),\qquad(7)$$

where ($\vec{\lambda}$ are Gell-Mann's matrices with $\mathrm{tr}\lambda_a\lambda_b = 2\delta_{ab}$)

$$\Phi(x) \equiv \frac{\phi^{(0)}}{\sqrt{3}} + \frac{\vec{\lambda}}{\sqrt{2}}\vec{\phi} = \begin{pmatrix} \frac{\pi^0}{\sqrt{2}} + \frac{\eta_8}{\sqrt{6}} + \frac{\eta_1}{\sqrt{3}} & \pi^+ & K^+ \\ \pi^- & -\frac{\pi^0}{\sqrt{2}} + \frac{\eta_8}{\sqrt{6}} + \frac{\eta_1}{\sqrt{3}} & K^0 \\ K^- & \bar{K}^0 & -\frac{2\eta_8}{\sqrt{6}} + \frac{\eta_1}{\sqrt{3}} \end{pmatrix}.\qquad(8)$$

The constant f_π, which is not fixed by symmetry requirements, is phenomenologically determined from the decay $\pi \to \mu\nu (f_\pi \simeq 93.3 MeV)$.

To lowest order in the number of derivatives, and in powers of $\mathcal{M}$ and external v_μ and a_μ fields, the most general effective Lagrangian invariant under local chiral transformations is given by

$$\mathcal{L}_{eff} = \frac{f_\pi^2}{4} \operatorname{tr}(D_\mu \tilde{U} D^\mu \tilde{U}^\dagger + \tilde{\chi} \tilde{U}^\dagger + \tilde{U} \tilde{\chi}^\dagger), \tag{9}$$

where

$$D_\mu \tilde{U} \equiv \partial_\mu \tilde{U} - i r_\mu \tilde{U} + i \tilde{U} l_\mu , \tag{10}$$

with r_μ and l_μ Hermitian 3×3 matrices in flavour space which under local chiral rotations transform as

$$l_\mu \equiv v_\mu - a_\mu \rightarrow g_L l_\mu g_L^\dagger + i g_L \partial_\mu g_L^\dagger ,$$
$$r_\mu \equiv v_\mu + a_\mu \rightarrow g_R r_\mu g_R^\dagger + i g_R \partial_\mu g_R^\dagger , \tag{11}$$

(hence $D_\mu \tilde{U} \rightarrow g_R D_\mu \tilde{U} g_L^\dagger$) ; and $\tilde{\chi}$ is a 3×3 matrix proportional to $\mathcal{M}(\mathcal{M} \rightarrow g_R \mathcal{M} g_L^\dagger)$

$$\tilde{\chi} \equiv 2 B_0 \mathcal{M} , \tag{12}$$

with B_0 a constant which, like f_π , is not fixed by symmetry requirements alone. Once special directions in flavour space are selected for the external fields or for the matrix $\mathcal{M}$, chiral symmetry is of course explicitly broken. For instance, to introduce electromagnetic interactions one should take $l_\mu = r_\mu = e Q A_\mu$, with $Q \equiv \operatorname{diag}(\frac{2}{3}, \frac{-1}{3}, \frac{-1}{3})$. The important point is that (9) then breaks chiral symmetry in exactly the same way as the fundamental Standard Model Lagrangian does.

Performing an appropriate chiral transformation, the quark mass matrix can be restricted to the form

$$\mathcal{M} = e^{i\theta/3} \operatorname{diag}(m_u, m_d, m_s) \equiv e^{i\theta/3} M , \tag{13}$$

with θ the full θ-vacuum angle in (4) . In the absence of the term (5) , the phase in (13) could be reabsorbed by the $U(1)_A$ transformation

$$\tilde{U} \rightarrow e^{i\theta/6} \tilde{U} e^{i\theta/6} . \tag{14}$$

In the presence of the $U(1)_A$ anomaly, and hence the term (5), this transformation generates new physical interactions. With (13) inserted in (9) , and (14) applied to the $\mathcal{L}_{U(1)_A}$ Lagrangian in (5) , the new form of the effective bosonic Lagrangian in the presence of the QCD θ-vacuum term is then

$$\mathcal{L}_{eff} = \frac{f_\pi^2}{4} \left\{ \operatorname{tr}(D_\mu \tilde{U} D^\mu \tilde{U}^\dagger + \tilde{\chi}^\dagger \tilde{U} + \tilde{U}^\dagger \tilde{\chi}) - \frac{a}{N_c} \{ \frac{i}{2}[\log(\det \tilde{U}) - \log(\det \tilde{U}^\dagger)] - \theta \}^2 \right\} \tag{15}$$

where now the matrix $\tilde{\chi}$ is real, positive and diagonal

$$\tilde{\chi} = \tilde{\chi}^\dagger = \mathrm{diag}(\chi_u^2, \chi_d^2, \chi_s^2). \tag{16}$$

If the term proportional to a/N_c were absent, we could take without loss of generality $<\tilde{U}> = 1$ and the diagonal entries χ_i^2 would correspond to the Goldstone boson masses:

$$
\begin{aligned}
\chi_u^2 &= m_{\pi^+}^2 + m_{K^+}^2 - m_{K^0}^2, \\
\chi_d^2 &= m_{\pi^+}^2 + m_{K^0}^2 - m_{K^+}^2, \\
\chi_s^2 &= m_{K^+}^2 + m_{K^0}^2 - m_{\pi^+}^2.
\end{aligned}
\tag{17}
$$

In this case, the constant B_0 introduced in eq. (12) directly relates the pseudoscalar masses to the current quark masses of the QCD Lagrangian:

$$B_0 = \frac{m_{\pi^+}^2}{m_u + m_d} = \frac{m_{K^+}^2}{m_u + m_s} = \frac{m_{K^0}^2}{m_d + m_s} = \frac{3 m_{\eta_8}^2}{m_u + m_d + 4 m_s}. \tag{18}$$

In the presence of the third term in eq. (15), $<\tilde{U}>$ cannot be set equal to the unit matrix, and therefore it is convenient, before applying this Lagrangian to calculate physical processes, to minimize the potential energy associated to $\mathcal{L}_{eff}$ in (15):

$$V(\tilde{U}) = -\frac{f_\pi^2}{4}\left\{ \mathrm{tr}(\tilde{\chi}^\dagger \tilde{U} + \tilde{U}^\dagger \tilde{\chi}) - \frac{a}{N_c}\{\frac{i}{2}[\log\left(\frac{\det \tilde{U}}{\det \tilde{U}^\dagger}\right)] - \theta\}^2 \right\}, \tag{19}$$

so as to fix $<\tilde{U}>$. With $\tilde{\chi}$ diagonal, $<\tilde{U}>$ can be restricted to be diagonal as well and of the form

$$<\tilde{U}> = \mathrm{diag}(e^{-i\varphi_u}, e^{-i\varphi_d}, e^{-i\varphi_s}). \tag{20}$$

The minimization conditions $\partial V/\partial \varphi_i = 0$ restrict the φ_i's to satisfy the Dashen [7] Nuyts [8] equations:

$$\chi_i^2 \sin \varphi_i = \frac{a}{N_c}(\theta - \sum_j \varphi_j), \qquad (i = u, d, s). \tag{21}$$

The φ_i's appearing in $\mathcal{L}_{eff}$ in eq. (15) can be reabsorbed in Hermitian matrices χ and H defined by

$$
\begin{aligned}
<\tilde{U}^\dagger> \tilde{\chi} &\equiv \chi + iH, \\
\tilde{\chi}^\dagger <\tilde{U}> &\equiv \chi - iH.
\end{aligned}
\tag{22}
$$

In fact, eqs. (20) fix H to be proportional to the unit matrix,

$$H = \frac{a}{N_c}(\theta - \sum_j \varphi_j) I \equiv \frac{a}{N_c}\bar{\theta} I. \tag{23}$$

The effective bosonic Lagrangian as a functional of $U(\phi)$ (eq. (6)), with $< U >= 1$, is then ($\bar{\theta} \equiv \theta - \sum_j \varphi_j$)

$$\mathcal{L}_{eff} = \frac{f_\pi^2}{4}\left\{\mathrm{tr}(D_\mu U D^\mu U^\dagger + \chi(U + U^\dagger)) - \frac{a}{N_c}\{\bar{\theta}^2 - \frac{1}{4}[\log(\frac{\det U}{\det U^\dagger})]^2\}\right.$$
$$\left. - i\frac{a}{N_c}\bar{\theta}\{\mathrm{tr}(U - U^\dagger) - \log(\frac{\det U}{\det U^\dagger})\}\right\}. \tag{24}$$

We are now in the position to discuss two salient physical features of this Lagrangian.

i) In the chiral limit $\chi \to 0$, the singlet η_1-particle acquires a mass from the third term induced by the $U(1)_A$-anomaly,

$$-\frac{1}{2}\frac{3a}{N_c}\eta_1(x)\eta_1(x). \tag{25}$$

ii) The last term in $\mathcal{L}_{eff}$ above generates strong CP-violating transitions between pseudoscalar particles. In particular it induces the phase-space allowed decays $\eta_8 \to \pi^+\pi^-, \pi^0\pi^0$ and $\eta_1 \to \pi^+\pi^-, \pi^0\pi^0$. The transition amplitudes for $\eta \to \pi\pi$ can be readily obtained from this effective Lagrangian. (Notice that in the effective Lagrangian formulation of eq. (24) tadpole-like diagrams have been eliminated via the correct vacuum alignment.) The result is

$$T(\eta \to \pi^+\pi^-) = (\cos\theta_P - \sqrt{2}\sin\theta_P)\frac{a}{N_c}\frac{\bar{\theta}}{\sqrt{3}f_\pi}, \tag{26}$$

where we have taken into account the $\eta_1 - \eta_8$ mixing

$$\eta = \eta_8 \cos\theta_P - \eta_1 \sin\theta_P,$$
$$\eta' = \eta_8 \sin\theta_P + \eta_1 \cos\theta_P. \tag{27}$$

The result in eq. (26) is in agreement with the earlier calculations in refs.[9] and [10].

We shall now discuss baryons and their interactions within the framework of chiral dynamics [11]. Here we shall adopt the standard non-linear representation of baryons and recall only the basics.

The wanted ingredient for a non-linear representation of the chiral group is the compensating $U(3)_V$ transformation $h(\phi, g)$ which appears under the action of the chiral $U(3)_L \otimes U(3)_R$ group on the left $\tilde{\xi}_L(\phi)$ and right $\tilde{\xi}_R(\phi)$ coset representatives ($g \equiv g_L \otimes g_R$):

$$\tilde{\xi}_L(\phi) \to g_L\tilde{\xi}_L(\phi)h^\dagger(\phi, g),$$
$$\tilde{\xi}_R(\phi) \to g_R\tilde{\xi}_R(\phi)h^\dagger(\phi, g). \tag{28}$$

In terms of $\tilde{\xi}_L(\phi)$ and $\tilde{\xi}_R(\phi)$, the unitary matrix $\tilde{U}(\phi)$ introduced in (6) and (7) is defined by the product

$$\tilde{U}(\phi) = \tilde{\xi}_R(\phi)\tilde{\xi}_L^\dagger(\phi). \tag{29}$$

The octet of baryon fields is then collected in a 3×3 matrix

$$B(x) = \begin{pmatrix} \frac{\Sigma^0}{\sqrt{2}} + \frac{\Lambda^0}{\sqrt{6}} & \Sigma^+ & p \\ \Sigma^- & -\frac{\Sigma^0}{\sqrt{2}} + \frac{\Lambda^0}{\sqrt{6}} & n \\ \Xi^- & \Xi^0 & -\frac{2}{\sqrt{6}}\Lambda^0 \end{pmatrix}, \qquad (30)$$

which under $U(3)_L \otimes U(3)_R$ transforms non-linearly

$$B \to h(\phi, g) \, B \, h^\dagger(\phi, g). \qquad (31)$$

We look for the most general $U(3)_L \otimes U(3)_R$ invariant effective Lagrangian one can write in terms of the matrices $B(x)$, $\bar{B}(x) \equiv B(x)^\dagger \gamma_0$, $\tilde{\xi}_L(\phi)$ and $\tilde{\xi}_R(\phi)$; and we wish to keep only the dominant terms which contribute to physical processes as leading powers in momenta and quark masses. The baryon-meson effective Lagrangian has a kinetic term

$$\mathcal{L}_{kin}^{(B)} = \mathrm{tr}(\bar{B}i\gamma^\mu D_\mu B) - M_B \mathrm{tr}(\bar{B}B), \qquad (32)$$

where M_B is a common mass term to all baryons, and D_μ denotes here the covariant derivative

$$D_\mu B = \partial_\mu B + [\Gamma_\mu, B], \qquad (33)$$

with

$$\Gamma_\mu \equiv \frac{1}{2}\{\tilde{\xi}_R^\dagger(\partial_\mu - ir_\mu)\tilde{\xi}_R + \tilde{\xi}_L^\dagger(\partial_\mu - il_\mu)\tilde{\xi}_L\}, \qquad (34)$$

and r_μ, l_μ the external fields introduced in (10) and (11). To the kinetic term in (32) we have to add possible interaction terms of $O(p)$ as well. These are

$$\mathcal{L}_{int}^{(B)} = -\frac{D}{2}\,\mathrm{tr}(\bar{B}\gamma^\mu\gamma_5\{\tilde{\xi}_\mu(\phi), B\})$$
$$-\frac{F}{2}\,\mathrm{tr}(\bar{B}\gamma^\mu\gamma_5[\tilde{\xi}_\mu(\phi), B])$$
$$+ S\,\mathrm{tr}(\tilde{\xi}_\mu(\phi))\,\mathrm{tr}(\bar{B}\gamma^\mu\gamma_5 B), \qquad (35)$$

with

$$\tilde{\xi}_\mu(\phi) \equiv i\{\tilde{\xi}_R^\dagger(\partial_\mu - ir_\mu)\tilde{\xi}_R - \tilde{\xi}_L^\dagger(\partial_\mu - il_\mu)\tilde{\xi}_L\}. \qquad (36)$$

Under $U(3)_L \otimes U(3)_R$ gauge transformations,

$$\tilde{\xi}_\mu(\phi) \to h(\phi, g)\,\tilde{\xi}_\mu(\phi)\,h^\dagger(\phi, g). \qquad (37)$$

The first two terms in (35) are the usual F and D couplings, which govern semilep-tonic hyperon decays. The third S coupling is specific to the axial flavour-singlet baryonic current.

Of special interest for our purposes are the possible terms generated by the explicit chiral symmetry breaking induced by quark mass terms in the underlying

QCD Lagrangian. The possible lowest $O(\mathcal{M})$ interactions induced in the effective meson-baryon Lagrangian are

$$\mathcal{L}_{\mathcal{M}}^{(B)} = -b_0 \operatorname{tr}(\tilde{\chi}_+) \operatorname{tr}(\bar{B}B) - b_1 \operatorname{tr}(\bar{B}\tilde{\chi}_+ B) - b_2 \operatorname{tr}(\bar{B}B\tilde{\chi}_+), \qquad (38)$$

where b_0, b_1 and b_2 are coupling constants with dimensions of an inverse mass, and $\tilde{\chi}_\pm$ is a shorthand notation for

$$\tilde{\chi}_\pm \equiv \tilde{\xi}_R^\dagger \tilde{\chi} \tilde{\xi}_L \pm \tilde{\xi}_L^\dagger \tilde{\chi}^\dagger \tilde{\xi}_R. \qquad (39)$$

Under $U(3)_L \otimes U(3)_R$ gauge transformations,

$$\tilde{\chi}_\pm \rightarrow h(\phi, g) \tilde{\chi}_\pm h^\dagger(\phi, g). \qquad (40)$$

With $\mathcal{M}$ restricted to the form in eq. (13) , and in the absence of the $U(1)_A$ anomaly, the phase in (13) could be reabsorbed by the same $U(1)_A$ transformation as in (14) i.e.,

$$\tilde{\xi}_L(\phi) \rightarrow e^{-i\theta/6} \tilde{\xi}_L(\phi) h^\dagger(\phi, \theta) \qquad \tilde{\xi}_R(\phi) \rightarrow e^{i\theta/6} \tilde{\xi}_R(\phi) h^\dagger(\phi, \theta). \qquad (41)$$

In the presence of the $U(1)_A$ anomaly, and hence the term (5), in the mesonic Lagrangian, this transformation generates new physical interactions between mesons, as we have seen previously, and new interactions as well between mesons and baryons, as we are going to see next. Again, before we proceed to analyze physical implications, it is convenient to rewrite the effective Lagrangian in a form compatible with the correct vacuum alignment.

From eqs. (22) and (29) it follows that

$$< \tilde{\xi}_R^\dagger > \tilde{\chi} < \tilde{\xi}_L > = < \tilde{\xi}_L > < \tilde{\xi}_R^\dagger > \tilde{\chi} = \chi + iH, \qquad (42)$$

where we have used the fact that $\tilde{\chi}$ is diagonal and , without loss of generality, $< \tilde{\xi}_L >$ and $< \tilde{\xi}_R >$ can be restricted to be diagonal as well. It is then convenient to introduce field matrices ξ_L and ξ_R, so that

$$\tilde{\xi}_L = < \tilde{\xi}_L > \xi_L \qquad \text{and} \qquad \tilde{\xi}_R = < \tilde{\xi}_R > \xi_R, \qquad (43)$$

with $< \xi_L > = < \xi_R > = 1$ and $U = \xi_R \xi_L^\dagger$. Furthermore, it is always possible to choose a coset representative such that

$$\xi_R = \xi_L^\dagger = \xi \quad ; \quad U = \xi^2. \qquad (44)$$

We then have

$$\tilde{\chi}_\pm = \chi_\pm + i\frac{a}{N_c}\bar{\theta}(U^\dagger \mp U), \qquad (45)$$

where

$$\chi_\pm \equiv \xi^\dagger \chi \xi^\dagger \pm \xi \chi \xi. \tag{46}$$

Inserting (45) in $\mathcal{L}_{\mathcal{M}}^{(B)}$ as given in (38), leads to CP non-conserving meson-baryon interaction terms modulated by the coupling $\frac{a}{N_c}\bar{\theta}$:

$$
\begin{aligned}
\mathcal{L}_{\mathcal{M}}^{(B)} = &- b_0 \operatorname{tr}(\chi_+) \operatorname{tr}(\bar{B}B) - b_1 \operatorname{tr}(\bar{B}\chi_+ B) - b_2 \operatorname{tr}(\bar{B}B\chi_+) \\
&- i\frac{a}{N_c}\bar{\theta} \left\{ b_0 \operatorname{tr}(U^\dagger - U)\operatorname{tr}(\bar{B}B) + b_1 \operatorname{tr}(\bar{B}(U^\dagger - U)B) \right. \\
&\left. + b_2 \operatorname{tr}(\bar{B}B(U^\dagger - U)) \right\}.
\end{aligned}
\tag{47}
$$

The first three terms in $\mathcal{L}_{\mathcal{M}}^{(B)}$ give contributions to baryon masses. From the experimentally known baryon mass splittings it is then possible to obtain the couplings b_1 and b_2, with the result (for $m_u = m_d = m$)

$$b_1 = \frac{M_\Xi - M_\Sigma}{4(m_K^2 - m_\pi^2)} \approx 0.14\,\text{GeV}^{-1} \quad , \quad b_2 = \frac{M_N - M_\Sigma}{4(m_K^2 - m_\pi^2)} \approx -0.28\,\text{GeV}^{-1}. \tag{48}$$

The term with b_0 gives an overall contribution to the baryon mass M_B, and therefore cannot be extracted from baryon mass splittings. The interesting terms for our purposes are the ones proportional to $\frac{a}{N_c}\bar{\theta}$. The interactions with the lowest possible number of fields, induced by these terms, are of the type

$$
\begin{aligned}
\mathcal{L}_{\bar{\theta}}^{(B)} = &-i\frac{a}{N_c}\bar{\theta}\frac{2i\sqrt{2}}{f_\pi} \left\{ \frac{n_f}{\sqrt{3}}b_0\,\eta_1 \operatorname{tr}(\bar{B}B) + b_1 \operatorname{tr}(\bar{B}\Phi B) + b_2 \operatorname{tr}(\bar{B}B\Phi) \right. \\
&\left. + O(\bar{B}B\phi\phi/f_\pi^2) \right\}.
\end{aligned}
\tag{49}
$$

We have now all the ingredients to do the chiral loop calculation of the neutron electric dipole moment. The possible Feynman diagrams which can generate a nEDM at the one-loop level are shown in fig. 1. The continuous line represents a baryon, the dashed line a meson, the wavy line the photon. The vertex with a dot is the CP-violating interaction induced by one of the couplings in eq. (49).

The normal CP-conserving $\bar{B}B\phi$ vertex interactions are those generated by the interaction Lagrangian $\mathcal{L}_{int}^{(B)}$ in (35), with ξ_μ in eq. (36) restricted to the term

$$\xi_\mu \doteq \frac{\sqrt{2}}{f_\pi}\left(\partial_\mu\Phi - ie\mathcal{A}_\mu[Q, \Phi]\right), \tag{50}$$

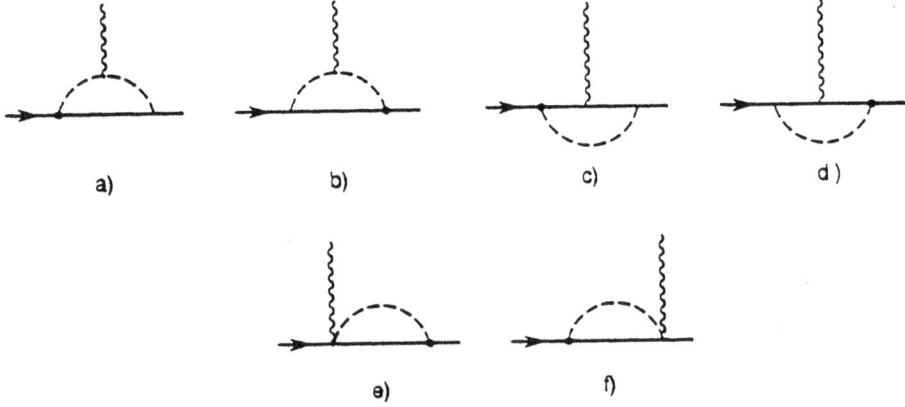

Fig.1   Feynman diagrams contributing to the nEDM at the one-loop level. The continuous line represents a baryon, the dashed line a meson, the wavy line the photon. The vertex with a dot is the CP-violating interaction induced by one of the couplings in eq. (49).

where Φ is the 3×3 matrix in (8), $\mathcal{A}_\mu$ the electromagnetic field and $Q = \text{diag}(\frac{2}{3}, \frac{-1}{3}, \frac{-1}{3})$ the electric charge matrix. The fact that the propagating pseudoscalar has to be charged restricts the CP-conserving $\bar{B}B\phi$ vertex interactions to those of F and D type only. There is no contribution from the S-coupling at the one-loop level.

We also require the $\phi\phi\gamma$ and $\bar{B}B\gamma$ vertices which follow from the first term in (24),

$$\mathcal{L}_{em} \doteq -ie\mathcal{A}_\mu(\phi^+\partial^\mu\phi^- - \phi^-\partial^\mu\phi^+) + e^2\mathcal{A}_\mu\mathcal{A}^\mu\phi^+\phi^-, \tag{51}$$

with $\phi = \pi, K$; and from the kinetic term $\mathcal{L}_{kin}^{(B)}$ in (32) , i.e.

$$\mathcal{L}_{em}^{(B)} = e\mathcal{A}_\mu \, \text{tr}\left(\bar{B}\gamma^\mu[\frac{1}{2}(\xi^\dagger Q\xi + \xi Q\xi^\dagger), B]\right). \tag{52}$$

The calculation is then rather straightforward. Only diagrams (a) and (b) give chiral logarithms, plus constant terms which we keep and higher order terms in the chiral expansion which are neglected. Diagrams (c) and (d) are suppressed by an additional baryon propagator; they don't give chiral logarithms, but produce constant terms which happen to cancel the constant terms from (a) and (b). Diagrams (e) and (f) cancel each other. The final result from the one-loop calculation is then

$$d_n^\gamma = \frac{a}{N_c}\bar{\theta}\,\frac{e}{16\pi^2 f_\pi^2}\left\{\frac{M_\Xi - M_\Sigma}{m_K^2 - m_\pi^2}(D{+}F)\log(\frac{M_N^2}{m_\pi^2}) + \frac{M_\Sigma - M_N}{m_K^2 - m_\pi^2}(D{-}F)\log(\frac{M_\Sigma^2}{m_K^2})\right\}. \tag{53}$$

The baryon mass in the chiral logarithm acts as an ultra-violet cut-off and should be considered as such. It is simply telling us that a complete calculation of d_n^γ to $O(p^2)$ must necessarily bring new local counterterms of the same chiral order, with coupling constants not fixed by symmetry arguments alone and probably rather hard to determine from phenomenology.

The main conclusion from our analysis is that the coefficient of the leading chiral logarithm ($\log m_\pi$) which contributes to the nEDM, as a result of strong CP-violation in the Standard Model, is unambiguously fixed by low energy phenomenology. The expression we find in eq. (53) has all the expected factors present: i) it vanishes in the large N_c limit; ii) it vanishes in the chiral limit where $\bar\theta \to 0$ [see eqs. (21) and (23)] ; iii) it is proportional to the couplings b_1 and b_2 [see eq. (48)] responsible for the baryon mass splitting; iv) it is proportional to the baryon D and F couplings; and v) it has the suppression factor $\frac{1}{16\pi^2 f_\pi^2}$ characteristic of a chiral loop.

Our result in (53) confirms the basic claims of the early work by the authors of ref. [9] and Di Vecchia [12].

In ref. [1] we have also identified the set of local couplings which can give contributions to the nEDM at the some $0(\mathcal{M})$ in ChPT. In principle, these couplings could be determined from a phenomenological comparison of data on $\gamma N \to \pi N$ and $\pi N \to \pi N$ reactions near threshold with ChPT predictions.

We have also calculated the electric dipole moment of the Λ, to lowest $O(\mathcal{M})$, with the result

$$d_\Lambda^\gamma = \frac{1}{2} d_n^\gamma. \tag{54}$$

In order to make a numerical estimate of d_n^γ, it is convenient to use the empirical fact that

$$\chi_u^2, \chi_d^2 \ll \chi_s^2, \frac{a}{N_c} \tag{55}$$

and use the approximate relation ($m_u \approx m_d = m \ll m_s$)

$$\frac{a}{N_c}\bar\theta \approx \frac{\theta}{\sum_i \frac{1}{\chi_i^2} + \frac{N_c}{a}} \approx \frac{1}{2}m_\pi^2\theta. \tag{56}$$

Using the values [13]

$$D = 0.61 \pm 0.04 \qquad \text{and} \qquad F = 0.40 \pm 0.03. \tag{57}$$

and the physical nucleon mass in the chiral logarithms, we get from the chiral loop expression in (53) the result

$$d_n^\gamma\big|_{Loop} = 3.3 \times 10^{-16}\,\theta\,e\,cm, \tag{58}$$

As an estimate of the error coming from the unknown contribution of the constant terms, we propose to vary the scale in the chiral logarithm between the value of the constituent quark mass, $M_Q \sim 320 \, MeV$, and the average mass scale $M_\Delta \sim 1500 \, MeV$ of the baryon decuplet. This gives as our final estimate

$$d_n^\gamma = (3.3 \pm 1.8) \times 10^{-16} \, \theta \, e \, cm \,. \tag{59}$$

From a comparison between this result and the experimental upper limit in eq. (3) , we conclude that

$$|\theta| < 5 \times 10^{-10} \,. \tag{60}$$

This is a much more stringent limit, of course, than the one obtained from $\eta \to \pi^+\pi^-$. The predicted branching ratio here (using $\theta_P = -20°$ for the $\eta_1 - \eta_8$ mixing angle) is

$$\mathrm{Br}(\eta \to \pi^+\pi^-) = 1.8 \times 10^2 \, \theta^2 \,; \tag{61}$$

while the present experimental upper limit is

$$\mathrm{Br}(\eta \to \pi^+\pi^-) < 1.5 \times 10^{-3} \,, \tag{62}$$

from which θ is limited to be $|\theta| < 3 \times 10^{-3}$.

The comparison between our result for the ΛEDM (54) and the present experimental upper bound (14) , $d_\Lambda^\gamma < 1.5 \times 10^{-16} \, e \, cm$ (95 % C.L.) , only limits θ to $|\theta| < 2$.

Acknowledgements

The Conference on Gauge Theories in honor of the 60th birthday of Martinus Veltman was a great success. I wish to thank the organizers for their kind hospitality.

References

1. A. Pich and E. de Rafael, *Nucl. Phys.* **B367** (1991) 313.
2. I.S. Altarev et al., *JETP Lett.* **44** (1987) 460.
3. K.F. Smith et al., *Phys. Lett.* **B234** (1990) 191.
4. N.F. Ramsey, private communication.
5. P. Di Vecchia and G. Veneziano, *Nucl. Phys.* **B171** (1980) 253.
6. E. Witten, *Ann. of Phys.* (N.Y.) **128** (1980) 363.
7. R. Dashen, *Phys. Rev.* **D3** (1971) 1879.
8. J. Nuyts, *Phys. Rev. Lett.* **26** (1971) 1604 ; **27** (1971) 361 (E).
9. R.J. Crewther, P. Di Vecchia, G. Veneziano and E. Witten, *Phys. Lett.* **B88** (1979) 123 ; **B91** (1980) 487 (E).
10. M.A. Shifman, A.I. Vainshtein and V.I. Zakharov, *Nucl. Phys.* **B166** (1980) 493.
11. S. Coleman, J. Wess and B. Zumino, *Phys. Rev.* **177** (1969) 2239
C. Callan, S. Coleman, J. Wess and B. Zumino, *Phys. Rev.* **177** (1969) 2247.

12. P. Di Vecchia, *Acta Phys.* Austriaca Suppl. **22** (1980) 477.
13. E. Jenkins and A.V. Manohar, Univ. California-San Diego preprint UCSD/PTH 91-05.
14. L. Pondrom et al. *Phys. Rev.* **D23** (1981) 814.

IS YOUR WEIGHT GAUGE DEPENDENT ?

Bernard DE WIT
Institute for Theoretical Physics,
University of Utrecht, Princetonplein 5
3584 CC Utrecht

ABSTRACT
In the presence of a continuous degeneracy of stationary points of the effective action the standard proofs of the gauge independence of the S-matrix become inapplicable. This explains the apparent lack of gauge independence of physical quantities in (perturbative) quantum gravity found in explicit calculations. We show that the S-matrix is still gauge independent when expressed in terms of physically relevant parameters.

As students in Utrecht in the late Sixties, when we started working on gauge theories, it was unavoidable that we had to acquaint ourselves with Tini Veltman's work on gauge theories. I still remember being struck by two remarkable features. First of all, Tini was not afraid of ghosts. It seemed that he, more than anybody else in the physics community, had accepted the idea that it was perfectly all right to live in an extended version of the physical Hilbert space, just because life was more convenient there. Secondly, he had developed a very personal style in dealing with Feynman diagrams and combinatorics. He freely introduced all sorts of new lines, vertices and even fish into the diagrams[1], was very experienced in cutting graphs, and was a renowned expert in extracting practical results from them. Occasionally, he would confess to us that he thought that diagrams were really much more fundamental than the underlying Lagrangian or Hamiltonian. In those days Tini belonged to a small minority of theorists who still took field theory seriously, and he was practically the only one working on gauge field theory with the ultimate goal of setting up a consistent quantum field theory for the weak interactions[2].

Looking back it is clear that Tini's methods for dealing with ghosts and diagrams were very instrumental in making the remarkable progress in gauge field theory that was about to take place in subsequent years. Ghosts and diagrams will also play a role in this lecture, which deals with the old problem of how to unambiguously extract physics results from the extended Hilbert space, populated by ghosts and phantoms and what have you. As is well known, this problem is solved by establishing the gauge independence of the S-matrix: one must set up a consistent procedure in order to restrict oneself to the physical subspace and extract the physically relevant information in an unambiguous way, irrespective of the gauge condition that is used.

The gauge independence of S-matrix elements in gauge theories has been proven by many authors (see, for instance, refs. 3-5). The proof makes use of the Ward identities for connected Green's functions and the kinematical structure of these functions. After a truncation of the propagator poles associated with the external lines corresponding to physical particles and an appropriate normalization of each of these lines (wave function renormalization), the result becomes independent of the gauge condition employed in the calculations. Of course, one should be careful and choose an admissible gauge condition, i.e., a gauge condition that breaks the gauge invariance at the perturbative level without inducing a spontaneously broken realization of the BRST invariance of the gauge-fixed action[6]. In principle the proofs apply to generic gauge theories.

Clearly, as long as all connected diagrams are well defined, at least in the context of some suitable regularization, the gauge independence of the (perturbative) S-matrix is beyond doubt. However, in certain cases there are diagrams that are not well defined, irrespective of the regularization scheme. What we have in mind here are so-called tadpole diagrams associated with massless fields: diagrams with a single external propagator line, which tend to diverge due to the singular behaviour of the massless zero-momentum propagator. Such diagrams are not only encountered in the calculation of scattering amplitudes, but also in various diagrammatic identities. For instance, in the Ward identity that governs the change of the S-matrix, massless ghost fields can give rise to such tadpole diagrams. Another example is quantum gravity, where tadpole diagrams can occur with a single external graviton line. Of course, we are not concerned here about fields that are massless just because of a special choice of the parameters or of the value of the fields about which one has chosen to expand the effective action. In those cases one can usually choose a different set-up, which would avoid the ambiguous terms. What concerns us here is the situation where the quantum effective action vanishes for certain fields taking arbitrary constant values. In that case these fields are obviously massless and the one-particle irreducible diagrams with zero-momentum external lines (corresponding to these fields) will vanish as well. Hence we are dealing with a continuous degeneracy in the stationary points of the effective action. For convenience we will remain in the context of a flat-space background, which is not always the situation one encounters in quantum gravity. However, we should stress that, in principle, our arguments do not depend on the precise nature of the degeneracy.

When the quantum effective action is degenerate for constant fields, then the vacuum-expectation values of these fields are arbitrary parameters, which are not constrained by the dynamics. Obviously the mass associated with these fields must be zero and there are no non-zero diagrams with a single line disappearing into the vacuum. Therefore ill-defined tadpole diagrams associated with these massless fields can be consistently suppressed. On the other hand there are extra free parameters corresponding to the vacuum-expectation values of the massless fields, that characterize the field configurations associated with the possible degenerate ground states. It is easy to write classical Lagrangians that have this property (i.e., Lagrangians with only kinetic terms or "velocity-dependent" potentials, such as non-linear sigma models), although usually this feature will not be preserved at the quantum level.

Whether or not different vacuum-expectation values will correspond to an inequivalent realization of the theory is another matter, which has to be studied separately. For gauge theories the ground states that are related by the action of the gauge group are strictly gauge equivalent and one should view the ground state as being non-degenerate. The existence of a continuous degeneracy of the effective action for field configurations associated with this ground state may be a feature that depends on the choice of the gauge condition. The classical action for gauge theories has an obvious degeneracy associated with the possibility of performing gauge transformations. In the quantum theory, most of this degeneracy will disappear. By its very nature the gauge condition breaks the gauge invariance, although it is by no means excluded that some harmless residual invariance remains. An example of such a residual invariance is the subgroup of rigid (i.e., space-time independent) gauge transformations, which is still manifest in many cases. Another example, which is more relevant for the purpose of this paper, is the following. Consider an abelian gauge field, A_μ, subject to standard gauge transformations $A_\mu \to A_\mu + \partial_\mu \Lambda$. Obviously, the Lorentz gauge, based on the gauge fixing term $F = \xi\, \partial_\mu A^\mu$, is not affected by gauge transformations satisfying $\partial^2 \Lambda = 0$. In particular, the theory must still be invariant under those transformations that shift the gauge potentials by a constant vector a_μ, i.e.,

$$A_\mu(x) \to A_\mu(x) + a_\mu. \tag{1}$$

These are the transformations corresponding to $\Lambda(x) = a_\mu x^\mu$. From this residual invariance it then follows that the part of the quantum effective action that depends exclusively on the gauge field and not on the matter fields must depend on the derivatives of the gauge field and not on the gauge field itself. Therefore the effective action vanishes for arbitrary constant gauge fields.*

In fact it is not so essential that (1) corresponds to a residual symmetry. As long as the gauge condition vanishes for constant potentials one may still expect a degeneracy of the effective action for these potentials. To prove this at the quantum level is not entirely straightforward. For instance, take the gauge condition $F(A) = f(A)\, \partial^\mu A_\mu$, where $f(A)$ is some arbitrary function of the gauge field A_μ. Although this gauge-fixing term still vanishes on the space of constant potentials, it is not, in general, invariant under (1). Therefore one can no longer invoke an invariance argument to show that the effective action is degenerate for constant gauge potentials. Indeed, an elementary one-loop calculation shows that the gauge-field part of the effective action no longer depends exclusively on derivatives of the gauge field. Nevertheless, it remains proportional to such derivatives, so that the effective action still vanishes for constant gauge potentials in accord with more intuitive expectation.

---

* The effective action is determined modulo an additive constant, which we choose such that the action vanishes for zero fields.

We were motivated to study the question of gauge independence in the presence of a continuous degeneracy of stationary points of the effective action by certain explicit calculations performed some time ago[7] in quantum gravity, which exhibited an apparent lack of gauge independence for physical quantities such as the mass. These one-loop calculations were performed in the context of dimensional regularization[8]. No explanation for this unusual phenomenon was offered, but the authors of ref. 7 claimed that it was still possible to regain gauge independence by exploiting certain symmetries of the gauge-fixed action, which were assumed to be typical for gravity. Later on the phenomenon was analyzed in ref. 9, where the analogy with spontaneously broken gauge theories was emphasized in an approach based on the effective action and one-particle irreducible graphs. Just recently Hari Dass and I argued that the degeneracy of the effective action was responsible for this apparent lack of gauge independence. We showed that in the degenerate case there is indeed a qualitative difference with regard to the way in which the gauge independence of the S-matrix is realized as compared to the standard situation[10]. We then established that the phenomenon is not restricted to quantum gravity, but is generic to all theories with a continuous degeneracy of stationary points of the effective action.

In this lecture I shall discuss the work of ref. 10 and derive a Ward identity, in a diagrammatic way just as in ref. 4, that governs the change of the S-matrix under a change of gauge, paying proper attention to the troublesome tadpole-like diagrams that may be present when the effective action exhibits the degeneracy discussed above. Our starting point is the standard quantized action in the presence of external sources J^i,

$$\mathcal{L}_{\mathrm{qu}} = \mathcal{L}_{\mathrm{inv}} - \tfrac{1}{2}(F^\alpha)^2 + ib^\alpha \frac{\partial \,\delta F^\alpha}{\partial \xi^\beta} c^\beta + J^i \,\phi^i, \tag{2}$$

where δF^α is the variation of the gauge-fixing term under infinitesimal gauge transformations with parameters ξ^α. Furthermore c^α and b^α are the (real) ghost and anti-ghost fields. We use a standard notation where the indices i and α refer not only to the type of field or gauge group generator, but also incorporate their dependence on space-time. The statistics of the ghost and anti-ghost fields is always opposite to the statistics of the corresponding gauge fields; for convenience, we restrict ourselves to bosonic symmetries, so that the (anti-)ghost fields are anticommuting. The infinitesimal gauge transformations of the fields ϕ^i are

$$\delta\phi^i = \left\{ t^i_\alpha + s^i_\alpha(\phi) \right\} \xi^\alpha, \tag{3}$$

where the quantities t^i_α and $s^i_\alpha(\phi)$ denote the field-independent and field-dependent term. It is straightforward to generalize this notation and include composite operators $Q(\phi)$, which then transform under infinitesimal gauge transformations according to

$$\delta Q(\phi) = \left\{ t^Q_\alpha + s^Q_\alpha(\phi) \right\} \xi^\alpha. \tag{4}$$

The quantities t^Q_α and $s^Q_\alpha(\phi)$ appear as extra operators in the Ward identity that couple to the ghost fields. The diagrammatic form of this identity, which applies to connected diagrams, is[4]

$$= 0, \qquad (5)$$

where the summation over J indicates that we have to sum over all possible attachments of the ghost line to the external sources. This identity can be directly generalized to more complicated operators coupling to external sources. When the external sources are taken on the mass shell, the last two sets of diagrams cancel. Hence we have

$$= 0. \qquad (6)$$

To establish the gauge independence of S-matrix elements one considers the Ward identity (5) in the presence of an extra source coupled to some operator $(\Delta F)^\beta$. The corresponding identity takes the form

$$+ \sum_J \left[\begin{array}{cc} b^\alpha \quad (\Delta F)^\beta \\ \\ iJ^i\, t^i_\alpha\, c^\alpha \end{array} + \begin{array}{cc} b^\alpha \quad (\Delta F)^\beta \\ \\ iJ^i\, s^i_\alpha\, c^\alpha \end{array} \right] = 0. \qquad (7)$$

The vertices in the second and third diagram originate from the infinitesimal gauge transformation of ΔF, which we defined by

$$\delta(\Delta F)^\alpha = \{t^\alpha_\beta + s^\alpha_\beta(\phi)\}\xi^\beta. \qquad (8)$$

Subsequently, one joins the two upper vertices in the diagrams (7), sums over α and β (which implies both a summation over the gauge group generators and an integration over the momentum associated with the newly formed loop, such that the net momentum flowing into the combined vertex is equal to zero). The result reads

$$\begin{array}{ccc} F^\alpha\,(\Delta F)^\alpha & ib^\alpha\, t^\alpha_\beta\, c^\beta & ib^\alpha\, s^\beta_\gamma\, c^\gamma \\ \\ J \quad J \quad J & + \qquad J \quad J \quad J & + \qquad J \quad J \quad J \end{array}$$

$$+ \sum_J \left[\begin{array}{c} b^\alpha\,(\Delta F)^\alpha \\ \\ iJ^i\, t^i_\alpha\, c^\alpha \end{array} + \begin{array}{c} b^\alpha\,(\Delta F)^\alpha \\ \\ iJ^i\, s^i_\alpha\, c^\alpha \end{array} \right] = 0. \qquad (9)$$

The procedure of joining the vertices leads to new operators. For the first three diagrams (9) these operators coincide with the terms generated from the quantum action (2) by a change of the gauge-fixing term equal to

$$F^\alpha \longrightarrow F^\alpha + (\Delta F)^\alpha. \qquad (10)$$

Taking into account the proper minus sign for a closed ghost loop, the sum of the operators corresponding to the first three graphs is equal to minus the change of the quantum action (2) under a gauge variation (10). Therefore (9) expresses the change of the connected diagrams under (10) in terms of the last two types of graphs containing the ghost line attachments to the external sources. However, certain contributions are still missing in (9). In the first graph of (9), before the actual joining of the two vertices, the operators F^α and $(\Delta F)^\alpha$ are part of the same connected diagram. There is an alternative situation, which also leads to diagrams that contribute to the gauge variation of the connected diagrams, where a diagram is formed by joining two separate connected diagrams, one containing F^α and the other one $(\Delta F)^\alpha$. Hence one has the following diagrams,

$$\tag{11}$$

In principle, those diagrams should be added to the diagrams (9), in order to describe completely the change of the connected diagrams. However, because the diagrams (11) vanish on the mass shell by virtue of (6), they will be ignored throughout this paper.

Hence we conclude that, modulo diagrams of type (11) that vanish when the external sources are on the mass shell, $\Delta\Gamma$, the change of the connected diagrams induced by a change of gauge (10) is expressed in terms of the last two sets of diagrams (9), where the ghost lines are attached to the sources J. On the mass shell each of these diagrams leads to a factor Δ_J, which depends on the type of source, multiplying the full amplitude, i.e.,

$$\Delta\Gamma = \sum_J \Delta_J \, \Gamma, \tag{12}$$

where Γ denotes the sum of all connected diagrams with a given configuration of external lines. In the proper normalization of the S-matrix the factors Δ_J are precisely cancelled by the change of the wave-function renormalization constants under the gauge change (10). As these features are standard and do not play a special role in our discussion, we refrain from giving further details and refer to refs. 3-5. We should stress, however, that it is assumed throughout this analysis that the propagators exhibit at most single poles.

This completes the standard proof of the gauge-independence of the S-matrix, where hitherto we have assumed that no ambiguities arise in the definition of the various amplitudes, at least in a regularization scheme such as dimensional regularization[8].

However, as already pointed out, such ambiguities may arise when dealing with tadpole diagrams corresponding to massless fields. Such tadpole diagrams may be formed in (9) upon joining the two upper vertices. If the top of the diagram is connected to the main body that contains all external source attachments via a single propagator line, then the momentum flowing through this line is forced to zero when the two upper vertices are joined. This is so because joining the two operators implies that we adjust the momenta flowing to the each of these operators, such that the total momentum flowing into the combined vertex vanishes. Subsequently we integrate over the momentum variable associated with the new closed loop, so that the result corresponds to the insertion of a space-time integral of a local operator. The diagrams of this type that contribute to (9) are

$$(13)$$

(a)   (b)   (c)   (d)   (e)

Whenever (some of) the ghosts and some of the ordinary fields corresponding to the propagator lines that connect the upper and the lower part of the graphs (13) are massless, we have diagrams that are ill-defined. This happens in the situation sketched previously, where the effective action has a continuous degeneracy of stationary points. Of course, we did not discuss the ghost-dependent terms in the effective action, but the ghost mass is always related to the mass of some of the ordinary fields. This follows, for instance, from a special case of the Ward identity (5),

$$= 0. \qquad (14)$$

As explained previously, in the case where the effective action is degenerate for certain fields taking arbitrary constant values, it is not meaningful to include the ambiguous tadpole graphs corresponding to these fields. Instead their vacuum-expectation values play the role of arbitrary parameters that characterize the field configuration corresponding to one of the possible (degenerate) ground states. Obviously, when we suppress the ambiguous tadpole diagrams from the connected diagrams, we should also suppress their variation under (10), which coincides with the first three diagrams (13). Let us first rewrite these graphs by applying the Ward identity (9) to the top part of the diagrams (13),

$$F^\alpha \left(\Delta F\right)^\alpha \qquad ib^\alpha \, t^\alpha_\beta \, c^\beta \qquad ib^\alpha \, s^\alpha_\beta \, c^\beta$$

$$b^\alpha \left(\Delta F\right)^\alpha \qquad b^\alpha \left(\Delta F\right)^\alpha$$

$$it^i_\alpha \, c^\alpha \qquad is^i_\alpha \, c^\alpha \tag{15}$$

For those diagrams of the last line of (15) that can be divided into two diagrams by cutting only one internal (ghost) line, we can make use of the identity (14). Diagrams that cannot be divided in this way will be called irreducible; such diagrams do not contain a massless (ghost) propagator pole associated with the external line. It is convenient to introduce the following notation for the irreducible graphs,

$$(\Delta T)_\alpha \equiv \qquad , \qquad (\Delta S)^i \equiv \qquad . \tag{16}$$

$$c^\alpha \qquad \qquad is^i_\alpha \, c^\alpha$$

Observe that we have truncated the ghost propagator from the diagrams contributing to $(\Delta T)_\alpha$ and that the diagrams contributing to $(\Delta S)^i$ are irreducible. Hence both diagrams are well defined in the limit of vanishing external momentum. The irreducible character of the diagrams (16) is indicated by using a double boundary circle.

Combining these steps leads to the following result for the sum of the first three diagrams (13),

where the index i is kept fixed for the moment. Note that the second diagram on the right-hand side of (17) no longer contains a (full) propagator associated with the field ϕ^i.

Now we return to the identity (9). The first three graphs describe the effect of a change of gauge (10) on the connected Green's functions which include the attachments of ambiguous tadpole diagrams associated with the massless fields. As we intend to suppress the tadpole diagrams associated with massless fields from the connected Green's functions, we should exclude the diagrams corresponding to their variation, i.e., the diagrams (13a-c) with a massless propagator. Hence we treat these diagrams separately and rewrite them with the aid of (17), where the index i will now be replaced by $\bar{\imath}$, to indicate that we restrict ourselves to the massless fields. Then there are the tadpole diagrams (13d-e), which contribute to the last two diagrams of (9). These diagrams can be rewritten by using the Ward identity (5) for the lower part of the diagram; the top part of the diagram corresponds precisely to $(\Delta T)_\alpha$, as defined in (16). Putting these elements together then leads to the following form of (9),

$$+\ (\Delta T)_\alpha \left[\quad - \quad \right] - (\Delta S)^{\bar{\imath}} \quad = 0\,.$$

where we sum over the indices $\bar{\imath}$ associated with the massless fields. The double boundary circles in the first line of (18) indicate that we suppressed the troublesome tadpole diagrams (13a-c). Therefore the first line coincides with minus the change of the connected Green's function without ambiguous tadpole diagrams attached (again modulo diagrams of type (11), which vanish on the mass shell). The double boundary circles in the second line indicate that we suppressed the diagrams (13d-e). Just as before, on the mass shell, these diagrams yield a multiplicative factor Δ_J, one for each external line (cf. (12)). Finally, the diagrams in the third line are such that the troublesome tadpole diagrams with a massless propagator are absent, so that equation (18) is well defined.

We now proceed and take the external sources on the mass shell. In that case the second diagram in the third line of (18) vanishes by virtue of (6), while we know that the singularity in the first diagram associated with the massless propagator labelled by $\bar{\imath}$ must still cancel. We would like to show that this first diagram is in fact proportional to the third diagram,

$$\frac{\partial \hat{\Gamma}}{\partial v^{\bar{\imath}}} \equiv \qquad \qquad . \tag{19}$$

Here $\hat{\Gamma}$ denotes the sum of the connected diagrams with a given configuration of external lines without any ambiguous tadpole attachments associated with the massless fields $\phi^{\bar{\imath}}$; $v^{\bar{\imath}}$ are the vacuum-expectation values of these fields, which, as explained before, play the role of extra parameters characterizing the degenerate ground states. The connected diagrams are evaluated for a given value of the parameters $v^{\bar{\imath}}$.

Although the massless propagator poles cancel in the diagrams of the third line of (18), their presence is important when evaluating the non-singular remainder. In order to see this, one must reconsider the joining of the vertices, which causes the momentum of the intermediate propagator line to vanish.* The result is that the contribution of the third line in (18) will involve both terms that are proportional to (19), where the momentum flowing through the top line of the diagram is zero, and terms proportional to the derivative of (19) with respect to the momentum of the top line. However, we will show that the derivative term is suppressed if one makes the additional assumption that the nature of the degeneracy of the effective action is not affected by the change of gauge (10). This is not a severe restriction; if the degeneracy were (partly) lifted by a change of gauge, then it would be best to proceed from a class of gauge conditions with a lower degree of degeneracy as a limiting case. Of course, a change in the nature of the degeneracy is a subtle matter, which often signals a phase transition. However, in the present context the phenomenon is only relevant to the gauge-dependent sector of the theory.

Assuming that the degeneracy is not affected by the change of gauge, the singular behaviour of the massless external propagator in the first three diagrams in (15) is cancelled, because the truncated version of these graphs (i.e., the graphs without the (full) external propagator) represents the change under (10) of the first derivative

---

* In ref. 10 we postponed joining the vertices until the end by introducing a limiting procedure that replaces the direct joining of two operators X^α and Y^β by

$$X^\alpha Y^\alpha \rightarrow [X^\alpha Y^\alpha]_{\mathcal{F}} \equiv \mathcal{F}_{\alpha\beta} X^\alpha Y^\beta,$$

where the test function $\mathcal{F}_{\alpha\beta}$ is such that a finite amount of momentum flows into the $X^\alpha Y^\alpha$ vertex. In this way the singular behaviour of troublesome tadpole-like diagrams is avoided and the calculations can be carried out without ever encountering ill-defined expressions. At the end one then takes the limit $\mathcal{F}_{\alpha\beta} \rightarrow \delta_{\alpha\beta}$, which corresponds to the zero-momentum limit.

of the effective action with respect to a massless constant field. As this derivative vanishes for constant values of these fields (because of the degeneracy) and the nature of the degeneracy is assumed to remain the same after a change of the gauge condition, it must remain zero. It then follows from (15) that the divergency caused by massless ghost propagators in the fourth diagram of (15) is also suppressed. This suppression suffices to prove that the on-shell contribution from the third line in (18) will be proportional to the diagram (19) and not to its derivative with respect to the external momentum of the top line. Hence we find that the standard equation (12) for the change of the connected on-shell Green's functions takes a modified form,

$$\Delta \hat{\Gamma} = \sum_J \Delta_J \hat{\Gamma} + (\Delta \phi)^{\bar{\imath}} \frac{\partial \hat{\Gamma}}{\partial v^{\bar{\imath}}}, \tag{20}$$

where the expression for $(\Delta \phi)^{\bar{\imath}}$ follows from explicitly evaluating the third line of (18). In simple cases one finds that $(\Delta T)_\alpha$ is equal to zero, which shows that the first three diagrams of (15) are indeed non-singular, so that the nature of the degeneracy remains the same after a choice of gauge. In that case we have $(\Delta \phi)^{\bar{\imath}} = (\Delta S)^{\bar{\imath}}$.

It is straightforward to derive the corresponding condition for the S-matrix. Introducing parameters ξ^a to parametrize a variety of gauge-fixing terms, the S-matrix satisfies the condition

$$\left\{ \frac{\partial}{\partial \xi^a} - \beta^{a\,\bar{\imath}}(\lambda, \xi^a, v^{\bar{\imath}}, \epsilon) \frac{\partial}{\partial v^{\bar{\imath}}} \right\} S(\lambda, \xi^a, v^{\bar{\imath}}, \epsilon) = 0. \tag{21}$$

Here λ generically denotes the (bare) parameters of the gauge-invariant Lagrangian and ϵ the cut-off; the functions $\beta^{a\,\bar{\imath}}$ follow from $(\Delta \phi)^{\bar{\imath}}$. In the standard case without a continuous degeneracy of stationary points of the effective action, the second term is absent and one finds that the regularized S-matrix elements are independent of the gauge parameters ξ^a. In the degenerate case, the second term is present and seems to give rise to a lack of gauge independence. This is the phenomenon noted in ref. 7 for quantum gravity (see also, ref. 9), whose origin was explained in terms of the degeneracy in ref. 10. However, the correct interpretation of this phenomenon is not that gauge independence is violated. According to (21) the S-matrix elements still depend on fewer parameters than explicitly indicated and, in fact, this equation suffices to prove that the S-matrix elements are gauge independent, provided they are expressed in terms of an independent set of physical parameters. Those are the parameters that can be extracted from S-matrix elements, such as masses, coupling constants, scattering lengths, etc.. This amounts to replacing the original parameters λ and $v^{\bar{\imath}}$ by physically relevant parameters G which are themselves S-matrix elements. Each of these parameters depends on λ, ξ^a, $v^{\bar{\imath}}$ and ϵ, and satisfies (21). According to (21), the S-matrix elements expressed in terms of G, ξ^a and ϵ no longer depend on the gauge-fixing parameters ξ^a.

It is clear that in the presence of a continuous degeneracy of stationary points of the effective action, the gauge independence is realized in a qualitatively different way.

In the usual situation S-matrix elements are independent of the gauge condition when we keep the (bare) parameters of the Lagrangian constant. Hence, the expression for a physical mass or a coupling constant is a function of the parameters of the Lagrangian, but does not depend on the gauge parameters. On the other hand, the degenerate case gives rise to S-matrix elements that do depend on the gauge parameters as well as on the parameters of the Lagrangian. Moreover they depend on parameters v^i that characterize the degenerate field configurations corresponding to the ground state. In actual calculations (like in quantum gravity) these parameters are often implicitly fixed. In spite of this complication, one finds that gauge independence is regained after expressing the results of the calculation in terms of physical parameters. Although this form of gauge independence is thus weaker than in the usual situation, it is sufficient for all practical purposes.

In the presence of a continuous degeneracy of stationary points of the effective action the number of parameters occurring in the S-matrix is larger than the number of parameters occurring in the Lagrangian. It should be emphasized that this enlargement of the parameter space is a phenomenon in its own right and has nothing to do with the issue of the gauge-independence of the S-matrix. In practice some of these parameters could be physically irrelevant, because the space of degenerate ground states may have certain invariances resulting in a physical equivalence of the corresponding theories. Even when symmetry arguments are lacking to relate the different ground states, it may still be possible to absorb these extra parameters into a smaller number of physically relevant parameters.

The latter phenomenon takes place in quantum gravity. In that case (for simplicity we assume that the ground state corresponds to flat space-time) the metric associated with the ground state can be given modulo a multiplicative factor ρ,

$$g_{\mu\nu} = \rho\, \eta_{\mu\nu}. \tag{22}$$

This parameter ρ may be viewed as the vacuum-expectation value of the trace of $g_{\mu\nu}$, which is a massless field. Already at the classical level, physical quantities will depend on the parameters in the Lagrangian and on the parameter ρ. For instance, consider gravity coupled to a scalar field described by the Lagrangian

$$\mathcal{L} = -\kappa^{-2}\sqrt{g}\,R - \tfrac{1}{2}\sqrt{g}\left\{g^{\mu\nu}\,\partial_\mu\phi\,\partial_\nu\phi + m^2\,\phi^2\right\}. \tag{23}$$

Expanding about the background (22), one finds that the physical graviton-graviton coupling constant, the (static) graviton-scalar coupling and the scalar mass all depend on ρ in the following way (for arbitrary space-time dimension d),

$$
\begin{aligned}
K_{ggg}^{\text{phys}} &\propto \rho^{\frac{1}{4}(2-d)}\kappa\,, \\
K_{g\phi\phi}^{\text{phys}} &\propto \rho^{\frac{1}{4}(6-d)}\kappa\, m^2\,, \\
M^{\text{phys}} &\propto \rho^{\frac{1}{2}}\, m\,.
\end{aligned}
\tag{24}
$$

Consider now two different scalar fields, ϕ_1 and ϕ_2 with (Lagrangian) masses m_1 and m_2, respectively, and calculate the one-graviton exchange amplitude in the static limit. The invariant amplitude again depends on ρ and is proportional to

$$\mathcal{M}(1+2 \to 1+2) \propto \rho^{\frac{1}{2}(6-d)} \kappa^2 m_1^2 m_2^2. \tag{25}$$

However, the explicit ρ-dependence disappears when expressing this result in terms of physical parameters, Newton's constant K_{ggg}^{phys} and the masses M_1^{phys} and M_2^{phys}. After correctly normalizing the incoming and outgoing particle states, one finds Newton's law with the $1/r$ potential and the proper coefficient. The latter is defined in terms of the physical parameters and not the parameters that appear in the Lagrangian (23). Although all these quantities depend on ρ, there is no explicit ρ-dependence left.

This example illustrates how the extra parameters associated with a degeneracy of the effective action can still be physically irrelevant. For gravity this phenomenon is not unexpected. One can always change the length scale by performing a special coordinate transformation $x^\mu \to a\, x^\mu$. This transformation will change the value of ρ in (22). In order to write down expressions for dimensionful quantities, the length scale, and thus the value of ρ, has to be fixed. In quantum gravity the issue is more subtle, because the full effective action is not invariant under general coordinate transformations. We refer to ref. 10 for a further discussion of this point.

I thank Jos Vermaseren for introducing me to the art of computer aided diagrammar.

References

1. M. Veltman, Nucl. Phys. **B7** (1968) 637; **B21** (1970) 288.

2. M. Veltman, Phys. Rev. Lett. **17** (1966) 562; *Relation between the practical results of current algebra techniques and the originating quark model*, Lectures at the Copenhagen Symposion, July 1968, reprinted in this volume.

3. E.S. Fradkin and I.V. Tyutin, Phys. Rev. **2** (1970) 2841.

4. G. 't Hooft and M. Veltman, Nucl. Phys. **B50** (1972) 318.

5. B.W. Lee and J. Zinn-Justin, Phys. Rev. **D7** (1973) 1049.
   E.S. Abers and B.W. Lee, Phys. Rep. **9** (1973) 1.

6. B. de Wit, Phys. Rev. **D12** (1975) 1843.
   B. de Wit and N. Papanicolaou, Nucl. Phys. **B113** (1976) 261.

7. I. Antoniadis, J. Iliopoulos and T.N. Tomaras, Nucl. Phys. **B267** (1986) 497.

8. G. 't Hooft and M. Veltman, Nucl. Phys. **B44** (1972) 189.
   C.G. Bollini and J.J. Giambiagi, Phys. Lett. **40B** (1972) 566; Nuovo Cim. **12B** (1972) 189.

9. D. Johnston, Nucl. Phys. **B293** (1987) 229.

10. B. de Wit and N.D. Hari Dass, *Gauge independence in quantum gravity*, preprint THU-91/12; ITFA 91-18, submitted to Nuclear Physics B.

LBL-30801; UCB-PTH-91/26
June 11,1991

EFFECTIVE THEORIES AND THRESHOLDS IN PARTICLE PHYSICS

MARY K. GAILLARD

Department of Physics, University of California

and

Physics Division, Lawrence Berkeley Laboratory, 1 Cyclotron Road,

Berkeley, California 94720

Abstract

The role of effective theories in probing a more fundamental underlying theory and in indicating new physics thresholds is discussed, with examples from the standard model and more speculative applications to superstring theory.

1. Introduction

Effective field theories have proven to be a useful phenomenological tool in elementary particle physics. They serve as probes of the underlying symmetries and structure of more fundamental theories, and the very limitations on their domain of validity can point to thresholds for new physics. I will first illustrate these points with examples from the Standard Model. The large Higgs mass limit of the Standard Model provides both a theoretical laboratory for checking the validity of an effective theory and also as a model for possible physics scenarios at the SSC/LHC.

Finally I will consider the Standard Model itself as an effective four-dimensional field theory that is the low energy limit of ten-dimensional superstring theory. This entails the study of four-dimensional effective supergravity theories that emerge as limits of the string theory at scales μ just

*Invited talk at the Conference on Gauge Theories–Past and Future, in Honor of the 60th birthday of M.J.G.Veltman, Ann Arbor, Michigan, May 16–18, 1991. This work was supported in part by the Director, Office of Energy Research, Office of High Energy and Nuclear Physics, Division of High Energy Physics of the U.S. Department of Energy under Contract DE-AC03-76SF00098 and in part by the National Science Foundation under grant PHY–90–21139.

below the string and/or compactification scales, and that should reduce to the Standard Model at still lower scales: $\mu \ll M_{Pl}$, where $M_{Pl} = (8\pi G_N)^{-\frac{1}{2}} \simeq 1.8 \times 10^{18} GeV$ is the reduced Planck mass. In addition, attempts to make the connection between superstrings and observed particle physics must be able to account for the origin of supersymmetry (SUSY) breaking; this motivates the study of effective lagrangians for gaugino condensation. The hope is to find a formulation that generates the observed large hierarchy of scales.

2. Effective Theories in the Standard Model

2.1. Fermi Theory

Low energy weak interactions are well described by the Fermi la-grangian:

$$\mathcal{L}_{\text{tree}} = 2\sqrt{2}G_F(\bar{\psi}_L\gamma^\mu\psi_L)(\bar{\psi}_L\gamma_\mu\psi_L), \tag{1}$$

which is now understood as the low energy limit of intermediate boson (W) exchange, with the identification

$$\sqrt{2}G_F = \frac{g^2}{4m_W^2} = \frac{\pi\alpha_2}{m_W^2}, \tag{2}$$

where $\alpha_2 = g^2/4\pi$ is the $SU(2)_L$ coupling constant. If we use (1) to calculate the one loop contribution to the four fermion coupling we get a quadratically divergent contribution

$$\mathcal{L}_{1-\text{loop}} \sim \frac{\sqrt{2}G_F\Lambda^2}{8\pi^2}\mathcal{L}_{\text{tree}}. \tag{3}$$

Evaluating instead the same coupling at the one-loop level of the renormal-izable Yang-Mills theory, and then taking the limit of low external momenta $|p|^2 \ll m_W^2$, gives

$$\mathcal{L}_{1-\text{loop}} \sim \frac{\alpha_2}{8\pi}\mathcal{L}_{\text{tree}}, \tag{4}$$

which, using (2), is the same as (3), provided we make the identification

$$\Lambda \to m_W. \tag{5}$$

Similarly, one-loop corrections in the effective theory (1) include logarithmi-cally divergent terms, for example, an 8-fermion coupling, that agree, after the substitution (5), with those calculated in the low energy limit of the Standard Model.

The effective theory defined by (1) provides a good description of weak interactions for energies $E^2 G_F \ll 1$, and the loop expansion converges if the cut-off satisfies $\Lambda^2 G_F < 1$. Before it was understood that the underlying physics of the Fermi theory was a Yang-Mills theory, these observations pointed to a new physics threshold $\Lambda < G_F^{-1} \approx 300 GeV$, a threshold that we now associate with the mass m_W of the intermediate boson. Veltman was one of the first people to recognize[1] the importance of Yang-Mills theories in this context.

When flavor changing four-fermion couplings are included in $\mathcal{L}_{\text{tree}}$, a cut-off less than a few GeV is required for consistency with observation; this led to the prediction[2], later made more precise by calculations[3] within the renormalizable Yang-Mills theory, of the charmed quark mass.

2.2. Pion Chiral Dynamics

As another example, the low energy physics of pions is described by an effective lagrangian where the pion field can be viewed as an interpolating field for the quark bilinear field operator:

$$\bar{q}\frac{\vec{\tau}}{2}\gamma_5 q \Rightarrow \vec{\pi}. \tag{6}$$

In this case we do not know how to take an analytic limit of the underlying QCD theory to obtain the effective pion theory. Rather the low energy limit of the latter is dictated by symmetries and their quantum anomalies:

$$\mathcal{L}_{eff} = \frac{1}{2}\partial_\mu \pi^i \partial^\mu \pi^j \left(\delta_{ij} + \frac{\pi_i \pi_j}{f_\pi^2 - \pi^2} \right) + \frac{\alpha}{6\pi}\frac{\pi^0}{f_\pi} F_{\mu\nu}\tilde{F}^{\mu\nu} + \cdots. \tag{7}$$

The first term in (7) is the unique two-derivative term that respects the chiral $SU(2)_L \otimes SU(2)_R$ symmetry of QCD with two massless quarks. It defines a nonrenormalizable effective theory for which the loop expansion series converges if the cut-off, $\Lambda < 4\pi f_\pi \sim GeV$, lies below the observed resonance mass scale. In fact, for a suitable choice of cut-off the one-loop corrections in this effective theory reproduce, for example, the low energy tail of the ρ resonance. The second term in (7), which induces the neutral pion decay $\pi^0 \to \gamma\gamma$, arises from the chiral anomaly[4] present in quark QED. Both terms will have analogues in the effective theory for gaugino condensation to be discussed below.

3. Is the Standard Model an Effective Theory?

In the above examples, the notion that the cut-off should indicate a scale of new physics is related to the unacceptability of fine tuning. For example, one could absorb the correction (3) (along with the leading divergent corrections in higher loop order) into the definition of the Fermi constant G_F. Since in the effective nonrenormalizable theory new (e.g., the 8-fermion coupling) terms are only log divergent, the limit on the cut-off would be much less stringent. However this would require arranging cancellations among large corrections to produce a very small number. In the spirit of avoiding fine tuning, we can point to two fine tuning problems in the standard model that might suggest new physics thresholds.

3.1. The Strong CP Problem

The QCD lagrangian in the Standard Model takes the form

$$\mathcal{L}_{QCD} = -\frac{1}{4}F_{\mu\mu}F^{\mu\nu} + \frac{\alpha_3}{8\pi}\theta\tilde{F}_{\mu\mu}F^{\mu\nu} + \bar{q}i\,\not{D}q + (\bar{q}_L M q_R + h.c.), \quad (8)$$

where α_3 is the QCD coupling constant, M is the quark mass matrix and the parameter θ violates P and CP. As discussed at this meeting by Eduardo de Raphael, the experimental limit on the neutron electric dipole moment sets a stringent bound: $\theta < 10^{-9}$, if we work in a basis where the quark mass matrix is hermitian $M = M^{\dagger}$. When radiative corrections from the CP violating weak sector are included, the quark mass matrix acquires a logarithmically divergent, nonhermitian correction. Rediagonalization of M then induces a correction[5] $\delta\theta$ to θ, that in the standard model is divergent first in 7-loop order:

$$\delta\theta_{SM} = \delta_{\inf} + \delta_{\text{finite}},$$

$$\delta_{\text{finite}} \sim \left(\frac{\alpha_3}{\pi}\right)^4 \left(\frac{\alpha_2}{\pi}\right)^2 \frac{m_c^2 m_s^2}{m_W^4}\theta_{CKM} \sim 10^{-16},$$

$$\delta_{\inf} = \frac{\alpha_1}{\pi}\left(\frac{\alpha_2}{\pi}\right)^6 \frac{m_t^4 m_b^4 m_c^2 m_s^2}{m_W^{12}}\theta_{CKM}\ln(\Lambda/m_t) < 2 \times 10^{-28} \quad (9)$$

assuming $m_t < 200 GeV$ and $\Lambda < M_{Pl}$. Here $\theta_{CKM} = s_1^2 s_2 s_3 \sin\delta$ is the usual CP violating parameter of the Cabibbo-Kobayashi-Maskawa matrix.[6] Although the contribution (9) increases when one includes additional couplings, such as in $SU(5)$ grand unification, it is clear that the strong CP "problem" does not provide useful information on possible new thresholds.

3.2. The Gauge Hierarchy Problem

In the standard model the renormalized Higgs mass is determined as

$$m_H^2 = \frac{\lambda}{8}(TeV)^2 = m_H^2(\text{tree}) + a\frac{g^2}{16\pi^2}\Lambda^2 + \cdots, \qquad (10)$$

where λ is the renormalized Higgs self-coupling constant, and a is a numerical coefficient of order unity. If the Higgs sector is weakly coupled, $\lambda < 1$, absence of fine tuning suggests a new threshold at a scale $\Lambda < 3TeV$, which is the well-known "second threshold", first emphasized by Veltman.[7]

A priori, there is nothing sacred about weak coupling. If we allow λ, and hence m_H, to become arbitrarily large, the Higgs sector becomes strongly coupled. At scales well below the Higgs mass m_H, the strong self-couplings of the three eaten Goldstone bosons φ^i of the Higgs sector manifest themselves as strong self-couplings among the longitudinally polarized intermediate bosons, $W^\pm, Z$. More precisely the S-matrix for the eaten Goldstones is equivalent,[8] up to corrections of order m_W^2/E_W^2, to that for the longitudinally polarized bosons. A recent alternative proof[9] by Hélène Veltman of this "equivalence theorem"[10] has resolved questions[11] that had been raised about its validity. To the extent that the linear σ-model is equivalent to the linear one (a possible discrepancy at the two loop level has been pointed out by van der Bij and M. Veltman[12]), the effective lagrangian for this system is identical to the QCD lagrangian for low energy pions, Eq.(7), with the substitutions $\pi \to \varphi$ and $f_\pi \to v$, where $v \sim \frac{1}{4}TeV$ is the vacuum expectation value of the Higgs field:

$$\mathcal{L}_{eff} = \frac{1}{2}\partial_\mu\varphi^i\partial^\mu\varphi^j\left(\delta_{ij} + \frac{\varphi_i\varphi_j}{v^2 - \varphi^2}\right)\left(1 - \frac{\Lambda^2}{8\pi^2v^2}\right) + \cdots. \qquad (11)$$

Eq.(11) gives a valid description of strong W, Z interactions over an energy range $m_W^2 \ll E^2 \ll \Lambda^2$, where Λ is the ultraviolet cut-off for the effective theory, and I have displayed the quadratically divergent part of the one-loop correction, which could be reabsorbed as a renormalization:

$$\varphi_{Ren} = Z^{\frac{1}{2}}\varphi, \quad v_{Ren} = Z^{\frac{1}{2}}v \approx 250GeV, \quad Z = 1 - \frac{\Lambda^2}{16\pi^2v^2} + \cdots.$$

This shows that, in order to avoid fine tuning, a new physics threshold $\Lambda < 4\pi v \sim 3TeV$ is indicated even in the strongly interacting limit of the electroweak sector.

One popular scenario for this new physics is technicolor;[13] in this case the strongly coupled part of the electroweak sector closely resembles the pion sector of QCD, including the resonance region, with a scaling in energy by a factor $v/f_\pi \approx 2800$. However there is no explicit realization of this scenario without phenomenological problems.

Veltman realized some time ago[14] that the cancellations among fermions and bosons in a supersymmetric theory[15] would damp the quadratic divergence in (10), provided the fermion-boson mass gap is not too large. It has proven difficult to construct a phenomenologically viable renormalizable theory with spontaneously broken supersymmetry (SUSY), but quite easy to accommodate explicit soft SUSY breaking, in the form of scalar and gaugino masses and trilinear scalar self-couplings. The scale parameter that determines the size of these effects plays the role of the cut-off in (10). In fact, the most recent measurements of the Standard Model gauge constants indicate that their unification is ruled out in the minimal Standard Model, while unification within the minimal supersymmetric extension of the Standard Model fits the data well, with a SUSY mass gap of about a TeV.[16]

In this context one must still understand why the mass gap is as small as it must be to conform to the data with no fine tuning. A favorite hypothesis is that local supersymmetry, in the form of a nonvanishing gravitino mass $m_{\tilde{G}}$, is broken spontaneously in a "hidden sector" of a (nonrenormalizable) supergravity theory, which may in turn be the low energy limit of a (finite) superstring theory. It is then the task of the superstring theorists to predict the correct scale for the SUSY mass gap.

4. The Heterotic Superstring

In the superstring scenario, one starts from a string theory[17] in ten dimensions with an $E_8' \otimes E_8$ gauge group, and ends up[18] in four dimensions with an effective $N = 1$ superstring theory with gauge group ($G' \in E_8'$) $\otimes$ ($SU(3) \otimes SU(2) \otimes U(1) \in G \in E_8$). G is the gauge group of a SUSY Yang-Mills theory coupled to matter, including the quarks, leptons and Higgs particles of the Standard Model and their superpartners. G' is the gauge group of a "hidden" SUSY Yang-Mills sector, that has only gravitational strength couplings to observed matter. A popular candidate mechanism for

SUSY breaking is gaugino condensation[19] in the hidden sector, which is assumed to be asymptotically free and infrared enslaved, so that the SUSY Yang-Mills theory becomes confined at some scale Λ_c where gaugino condensation occurs:

$$< \bar{\lambda}\lambda >_{hid} \sim \Lambda_c^3. \tag{12}$$

An additional source of SUSY breaking could be[20] the (quantized) vev of the field strength H_{lmn} of ten-dimensional supergravity.

$$\int dV^{lmn} < H_{lmn} >= 2\pi n \neq 0, \quad l, m, n = 4, \ldots, 9,$$

$$H_{LMN} = \nabla_L B_{MN}, \quad L, M, N = 0, \ldots, 9. \tag{13}$$

When both sources of SUSY breaking are present, it is possible[20] to have "local" SUSY breaking, in the sense that the gravitino acquires a mass: $m_{\tilde{G}} \neq 0$, with a vanishing cosmological constant at the classical level of the effective theory. Supersymmetry breaking should be communicated by radiative corrections to the observable sector, resulting in a SUSY mass gap, i.e., "global" SUSY breaking.

4.1. The Effective Supergravity Theory

The particle spectrum of the effective four dimensional field theory includes the gauge supermultiplets W^a, the matter chiral multiplets Φ^i and the supergravity multiplet. In addition there is a gauge singlet chiral multiplet S, with a scalar component Res that is the "dilaton" field whose vev determines the value of the gauge coupling constant at the unification scale:

$$< \text{Res} >= g^{-2}, \tag{14}$$

as well as singlet chiral multiplets T_α, called "moduli", whose scalar components t_α are related to the structure of the compact manifold. In the case of a single modulus[21] the unification (or compactification) scale is determined as

$$\Lambda_{GUT}^2 = \frac{M_{Pl}^2}{< \text{ResRet} >} + O(< |\varphi^i|^2 >) \tag{15}$$

where φ^i is the (complex) scalar component of the superfield Φ^i.

The effective theory just below the compactification scale is an $N = 1$ supergravity theory. In the Kähler covariant superfield formulation[22] of supergravity, the lagrangian takes the simple form

$$\mathcal{L} = \mathcal{L}_E + \mathcal{L}_{pot} + \mathcal{L}_{YM}. \tag{16}$$

The first term

$$\mathcal{L}_E = -3 \int d^2\Theta \mathcal{E}\mathcal{R} + h.c. \tag{17}$$

is the generalized Einstein term. It contains the pure supergravity part as well as the kinetic energy terms for the chiral supermultiplets. The second term:

$$\mathcal{L}_{pot} = \int d^2\Theta \mathcal{E}e^{K(Z,\bar{Z})/2}W(Z) + h.c., \tag{18}$$

contains the Yukawa couplings and the scalar potential, and the third term

$$\mathcal{L}_{YM} = \frac{1}{4} \int d^2\Theta \mathcal{E}f_b^a(Z)W_\alpha^b W_a^\alpha + h.c. \tag{19}$$

is the Yang-Mills lagrangian. The above lagrangian is invariant under a Kähler transformation, which is a redefinition of the Kähler potential $K(Z,\bar{Z}) = K(Z,\bar{Z})^\dagger$ and of the superpotential $W(Z) = \overline{W}(\bar{Z})^\dagger$ by a holomorphic function $F(Z) = \bar{F}(\bar{Z})^\dagger$ of the chiral supermultiplets $Z = \Phi^i, S, T_\alpha$:

$$K \rightarrow K' = K + F + \overline{F}, \quad W \rightarrow W' = e^{-F}W. \tag{20}$$

Since this transformation changes $e^{K/2}W$ by a phase that can be compensated by a phase transformation of the integration variable Θ, the theory defined above is classically invariant[23,22] under Kähler transformations provided one transforms the superfields $\mathcal{R}$ and W_α^a by a compensating phase; for example the Yang-Mills superfield transforms as:

$$W_\alpha^a \rightarrow e^{-i\mathrm{Im}F/2}W_\alpha^a. \tag{21}$$

This last transformation, which implies a chiral rotation on the left-handed gaugino field λ_L^a:

$$\lambda_\alpha^a \rightarrow e^{-i\mathrm{Im}F/2}\lambda_\alpha^a, \tag{22}$$

is anomalous at the quantum level, a point that will be important in the discussion below. (Here a is a gauge index and α is a Dirac index.).

The theory is completely specified by the field content, the gauge group and the three functions K, W and f of the chiral superfields. In theories from superstrings one has $f^a_b(Z) = \delta^a_b S$, resulting in the identification (14). The Kähler potential depends on the dilaton and the moduli fields in such a way that the compactification scale (15) is determined by the vev:

$$\frac{\Lambda_{GUT}}{M_{Pl}} = (2g)^{\frac{2}{3}} < e^{K/6} > . \tag{23}$$

4.2. The Effective Lagrangian for Gaugino Condensation

In order to incorporate supersymmetry breaking, we include an effective potential for gaugino condensation that is constructed by the introduction of a composite superfield operator[24] U as an interpolating field for the Yang-Mills composite operator:

$$\frac{1}{4} W^a_\alpha W^\alpha_a \Rightarrow U = e^{K/2}\widetilde{W}(H). \tag{24}$$

Here H is a chiral supermultiplet that represents the lightest bound state of the confined SUSY Yang-Mills sector, in the same way that the pion is an interpolating field for the composite quark operator, Eq.(6), in low energy QCD. Kähler invariance requires

$$\widetilde{W}(H) \rightarrow e^{-F}\widetilde{W}(H) \tag{25}$$

under (20).

Just as the symmetries of the Standard Model and their quantum anomalies uniquely determine the low energy pion lagrangian (7), the symmetries of the effective supergravity theory determine the effective supergravity lagrangian for the bound state supermultiplet H. In addition to the chiral anomaly related to the transformation (21), (22) under Kähler transformations, there is a conformal anomaly associated with a rescaling of the cut-off (15), (23) under (20):

$$\Lambda_{GUT} \rightarrow e^{\mathrm{Re}F/3}\Lambda_{GUT}. \tag{26}$$

The effective lagrangian for gaugino condensation is defined by[25,26]

$$\mathcal{L}^{eff}_{pot} \equiv \int d^2\Theta \mathcal{E} e^{K/2} W(H,S) = \int d^2\Theta \mathcal{E} e^{K/2}\widetilde{W}(H) 2b_0 \lambda \ln(H/\mu) + h.c.$$

$$= \int d^2\Theta \mathcal{E} e^{K/2} 2b_0 \lambda e^{-3S/2b_0} H^3 \ln(H/\mu) + h.c., \qquad (27)$$

where b_0 determines the β-function for the confined Yang-Mills theory:

$$\frac{\partial g}{\partial \ln \mu} = -b_0 g^3,$$

and λ and μ are constants of order unity. The H-superfield kinetic energy term is determined by the Kähler potential[26,27]:

$$K = -\ln(S + \bar{S}) - 3\ln(T + \bar{T} - |\Phi|^2 - |H|^2). \qquad (28)$$

Then under a Kähler transformation (20), (25), with $H \to e^{-F/3}H$, the lagrangian (27) undergoes the shift

$$\delta\mathcal{L}_{pot}^{eff} = -\frac{2b_0}{3} \int d^2\Theta \mathcal{E} F(Z) U + h.c.$$

$$= \sqrt{-\det g} \frac{b_0}{3} (\mathrm{Re}F(z)F^{\mu\nu}F_{\mu\nu} + \mathrm{Im}F(z)\tilde{F}^{\mu\nu}F_{\mu\nu} + \cdots), \qquad (29)$$

which correctly reproduces the known variations under the trace and conformal anomalies.[24] Note that in this formalism the anomaly is reflected in the interaction term (27), but the H kinetic energy term, defined by (28), respects the classical symmetry of the theory.

If we now solve for the vacuum value of the scalar component h of the superfield H:

$$< h >= \mu e^{-\frac{1}{3}}, \qquad (30)$$

and integrate out the H-supermultiplet, which at the classical level of the theory defined by (27) and (28), amounts to fixing H at its ground state value (30), we obtain an effective theory for Φ^i, S, T and the observable-sector Yang-Mills fields that is defined by (28) with $H = < h >$ and by the superpotential

$$W(Z) = c_{ijk}\Phi^i\Phi^j\Phi^k + \tilde{c} + \tilde{h}e^{-3S/2b_0}, \qquad \tilde{h} = -\frac{2b_0}{3}\lambda\mu^3 e^{-1}, \qquad (31)$$

where the constant $\tilde{c}$ is proportional to the *vev* (13). This is precisely the effective theory obtained earlier by Dine *et al.*[20] using arguments based on a nonanomalous chiral $U(1)$ symmetry:

$$\lambda_\alpha^a \to e^{3i\beta/2}\lambda_\alpha^a, \qquad s \to s + 2b_0 i\beta. \qquad (32)$$

The effective theory defined by (31) has a positive semi-definite potential which vanishes at the minimum. If $\bar{c} = 0$, the vacuum energy is minimized for $\bar{h} = 0$ ($< H >= 0$) or $< s > \rightarrow \infty$ ($g = 0$), that is, condensation does not occur and supersymmetry remains unbroken. For $\bar{c} \neq 0$ the effective theory has the following properties at the classical level[20] and at the one-loop[28] level: the cosmological constant vanishes, the gravitino mass $m_{\tilde{G}}$ can be nonvanishing, so that local supersymmetry is broken, in which case the vacuum is degenerate, and there is no manifestation of SUSY breaking in the observable sector. Nonrenormalization theorems for supergravity, together with a classical $SL(2, \mathcal{R}) \otimes U(1)$ symmetry of this effective theory, indicate[27] that these results will persist to all orders of the effective theory defined by (31).

The $SL(2, \mathcal{R}) \otimes U(1)$ symmetry effects a Kähler transformation (20), and hence is broken by anomalies at the quantum level. The effects of anomalies can be made manifest in the effective low energy theory if we first integrate out the H supermultiplet at the one-loop level. Then soft SUSY breaking in the form of gaugino masses appears at the one loop level of the effective low energy theory; more precisely these terms arise from diagrams with one H loop and one "light"particle loop. Evaluating this contribution requires first fully determining the one-H-loop effective lagrangian and performing the appropriate wave function renormalizations and the Weyl transformation needed to recast the renormalized Einstein curvature term[29] in canonical form.

Including loop corrections from the H-sector, one finds[26] that masses are generated for the gauginos of the observable sector that are of order

$$m_{\tilde{g}} \sim \frac{m_{\tilde{G}} m_H^2 \Lambda_c^2}{(4\pi M_{Pl})^4} < 4 \times 10^{-15} M_{Pl} \sim 7 TeV,$$

$$\text{for } m_{\tilde{G}} < m_H \sim \Lambda_c < 10^{-2} M_{Pl}, \tag{33}$$

where m_H is the mass of the H-supermultiplet. The factor $(4\pi)^{-4}$ appears in (33) because the effect arises first at two-loop order in the effective theory, the factor $m_{\tilde{G}}$ is the necessary signal of SUSY breaking, the factor m_H^2 is the signal of the anomalous breaking of $SU(2, \mathcal{R}) \otimes U(1)$, and Λ_c^2 is the effective cut-off. This last factor arises essentially for dimensional reasons: the couplings responsible for transmitting the knowledge of symmetry breaking to the observable sector are nonrenormalizable interactions with dimensionful coupling constants proportional to M_{Pl}^{-2}. Note that the ground state equations give

$$m_{\tilde{G}} = <e^{K/2}W> \approx \frac{\lambda\mu^3\Lambda_c^3}{2eg^4M_{Pl}^2}, \quad \mu\Lambda_c \sim \left(\frac{\bar{c}e}{2\lambda}\right)^{\frac{1}{3}}\Lambda_{GUT}, \tag{34}$$

so it is not possible to generate a hierarchy of more than a few orders of magnitude between $m_{\tilde{G}}$ and Λ_{GUT} if $\tilde{c}$ is quantized as in (13). However this initial small hierarchy is enough to generate a viable gauge hierarchy if observable SUSY breaking is sufficiently suppressed, as in (33), relative to local SUSY breaking. For example, recent LEP data[16] suggest $\Lambda_{GUT} \sim 10^{16}GeV$, $g^{-2} \sim 2$, so for a hidden E_8 gauge group ($b_0 = .56$) we get $\Lambda_c \sim .6\Lambda_{GUT} \sim 3 \times 10^{-3}M_{Pl}$.

4.3. Restoration of Space-Time Duality

In the formalism presented above, the classical $SL(2,\mathcal{R})\otimes U(1)$ symmetry is broken by anomalies to a Peccei-Quinn type $U(1)$ symmetry: $T \to T + i\gamma$. However the discrete subgroup $SL(2,\mathcal{Z})$ of $SL(2,\mathcal{R})$ is known[30] to be an exact symmetry to all orders in string perturbation theory. Similar symmetries are present in more general string compactifications.

This so-called "modular invariance", which includes the "duality" inversion $R \to R^{-1}$ of the radius of compactification, is restored by adopting, instead of (27), the effective lagrangian[31]

$$\mathcal{L}_{pot}^{eff} = \int d^2\Theta \mathcal{E}e^{K/2}2b_0\lambda e^{-3S/2b_0}H^3\ln(H\eta^2(T)/\mu) + h.c., \tag{35}$$

where $\eta(T)$ is the Dedekind η-function. This is the unique function of the chiral superfields that has the required analyticity and $SL(2,\mathcal{Z})$ transformation properties.[32,31] For different compactifications it will be replaced by different moduli-dependent functions.[33,34] This additional contribution to the Yang-Mills wave function renormalization can be understood[35] as arising from finite threshold corrections[36,33] to the leading log approximation that arise from heavy string mode loops, and is closely related to the anomalous quantum correction due to the (nonrenormalizable) coupling of the Kähler connection,

$$\Gamma_\mu = \frac{1}{4}(K_i\partial_\mu z^i - K_{\bar{\imath}}\partial_\mu z^{\bar{\imath}}) = -\frac{1}{4}\left[\frac{\partial_\mu(s-\bar{s})}{s+\bar{s}} + 3\frac{\partial_\mu(t-\bar{t})+\varphi^i\partial_\mu\varphi^{\bar{\imath}}-\varphi^{\bar{\imath}}\partial_\mu\varphi^i}{t-\bar{t}-|\varphi|^2}\right],$$

to the axial $U(1)$ current.[27,37,35] This ABBJ-type anomaly[4] induces a coupling of the moduli to the fermion axial current and hence to the gauge field strength $F_{\mu\nu}\tilde{F}^{\mu\nu}$, in analogy with the $\pi\gamma\gamma$ coupling in (7).

Whether or not this "corrected" effective lagrangian, or its generalization to more realistic compactifications, can produce as promising a result for phenomenology as the one in (33) remains to be seen.[38]

Acknowledgement. This work was supported in part by the Director, Office of Energy Research, Office of High Energy and Nuclear Physics, Division of High Energy Physics of the U.S. Department of Energy under Contract DE-AC03-76SF00098 and in part by the National Science Foundation under grant PHY-90-21139.

References

1. M. Veltman, *Nucl. Phys.* **B7**: 647 (1968) and **B21**: 288 (1970).

2. S.L. Glashow, J. Iliopoulos and L. Maiani, *Phys. Rev.* **D2**: 1285 (1970).

3. M.K. Gaillard and B.W. Lee, *Phys. Rev.* **D10**: 897 (1974).

4. J.S. Bell and R. Jackiw, *Nuovo Cim.* **60A**: 47 (1969); S. Adler, *Phys. Rev.* **177**: 2426 (1969); W.A. Bardeen, *Phys. Rev.* **184**: 1848 (1969).

5. J. Ellis and M.K. Gaillard, *Nucl. Phys.* **B150**: 141 (1979).

6. N. Cabibbo, *Phys. Rev. Lett.* **10**: 531 (1963); M. Kobayashi and K. Maskawa, *Prog. Theor. Phys.* **49**: 652 (1973).

7. M. Veltman, *Acta Phys. Pol.* **B8**: 475 (1977).

8. M.S. Chanowitz and M.K. Gaillard, *Nucl. Phys.* **B261**: 379 (1985); G.J. Gounaris R. Kögeler and H. Neufeld, *Phys. Rev.* **D34**: 3257 (1986).

9. H. Veltman, *Phys. Rev.* **D41**: 2294 (1990).

10. J.M. Cornwall, D.N. Levin and G. Tiktopoulos, *Phys. Rev.* **D10**: 1145 (1974). C.E. Vayonakis, *Lett. Nuovo Cimento* **17**: 383 (1976); B.W. Lee, C. Quigg and H. Thacker, *Phys. Rev.* **D16**: 1519 (1977).

11. Y. Yao and C. Yuan, *Phys. Rev.* **D38:** 2237 (1986); K. Aoki, *Proc. Tsuduba Meeting* QCD 161: **M376:** 20 (1987); M. Veltman and F. Yndurain, *Nucl. Phys.* **B325:** 1 (1989); J. Bagger and C. Schmidt, *Phys. Rev.* **D41:** 264 (1990).

12. J. van der Bij and M. Veltman, *Nucl. Phys.* **B231:** 205 (1984).

13. S. Weinberg, *Phys. Rev.* **D13:** 974 (1976); L. Susskind, *Phys. Rev.* **D16:** 1519 (1977).

14. M. Veltman, *Acta Phys. Pol.* **B12:** 437 (1981).

15. J. Wess and B. Zumino, *Nucl. Phys.* **B70:** 39 (1974)

16. For recent analyses, see P. Langacker and M. Luo, Univ. of Penn. preprint UPR-0466T (1991); U. Amaldi, W. de Boer and H. Fürstenau, CERN-PPE/91-44 (1991).

17. M. Green and J. Schwarz, *Phys. Lett.* **B149:** 117 (1984); D. Gross, J. Harvey, E. Martinec and R. Rohm, *Phys. Rev. Lett.* **54:** 502 (1985).

18. P. Candelas, G. Horowitz, A. Strominger and E. Witten, *Nucl. Phys.* **B258:** 46 (1985).

19. H.P. Nilles, *Phys. Lett.* **115B:** 193 (1982).

20. M. Dine, R. Rohm, N. Seiberg and E. Witten, *Phys. Lett.* **156B:** 55 (1985).

21. E. Witten, *Phys. Lett.* **155B:** 151 (1985).

22. P. Binétruy, G. Girardi, R. Grimm and M. Muller, *Phys. Lett.* **189B:** 83 (1987).

23. E. Cremmer, S. Ferrara, L. Girardello, and A. Van Proeyen, *Nucl. Phys.* **B212:** 413 (1983).

24. G. Veneziano and S. Yankielowicz, *Phys. Lett.* **113B:** 231 (1982).

25. T.R. Taylor, *Phys. Lett.* **164B:** 43 (1985).

26. P. Binétruy and M.K. Gaillard, *Phys. Lett.* **232B:** 83 (1989).

27. P. Binétruy and M.K. Gaillard, Annecy-CERN preprint LAPP-TH-273/90, CERN-TH.5727/90 (1990), to be published in *Nuclear Physics*.

28. P. Binétruy, S.Dawson, M.K. Gaillard and I. Hinchliffe, *Phys. Lett.* **192B**: 377 (1987), and Phys. Rev. **D37**: 2633 (1988).

29. G. 't Hooft and M. Veltman, *Ann. Inst. Henri Poincaré* **20**: 69 (1974); P. Binétruy and M.K. Gaillard, *Nucl. Phys.* **B312**: 341 (1989): P. Binétruy and M.K. Gaillard, *Phys. Lett.* **220B**: 68 (1989).

30. A. Giveon, N. Malkin and E. Rabinovici, *Phys. Lett.* **220B**: 551 (1989); E. Alvarez and M. Osorio, *Phys. Rev.* **D40**: 1150 (1989).

31. S. Ferrara, N. Magnoli, T. Taylor and G. Veneziano, *Phys. Lett.* **B245**: 409 (1990).

32. A. Font, L.E. Ibáñez, D. Lüst and F. Quevedo, *Phys. Lett.* **B245**: 401 (1990).

33. L. Dixon, V. Kaplunovsky and J. Louis, *Nucl. Phys.* **B355**: 649 (1991).

34. S. Ferrara, C. Kounnas, D. Lüst, and F. Zwirner, CERN-TH.6090/91, ENS-LPTENS-91/14 (1991).

35. V. Kaplunovsky and J. Louis, SLAC-PUB to appear, and J. Louis, SLAC-PUB-5527 (1991), to appear in the proceedings of the 2nd International Symposium on Particles, Strings and Cosmology.

36. V.S. Kaplunovsky, *Nucl. Phys.* **B307**: 145, (1988).

37. J.P. Derendinger, S. Ferrara, C. Kounas and F. Zwirner, CERN-TH.6004/91 (1991).

38. T. Taylor, *Phys.Lett.* **B252**: 59 (1990); P. Binétruy and M.K. Gaillard, *Phys. Lett.* **253B**: 119 (1991); M. Cvetič, A. Font, L.E. Ibáñez, D. Lüst and F. Quevedo, CERN-TH.5999/91, UPR-0444-T (1991).

Neutrino Mass Matrices with Minimal Polynomials of Second or Third Degree in the Three-Family Model

Cecilia Jarlskog

Department of Physics, University of Stockholm

Stockholm, Sweden

ABSTRACT

After some historical notes on the neutrino theory of the light I study the structure of some neutrino mass matrices in the framework of the three-family electroweak model with three right-handed neutrinos. I consider mass matrices which have the minimal polynomials of the second or third degree. These mass matrices are "simple" in the sense that they correspond to a large degree of degeneracy in the neutrino masses.

1 Some early papers on the neutrino theory of the light

The neutrino was introduced by Wolfgang Pauli in a letter written in the his usual colourful style and dated fourth of December 1930. The letter was addressed to the participants in a conference in Tübingen on radioactivity.

From the very beginning the neutrino concept enjoyed a great deal of popular support. In 1930's many papers were written on neutrinos. Some of them turned out to be both relevant and very important (for some references see, for example, [1]). However, there were also several very interesting papers, full of original ideas and written by great physicists, which turned out to be wrong (as far as we know). I feel that some such "wrong" papers can be quite instructive and certainly do deserve attention. Here I would like to mention a few of the wrong papers on neutrinos which had to do with the neutrino theory of light.

The concern about the proliferation of particles, forces, etc. has a long history. In 1930's, which concern me here, the question was again raised by, for example, Max Born and Leopold Infeld [2] who distinguished between two opposite philosophies in physics, namely what they called the "unitarian standpoint" and the "dualistic standpoint". In the unitarian standpoint one assumes only one physical entity, the electromagnetic field. Particles of matter were considered as singularities of the radiation field. In the dualistic point of view the radiation and the particles of matter were intrinsically "different agencies". For an example of a unitarian theory see, for example, Wentzel's article "Zur Frage der Äquivalenz von Lichtquanten und Korpuskelpaaren" [3].

In a somewhat different version of the unitarian approach one considered the particles of matter as being the "real entities" and constructed the radiation (in our modern language, the gauge fields) out of them. With the neutrino on the scene there seemed to be a natural way to do so. L. de Broglie suggested that the photon is made up of a pair of neutrinos [4]. This suggestion came from a very prominent scientist. De Broglie had indeed electrified the world of physics in the 1920's by suggesting the very successful hypothesis of the wave nature of particles of matter and had paved the way to the discovery of wave mechanics. Was he again leading to a great new world by suggesting that the photon be a bound state of a neutrino pair? Some of the great theorists of that time got very excited about the idea and many papers were written in the span of a few years.

Pascual Jordan seems to have been one of the most ardent adherents of the neutrino theory of light. Being mathematically very skilled, and among other things a champion of second quantization, it was for him a challenge to solve the mathematical problems raised by the new hypothesis. He published indeed many papers on the subject [5] with titles such as "Zur Neutrinotheorie des Lichtes", "Lichtquant und Neutrino", "Zur Herleitung der Vertauschungsregeln in der Neutrinotheorie des Lichtes"; "Beiträge zur Neutrinotheorie des Lichtes".

The neutrino theory of light was also taken very seriously by Ralph de L. Kronig, Groningen, who published several papers ("Zur Neutrinotheorie des Lichtes" I, II, and III) on the topic in the Dutch journal Physica [6]. Later these two authors also published a joint paper [7]. The papers by Jordan and Kronig were concerned with solving problems such as the construction of (composite) creation and annihilation operators for a photon which is made up of a pair of fermions; the commutation vs. anticommutation relations; and the thermodynamics of the neutrino-made light, etc.

In spite of the optimistic tone in most of the papers by the above authors it is fair to say that the neutrino theory of light never reached the status of a "serious" theory. Eventually it was realized that although the mathematics looked O.K. the dynamical problems were very serious. This was nicely expressed by O. Scherzer [8] in "Zur Neutrinotheorie des Lichts" who concluded his paper with the following revealing statement: "Die übereinstimmende Flugrichtung der beiden Neutrinos, die ein Lichtquant vertreten, bringt einige noch ungeklärte Schwierigkeiten mit sich". In other words (if the theory is constructed in more than one space dimension) what makes the two neutrinos fly in the same direction and behave as a single entity? The answer to this question was never found and in spite of the fact that more papers were written on the subject later on eventually the theory was abandoned.

From letters written by Pauli [9] one sees that he was very interested in and was closely following what was being done on the neutrino theory of light but found all the work that had been done "weak and unsatisfactory". For example, in a letter addressed to Kronig in 1935, he writes "Des ersteren Idee, das Neutrino mit dem Licht in Verbindung zu bringen ist sehr bestechend, aber die bisherigen Versuche, diese Idee durchzuführen, fand ich alle schwach und unbefriedigend".

2 Preliminaries on neutrino mass matrices

As we have seen in the above section the neutrinos are not easy to work with. Many papers written on them have not stood up to the passage of the time and I believe that the situation is not so different nowadays. Just imagine the present "wealth" of literature on various aspects of neutrino physics, such as the solar neutrino problem or the 17 KeV neutrino, etc. Even for a "neutrino physicist" it is hard if not impossible to keep track of all that is being published on neutrinos. There is no doubt that most of these papers will end up in the dust bin of history, being either irrelevant, uninteresting or at best interesting but wrong. Here I wish to present some models which most probably will share the same destiny, that is they will end up in history's dust bin. My reason for sticking to them is that they may be of some pedagogical interest since they belong to the class of the "simplest" models. I shall consider, in some detail, neutrino mass matrices which has the lowest degree (second) degree minimal polynomials. For such mass matrices it is easy to work out all the couplings and study them. I also take a first look at mass matrices with minimal polynomials of third degree. The point of these exercises is that such mass matrices give automatically a large degree of degeneracy for neutrino masses and that I find is interesting.

I study the neutrino mass matrices in the Electroweak Model [10] with three families. The framework considered is as follows:

1. The Standard Electroweak Model with three families of quarks and leptons together with just one Higgs doublet.

2. The neutrinos are taken to be massive due to the existence of three right-handed neutrinos.

3. The neutrino mass matrices are assumed to be real (i.e., no CP violation in the leptonic sector is allowed).

Thus the framework is a rather standard one and hardly requires any explanation or justification. The leptonic sector of the three-family model, in its minimal version, has the following particle/field content:

$$\begin{pmatrix} \nu_{eL} \\ e_L \end{pmatrix}, \quad \begin{pmatrix} \nu_{\mu L} \\ \mu_L \end{pmatrix}, \quad \begin{pmatrix} \nu_{\tau L} \\ \tau_L \end{pmatrix}, \quad e_R, \ \mu_R, \ \tau_R \tag{1}$$

Without lack of generality, the left-handed neutrinos are introduced in a basis where the charged lepton mass matrix is diagonal. In the extension of the minimal version considered here, there are also three right-handed neutrino singlets, viz.,

$$\nu_{1R}, \ \nu_{2R}, \ \nu_{3R} \tag{2}$$

These particles, having zero weak isospin and hypercharge, do not couple to the gauge bosons (W and Z). Therefore they are often referred to as sterile neutrinos. However, they are not completely sterile as they can couple to the Higgs boson and the left-handed neutrinos. The right-handed neutrinos can also have self-coupling of the kind $\overline{(\nu_{kR})^c}\nu_{jR}$, $j,k = 1,2,3$. Such couplings would also arise if Higgs singlets were added to the model.

Within the above framework the neutrino mass term in the Lagrangian of the three-family model is given by

$$L_{\nu-mass} = -\frac{1}{2} \left(\overline{(\nu_L)^c} , \ \overline{\nu_R} \right) \mathcal{M} \begin{pmatrix} \nu_L \\ (\nu_R)^C \end{pmatrix} + h.c. \tag{3}$$

where

$$\nu_L = \begin{pmatrix} \nu_{eL} \\ \nu_{\mu L} \\ \nu_{\tau L} \end{pmatrix} \qquad \nu_R = \begin{pmatrix} \nu_{1R} \\ \nu_{2R} \\ \nu_{3R} \end{pmatrix} \tag{4}$$

Furthermore $(\psi_X)^c \equiv C\tilde{\overline{\psi}}_X$, where C is the charge conjugation matrix (see, for example, Ref. [11]) and X denotes the chirality, X=L,R. The mass matrix, $\mathcal{M}$, is six by six. It is a real matrix (by assumption) and symmetric, because of the identity $\overline{\nu_j^c}\nu_k = \overline{\nu_k^c}\nu_j$. Thus

$$\mathcal{M} = \begin{pmatrix} 0 & A \\ \tilde{A} & M \end{pmatrix} \tag{5}$$

where the matrices **A** and **M** are three by three real matrices, the latter being symmetric. The matrix **A** is usually called the Dirac mass matrix and **M** is referred to as the Majorana mass matrix. The tilde denotes, throughout this paper, transposition. The **0** in the upper left corner of $\mathcal{M}$ is due to the fact that the coupling $\overline{\nu_L^c}\nu_L$ has $I = 1$ and, therefore, can not be generated by spontaneous symmetry breaking via the Higgs doublet of the Standard Model. Note that the fields in the Lagrangian, Eq. (3), are neither Majorana nor Dirac fields. They are simply unphysical (Weyl) fields with well-defined transformation properties under the Lorentz group. The diagonalization of the above mass matrix produces six Majorana neutrinos (a fact which is well known and trivial). "Degenerate" Majorana neutrinos may combine to form Dirac neutrinos. Here by "degenerate" one means that $|\lambda_j| = |\lambda_k|$ where the λ's refer to the eigenvalues of the mass matrix (see, for example [12], [13]).

3 Symmetries of the Mass Matrix $\mathcal{M}$

The characteristic equation as well as the Cayley-Hamilton equation (see section 4) of the mass matrix $\mathcal{M}$ are invariant under rotations defined by

$$\mathcal{M} \rightarrow \mathcal{M}' = X\,\mathcal{M}\,\tilde{X} \tag{6}$$

$$\tilde{X}X = X\tilde{X} = E$$

where E is the six by six identity matrix. However, a general rotation does not leave the form of the mass matrix invariant. The three by three null matrix (which signals the absence of self-coupling of the left-handed neutrinos as well as the absence of Higgs triplets) will no longer be zero, after the rotation. In order to preserve the form of the matrix only block-diagonal rotations are allowed, viz.,

$$X = \begin{pmatrix} X_1 & 0 \\ 0 & X_2 \end{pmatrix} \tag{7}$$

where X_1 and X_2 are arbitrary three by three orthogonal matrices. The transformations mediated by rotations in Eq.(7) do two things, viz., 1) they re-parametrize the mass matrix and 2) they introduce a non-trivial mixing matrix in the charged currents if X_1 is non-trivial, i.e., different from the identity matrix. First consider the re-parametrization of the matrix. Eq. (6) leads to a new mass matrix

$$\mathcal{M}' = \begin{pmatrix} 0 & A' \\ \tilde{A}' & M' \end{pmatrix} \tag{8}$$

where

$$\mathbf{A'} = \mathbf{X_1} \ \mathbf{A} \ \tilde{\mathbf{X}}_2$$
$$\mathbf{M'} = \mathbf{X_2} \ \mathbf{M} \ \tilde{\mathbf{X}}_2 \qquad (9)$$

The rotations given in Eq. (7) also redefine the left-handed as well as the right-handed fields. The new (primed) fields are related to the old (unprimed) ones via

$$\nu'_L = \mathbf{X_1} \ \nu_L$$
$$(\nu'_R)^c = \mathbf{X_2} \ (\nu_R)^c \qquad (10)$$

After re-writing the Lagrangian in terms of the new (primed) fields the rotation matrix $\mathbf{X_2}$ disappears. However $\mathbf{X_1}$ does. It shows up in the charged currents which no longer are diagonal but are now given by

$$\overline{l_L} \ \gamma^\lambda \ \nu_L \to \overline{l_L} \ \gamma^\lambda \ \tilde{X}_1 \ \nu'_L \qquad (11)$$

Here l denotes the vector in the leptonic flavour space which has the components e, μ and τ respectively.

4 Neutrino Mass Matrices with Minimal Polynomials of Second Degree

The eigenvalue equation for the neutrino masses is of the form

$$\Delta(\lambda) = 0$$
$$\Delta(\lambda) \equiv \lambda^6 + c_5\lambda^5 + c_4\lambda^4 + c_3\lambda^3 + c_2\lambda^2 + c_1\lambda + c_0 \qquad (12)$$

where λ denotes the eigenvalue and the c_j, $j = 0, .., 5$, are invariant coefficients (functions of traces of powers of $\mathcal{M}$). For a general matrix the expressions for some of these coefficients are very complicated. For the mass matrices of the Electroweak Model, it turns out that [14] one obtains much simpler expressions which I shall not quote here.

As is well known the matrix $\mathcal{M}$ itself satisfies the above eigenvalue equation (i.e., the equation obtained by replacing λ by $\mathcal{M}$, in Eq. 12). This is the so-called Cayley-Hamilton equation which here is of the sixth order.

In this paper, I shall be looking for such mass matrices which satisfy, in addition, an equation of a much lower degree (second or third). Such mass matrices correspond to having a high degree of degeneracy in the neutrino masses. Thus the equation we are looking for is of the form

$$\psi(\mathcal{M}) = 0$$
$$\psi(\mathcal{M}) \equiv \mathcal{M}^k + b_{k-1}\mathcal{M}^{k-1} + .. + b_0\mathbf{E} \tag{13}$$

where $\mathbf{E}$ denotes the (six by six) unit matrix This equation with the lowest possible value of k defines the minimal polynomial of the mass matrix. The minimal polynomial is unique and factorizes the characteristic equation (see, for example [15])

$$\Delta(\mathcal{M}) = f(\mathcal{M})\psi(\mathcal{M}) \tag{14}$$

where f is again a polynomial in $\mathcal{M}$ but of a degree which is less than or equal to k. The minimal polynomial, ψ, and the characteristic polynomial, Δ, have the same set of distinct eigenvalues. In other words ψ is of the form

$$\psi(\lambda) = (\lambda - \lambda_1)^{r_1}(\lambda - \lambda_2)^{r_2}...(\lambda - \lambda_j)^{r_j} \tag{15}$$

where $\lambda_1, .., \lambda_j$ are the distinct eigenvalues of the characteristic polynomial $\Delta(\lambda)$ and $r_1 + r_2 + .. + r_j \leq 6$, for the case under consideration. Unfortunately, however, the powers r_1, etc, do not in general tell us about the degree of degeneracy of the respective eigenvalue. In general, they leave open many possibilities as we shall see below.

We now ask the following question: What is the minimal polynomial of the lowest possible degree for the neutrino mass matrix in the three-family model? We start with the trial polynomial $\psi(\mathcal{M}) = \mathcal{M} + b_0\mathbf{E} = 0$. This equality cannot be satisfied except for the trivial case that b_0 as well as the mass matrix both vanish. The neutrinos are then massless, a case which does not interest us. We now try a minimal polynomial of second degree

$$\psi(\mathcal{M}) = \mathcal{M}^2 + b_1\mathcal{M} + b_0\mathbf{E} = 0 \tag{16}$$

Using Eq. (5) we find that the matrix $\mathbf{A}$ must then be proportional to an orthogonal matrix and that the matrix $\mathbf{M}$ has to be a multiple of the unit matrix

$$\mathbf{A} = \xi\,\mathbf{O}, \qquad \mathbf{M} = \mu\,\mathbf{E} \tag{17}$$

where **O** is an arbitrary (three by three) orthogonal matrix and ξ as well as μ are arbitrary real numbers. As before, we shall take these numbers to be non-zero. The minimal polynomial is given by

$$\mathcal{M}^2 - \mu\mathcal{M} - \xi^2\mathbf{E} = 0 \qquad (18)$$

The roots of this equation are given by the familiar-looking expression

$$\lambda_\pm = \frac{1}{2}(\ \mu\ \pm\ \sqrt{\mu^2 + 4\xi^2}\ ) \qquad (19)$$

Thus

$$\psi(\lambda) = (\lambda - \lambda_+)\ (\lambda - \lambda_-) \qquad (20)$$

We can, therefore, conclude that the characteristic equation has only two distinct roots. A priori the multiplicities of the two roots could be (5,1), (4,2), (3,3), (2,4) and (1,5) where the numbers in the parentheses denote the multiplicities of the two roots respectively. However a simple calculation shows that the answer is (3,3), i.e.,

$$\Delta(\lambda) = (\lambda - \lambda_+)^3\ (\lambda - \lambda_-)^3 \qquad (21)$$

In this (in some sense simplest) case there are three degenerate neutrinos with a common mass $|\lambda_-|$. The other three are also degenerate with a common mass λ_+. Since the degenerate neutrinos are indistinguishable the lepton mixing matrix is trivial (i.e., the unit matrix) in this case. The charged currents, up to an overall constant, are given by

$$C.C. = \bar{e}\Gamma_\lambda(ic_\alpha\nu_e - s_\alpha N_e)\ +\ \bar{\mu}\Gamma_\lambda(ic_\alpha\nu_\mu - s_\alpha N_\mu)\ +\ \bar{\tau}\Gamma_\lambda(ic_\alpha\nu_\tau - s_\alpha N_\tau) \qquad (22)$$

Here the quantities $\Gamma_\lambda = \gamma_\lambda(1 - \gamma_5)$ and

$$c_\alpha = \sqrt{\frac{\lambda_+}{\lambda_+ + |\lambda_-|}}, \qquad s_\alpha = \sqrt{\frac{|\lambda_-|}{\lambda_+ + |\lambda_-|}} \qquad (23)$$

Note that, due to degeneracy, the three parameters of the orthogonal matrix **O** have disappeared. There are only two parameters (the two masses) in this sector of the model. However the orthogonal matrix will manifest itself in the Higgs sector (see below).

If we take the limit $|\mu| >> |\xi|$ we obtain the so called see-saw model where each family has one light and one heavy neutrino and there is no mixing between families. In general it is not all that trivial to get a see-saw behaviour by just assuming that the entries in the Majorana mass matrix **M** should be large. One must require that each of the three eigenvalues of this matrix is large. Evidently, the assumption that the minimal polynomial be of lowest possible degree is a way to get the see-saw behavior without family mixing.

Computing the neutral currents we find again they only depend on the two masses λ_+ and $|\lambda_-|$

$$
\begin{aligned}
N.N. \;=\; & c_\alpha^2(\overline{\nu_e}\Gamma_\lambda\nu_e + \overline{\nu_\mu}\Gamma_\lambda\nu_\mu + \overline{\nu_\tau}\Gamma_\lambda\nu_\tau) \\
& + s_\alpha^2(\overline{N_e}\Gamma_\lambda N_e + \overline{N_\mu}\Gamma_\lambda N_\mu + \overline{N_\tau}\Gamma_\lambda N_\tau) \\
& + c_\alpha s_\alpha \left\{ i(\overline{\nu_e}\Gamma_\lambda N_e + \overline{\nu_\mu}\Gamma_\lambda N_\mu + \overline{\nu_\tau}\Gamma_\lambda N_\tau) + h.c. \right\}
\end{aligned}
\tag{24}
$$

The Higgs interactions with the neutrinos are much more involved. The individual elements of the orthogonal matrix $\mathbf{O}$ now appear explicitly in the couplings. We have

$$
L(\nu - H) = L_1 + L_2
\tag{25}
$$

where

$$
L_1 = \frac{-H\sqrt{|\lambda_- \lambda_+|}}{2v} \left\{ s_\alpha c_\alpha(\overline{\nu_e}, \overline{\nu_\mu}, \overline{\nu_\tau})(\mathbf{O} + \tilde{\mathbf{O}}) \begin{pmatrix} \nu_e \\ \nu_\mu \\ \nu_\tau \end{pmatrix} + s_\alpha c_\alpha(\overline{N_e}, \overline{N_\mu}, \overline{N_\tau})(\mathbf{O} + \tilde{\mathbf{O}}) \begin{pmatrix} N_e \\ N_\mu \\ N_\tau \end{pmatrix} \right\}
\tag{26}
$$

and

$$
L_2 = \frac{-H\sqrt{|\lambda_- \lambda_+|}}{2v} \left\{ i(\overline{\nu_e}, \overline{\nu_\mu}, \overline{\nu_\tau})\gamma_5(c_\alpha^2\tilde{\mathbf{O}} - s_\alpha^2\mathbf{O}) \begin{pmatrix} N_e \\ N_\mu \\ N_\tau \end{pmatrix} + i(\overline{N_e}, \overline{N_\mu}, \overline{N_\tau})\gamma_5(c_\alpha^2\mathbf{O} - s_\alpha^2\tilde{\mathbf{O}}) \begin{pmatrix} \nu_e \\ \nu_\mu \\ \nu_\tau \end{pmatrix} \right\}
\tag{27}
$$

Finally, it should be noted that although we have been focusing on the case of three families the above results are valid for any number of families, where there are as many right-handed as left-handed neutrinos.

In the next section we shall consider the case where the minimal polynomial of $\mathcal{M}$ is of third degree. We shall see that the above simplicity will be largely lost.

5 Neutrino Mass Matrices with Minimal Polynomial of Third Degree

Let us now go one step further and consider the case where the minimal polynomial of $\mathcal{M}$ is of third degree, viz.,

$$\psi(\mathcal{M}) = \mathcal{M}^3 + b_2\mathcal{M}^2 + b_1\mathcal{M} + b_0\mathbf{E} = 0 \tag{28}$$

Without lack of generality we may reparametrize the mass matrix by rotating it as in Eq. 6 choosing $\mathbf{X_2}$ such that the Majorana mass matrix $\mathbf{M}$ is diagonal

$$\mathbf{M} = diag.(M_1, \ M_2, \ M_3) \tag{29}$$

Eq. (28) gives the following restrictions

$$\begin{aligned}
\mathbf{AM\tilde{A}} + b_2\mathbf{A\tilde{A}} + b_0\mathbf{E} &= 0 \\
\mathbf{A}(\mathbf{\tilde{A}A} + \mathbf{M}^2) + b_2\mathbf{AM} + b_1\mathbf{A} &= 0 \\
(\mathbf{M} + b_2\mathbf{E})(\mathbf{\tilde{A}A} + \mathbf{M}^2) + \mathbf{\tilde{A}AM} + b_1\mathbf{M} + b_0\mathbf{E} &= 0
\end{aligned} \tag{30}$$

as well as the transposed equations. These equations look indeed very complicated. Therefore, I shall only consider a special class of solutions where all neutrino masses are taken to be non-zero.

If all neutrino masses are non-zero then the matrix $\mathbf{A}$ is non-singular. Thus one must take b_0 to be non-vanishing. Otherwise, the minimal polynomial as well as the characteristic equation will have zero eigenvalues which is not possible because $det\mathcal{M} = -(det\mathbf{A})^2$. In this case the above constraint equations simplify considerably. Only two independent constraint matrix equations are obtained. They read

$$\begin{aligned}
\mathbf{M} + b_2\mathbf{E} + b_0(\mathbf{\tilde{A}A})^{-1} &= 0 \\
\mathbf{\tilde{A}A} + \mathbf{M}^2 + b_2\mathbf{M} + b_1\mathbf{E} &= 0
\end{aligned} \tag{31}$$

Since $\mathbf{M}$ is diagonal we have that the matrix $\mathbf{\tilde{A}A}$ is also diagonal

$$\mathbf{\tilde{A}A} \equiv \Sigma \equiv diag(\sigma_1, \sigma_2, \sigma_3) \tag{32}$$

where σ_j, j=1,2,3, are three non-negative numbers. Thus the matrix $\mathbf{A}$, up to an overall sign is given by

$$\mathbf{A} = \mathbf{O} \ diag.(\sqrt{\sigma}, \sqrt{\sigma}, \sqrt{\sigma'}) \tag{33}$$

where O is again a general three by three orthogonal matrix. Solving for M from the first equation in (31) and substituting it into the second one we find

$$\Sigma^3 + b_1 \Sigma^2 + b_0 b_2 \Sigma + b_0^2 = 0 \tag{34}$$

This equation may be considered as a linear system of equations for three unknowns b_0^2, $b_0 b_2$ and b_1. Solving for these unknowns we find that if all three σ's are different $b_0^2 = -\sigma_1 \sigma_2 \sigma_3$ which is not possible because the σ's are positive. Thus there is no such minimal polynomial of third degree. Taking two of the σ's to be equal we have, for example

$$\Sigma = diag(\sigma, \sigma, \sigma') \tag{35}$$

We then obtain

$$
\begin{aligned}
b_0^2 &= \sigma \sigma'(\sigma + \sigma' + b_1) \\
b_0 b_2 &= -(\sigma^2 + \sigma'^2 + \sigma \sigma') - (\sigma + \sigma') b_1
\end{aligned}
\tag{36}
$$

A necessary condition for having a solution is that $b_1 > -(\sigma + \sigma')$. As an explicit example one can take, for example $b_2 = 0$. We then obtain

$$
\begin{aligned}
b_0 &= \pm \frac{\sigma \sigma'}{R} \\
b_1 &= -(\sigma + \sigma') + \frac{\sigma \sigma'}{R^2}
\end{aligned}
\tag{37}
$$

where $R = \sqrt{\sigma + \sigma'}$. The minimal polynomial is in this case given by

$$\psi(\lambda) = (\lambda - \lambda_1)(\lambda - \lambda_2)(\lambda - \lambda_3) \tag{38}$$

where for $b_0 = +\sigma \sigma'/R$ we have

$$\lambda_1 = \sigma/R, \qquad \lambda_2 = \sigma'/R, \qquad \lambda_3 = -R \tag{39}$$

For b_0 with the opposite sign we just have to change the signs of all three eigenvalues. For this special case we find that the Majorana mass matrix is given by

$$M = diag.(-\sigma'/R, -\sigma'/R, -\sigma/R) \tag{40}$$

Thus our mass matrix $\mathcal{M}$ has three distinct eigenvalues λ_1, λ_2 and λ_3. A simple computation then gives that

$$\Delta(\lambda) = (\lambda - \lambda_1)^2 (\lambda - \lambda_2)(\lambda - \lambda_3)^3 \tag{41}$$

In other words, there are two degenerate neutrinos with mass λ_1, one with mass λ_2 and three degenerate ones with mass λ_3. One may work out all the couplings as was done in the previous case. The calculations are trivial but a bit long. In this case the elements of the orthogonal matrix O do appear in the charged currents. The expressions obtained are more complicated and, therefore, will not be quoted here. This model does not exhibit see-saw-like behavior but has a different pattern of hierarchy. This and the phenomenological consequences of the above models will be presented elsewhere.

Acknowledgements

I am indebted to Prof. R. Akhoury for asking me to write an article for the Veltman Festschrift "Gauge Theories - Past and Future". I remember vividly that in 1969 I took part in a school of physics at Nordita where Tini Veltman was one of the speakers. My advisor Gunnar Källen had been killed in a plane accident the year before and thus I had to stand on my own feet. I took advantage of having met Tini and asked him what he believed one should work on to which he promptly replied "why don't you work on Yang-Mills theories". Alas, I did not understand him, nor did I understand his lectures on what we now call gauge theories. That was not the way one used to think and talk in those days. It is for me a great pleasure to dedicate this paper to him on the occasion of his sixtieth birthday.

I wish to thank Prof. Val Telegdi for his comments on the manuscript and for showing me several letters by Pauli where he clearly states his (sceptical) opinion on the neutrino theory of light.

This work has been supported by the Swedish National Research Council (NFR).

References

[1] C. Jarlskog, Nucl. Phys. A518 (1990) 129

[2] M. Born and L. Infeld, Proc. Roy. Soc. A144 (1934) 425.

[3] G. Wentzel, Zeit. f. Phys. 92 (1934) 337

[4] L. de Broglie, C. R. 195 (1932) 536, 577, 862; *ibid* 197 (1933) 1377; *ibid* 198 (1934) 135; *ibid* 199 (1934) 445, 1165;

L. de Broglie and J. Winter, C. R. 199 (1934) 813

[5] P. Jordan, Zeit. f. Phys. 93 (1935) 464; *ibid* 98 (1936) 759; *ibid* 99 (1936) 109; *ibid* 102 (1936) 243; *ibid* 105 (1937) 115, 229

[6] R. De L. Kronig, Physica 2 (1935) 491, 854, 968

[7] P. Jordan and R. De L. Kronig, Zeit. f. Phys. 100 (1936) 569

[8] O. Scherzer, Zeit. f Phys. 97 (1935) 725

[9] "Wolfgang Pauli, Scientific Correspondence", Vol. II (1930-1939), Eds. K. von Meyenn, A. Herman and V. Weisskop (Springer-Verlag 1985).

[10] S. L. Glashow, Nucl. Phys. 22 (1961) 579;

S. Weinberg, Phys. Rev. Lett. 19 (1967) 1264 ;

A. Salam, *in* Elementary Particle Theory, Ed. N. Svartholm (Almqvist and Wiksell, 1968).

[11] J. D. Bjorken and S. D. Drell, Relativistic Quantum Mechanics (McGraw-Hill, 1964) Chap. 5

[12] H. Umezawa, Quantum Field Theory (North-Holland, 1956) Chap. III

[13] L. Wolfenstein, Nucl. Phys. B186 (1981) 147

[14] C. Jarlskog and B.-Å Lindholm, Stockholm preprint USITP 91-17 (Sept. 1991) to appear in Zeit. Phys. C

[15] F. R. Gantmacher, Théorie des Matrices, Vol. 1 (Dunond, 1966)

QCD: SOME LIKE IT HOT
or
WHY THERE HAVE BEEN SMALL BANGS[*]

by
C.P.Korthals Altes
Centre Physique Theorique au C.N.R.S., Section 2
Campus de Luminy, F13288, Marseille, France

Abstract

We report on recent work in perturbative hot QCD. Amusingly, the surface tension between two pure gluonic $Z(N)$ vacua is calculable to two loop precision and we speculate on how to keep infrared divergencies at bay to any order in the surface tension. A conspiracy between quarks and gluons creates meta stable $Z(N)$ vacua, in which charge conjugation is broken spontaneously. We discuss briefly how our results lead to the occurrence of small bangs.

..avenir,passé,cosmogonie,néant. Je
suis le maître en fantasmagorie.

A.Rimbaud,La Nuit de l'enfer.

1. Introduction

Hot field theories have been relevant during a few seconds just after the Big Bang. It seems therefore hardly appropriate to bring the subject up on Tini's birthday, some 10^{10} years too late. Yet, I may have some excuse because of the planning of the heavy ion colliders, though controversy is still raging about the true thermal state of affairs in the collisions they will bring about.

---

[*] January 1992. To appear in the Festschrift for M.J.G.Veltman, University of Michigan, May 1991.

Here we will be concerned with hadrons at temperatures much higher than the deconfinement temperature of QCD. Under these circumstances the coupling will—at least naively—become small and the question was raised long ago whether perturbation theory was possible. The naivety became apparent at multiple loop order: quantities like free energy became infrared divergent, so it seemed that the usual picture of a gas of free quarks and gluons interacting only weakly at very high temperature is inadequate.

Much speculation has been going on on the "true" nature of the deconfined phase, unfortunately without much tangible results. Recently however, there has been a revival of interest in the small coupling expansion through the work of R.Pisarski, J.C.Taylor and others[1]. The work[2] described here is another example of how well perturbation theory works for a very specific observable: the surface tension in hot QCD. In fact, it works so well, that Tini will be mightily pleased, as one of the authors of this seminal paper on gauge theories: "Example of a gauge theory" of twenty years ago[3]. Still, in our case, the workings of the hot theory in more than two loops are fairly obscure, to put it mildly.

In particular, the occurrence of infrared divergencies is said to jeopardize the use of perturbation theory. However these divergencies will arguably occur in a very special sector of the theory, as will be done in the last section. At any rate, the current state of the art allows us to use our two loop calculation and speculate on the occurrence of bubbles of a false vacuum of quark-gluon-matter decaying through a series of small bangs into the true vacuum[4].

The set-up of this paper is as follows. In the next section we will fix notations, and remind the reader of the general ideas of what happens to QCD in its hot phase, and how to describe it. In section 3 we will discuss the surface tension and a novel approach to compute it in QCD, the gauge independence and the stability of the ground-state. In section 4 we look at the theory with quarks involved and the question of stability comes in again. Section 5 deals with the small bangs. An outlook closes the paper.

2. A bird's view of hot QCD

The starting point of our discussion will be the second quantized Hamiltonian H:

$$H = Tr \int d\vec{x} [g^2 (\vec{E})^2 + \frac{1}{g^2} (\vec{B})^2] \qquad (2.1)$$

The electric and magnetic field strengths $\vec{E}$ and $\vec{B}$ are written in $N x N$ matrix form and the gauge group is $SU(N)$. The trace is over colour, and the coupling g is the bare coupling. We'll leave out for a start the quark content: it will come in section 4.

This Hamiltonian can be used to compute thermodynamical quantities. For example the free energy one gets by summing over matrix elements between physical states:

$$\exp -\frac{F}{T} = \sum_{phys} exp -\frac{H}{T} \qquad (2.2)$$

One can convert this trace into a path integral. To do so, one introduces a fictitious time t and chops up the fictitious time lapse $\frac{1}{T}$ in front of H in little "time" bits. The physical states in 2.2 are gotten from the canonical ones by projecting over the gauge group which leads to the well known relation:

$$\sum_{phys} exp -\frac{H}{T} = \int D[A_0] D[\vec{A}] \exp -\frac{1}{g^2} S \qquad (2.3)$$

The potentials are to be periodic in the time direction, and the action S one gets from the usual Lagrange density by integrating over all of space and the "time" lapse $\frac{1}{T}$. This path integral delivers Feynman rules, good for a perturbative calculation. It can be put on a lattice to do a non-perturbative simulation.

The periodicity of the gauge potentials can be extended to periodicity up to a gauge transformation. Gauge transformations can be periodic modulo a centergroup transformation. Under the latter the Hamiltonian and the action are still invariant (Z(N) symmetry). But the Polyakov "loop" $p(\vec{x})$, defined as a path ordered exponential along a straight path in the "time" direction, is not. In other words

$$p(\vec{x}) = \frac{1}{N} Tr P \exp i \int\limits_0^{1/T} dt A_0(t, \vec{x}) \qquad (2.4)$$

feels the presence of the discontinuity, and serves as an order parameter if the symmetry gets spontaneously broken.

Is there any reason to believe that this symmetry will be spontaneously broken? Since the modulus of the Polyakov-loop is the exponential of minus the free energy of a single infinitely heavy quark, the answer is: yes, if at high enough temperature the free energy of a single external quark gets from infinite (confinement) to a finite value (deconfinement). If that is realized, the Polyakov loop acquires a non-zero value for high enough T.

This ends our brief introduction to hot gauge theory. I will now describe how one simulates the surface tension in hot QCD.

People have done the following simulation[7] with $SU(3)$: take an elongated box of size $L_t{}^2$ by L, L much larger than L_t (see fig.3). Eventually all box sizes will be taken to infinity.

Now we put hot QCD in this box and fix at one boundary in the elongated direction the value of the Polyakov loop (eqn(2.4)) to be 1, at the other boundary to be $\exp i\frac{2\pi}{3}$ (see fig.3). Configurations with the value of the loop being in Z(3), will be called "vacua of hot QCD". Justification follows in the next section below eqn(3.3). Because of the Z(3) symmetry the energy is degenerate at the two boundaries. Now one can compute the free energy of this box and compare it to the free energy of a box with the fixed values of the loop both 1, say. What is found at large enough temperature, is an excess free energy associated to the "turning around" of the value of the loop from the 1 to $\exp i\frac{2\pi}{3}$. In other words, formation of a domain wall between the two vacua at the ends of the box takes place.

How precisely the turning around goes, depends on the temperature and the results are shown in fig.1. So at very high T the average values of the loop follow very closely the border of the allowed region as we move from one end to the other in the elongated direction. The border of the allowed region will be parametrized by the parameter q through $tr \exp i2\pi q\sqrt{6}\lambda_8$ in an obvious notation

So q runs from 0 to 1 in our box. What happens for values from 1 to 2 and 2 to 3 is determined by the $Z(3)$ symmetry from the values between 0 and 1. The excess free energy is proportional to the cross section L_t^2 of the container and the proportionality constant is called the surface tension α. Its values from a simulation are shown in fig.2 as a function of temperature.

3. Effective action and the surface tension.

Let us now get to the perturbative calculation of the surface tension and properties of the interface profile. We will do so by first calculating an effective action Γ as a function of the averaged order parameter $< p(z) >$, z being the coordinate in the elongated direction of our box. For our purposes we define this effective action Γ as follows:

$$\exp -\frac{1}{T}L_t^2\Gamma = \int DA_\mu \prod_z \delta(p(z) - \frac{1}{L_t^2} \int dx\, dy p(\vec{x})) \exp -\frac{1}{g^2}S(A)$$

$$(3.1)$$

This definition tells us that Γ, up to a normalisation, is the probability that a given profile $p(z)$ appears, with the boundary conditions to be specified. The normalisation also appears with these boundary conditions.

So this is the quantity of interest: once we have it explicitly we obtain the interface by minimising Γ under the boundary condition described in detail above. We will be interested in temperatures T much higher than any hadronic scale. Then the coupling can be supposed to be small and perturbation theory should be applicable. Our gauge choice will be motivated by the physical context: we will work in a classical background field $A_0 = C(\vec{x}), A_i = 0$ and use this background field in the definition of the Polyakov loop, eqn(2.1). So the loop will be a function of C, which can be chosen to be time independent and diagonal, by a suitable gauge transformation. This choice will be kept in the final result for the effective action Γ. Γ will depend on the diagonal elements $C_i, i = 1...N$, and with the constraint $\sum_i C_i = 0$. Throughout backgound gauge fixing is used:

$$S_{gauge\ fixing} = \frac{1}{\xi} Tr(D_\mu Q_\mu)^2 \qquad (3.2)$$

Here we have written D_μ as the covariant derivative with respect to the background field C and Q_μ is the quantum fluctuation around C. An important property of this gauge fixing term is that the propagators for the gluons are infra red protected for all non diagonal quantum fluctuations. To be precise: we take the colour basis formed of the $N(N-1)$ matrices $E_{i,j}$ having entry zero everywhere except on the ith row and the jth column, where we have entry 1, and a set of $N-1$ diagonal traceless matrices; then it is easy to verify that the propagator that propagates such an off diagonal quantum $Q_{i,j}$ will have have a fourth component of momentum of the form $2\pi nT + C_i - C_j$. The integer valued n is due to the periodicity in the time direction. This form of the fourth component is due to our use of the *covariant* rather then the traditional derivative. When the integer n is set to zero, the background field $C_i - C_j$ will serve as a mass in the remaining *three* dimensional propagator.

So any infrared singularities in the $C \neq 0$ sector are due to the maximal Abelian subgroup of $SU(N)$. This statement, strictly speaking, is true for $SU(2)$; for $SU(3)$ the $C \neq 0$ sector is two dimensional and contains *lines*,where the background field mass vanishes for some, but not all of the off-diagonal fields.

We have computed the effective action to order g^2 under the assumption, that the background is very smoothly varying. Then we are allowed to make a gradient expansion in the result we get from computing the 1 and 2 loop graphs shown in fig.4. The gradient terms come from the classical action (remember our background field has only a time component and is diagonal, so only the gradient term survives in the classical action) and from the one-loop graph and give rise to the kinetic term in the effective potential. Since the gradient term is small, of order g as it turns out (see below eqn(3.6)), we do not need to go any further in the gradient expansion.

The constant potential has contributions from both one and two-loop graphs, to order g^2. The result can be given in terms of the differences $C_i - C_j$, appearing in all energies. To avoid clutter in our formulae, we will limit the discussion to the case where $C_1 = C_2 = ... = C_{N-1}$ and take $\frac{C_{N-1}-C_N}{2\pi T} = q$. One sees immediately that the summation over the discrete energies will leave us with a result periodic modulo 1 in q. This definition of q coincides for $N = 3$ with the one given at the end of section 2.

We are interested in computing the surface tension. So the result we give is obtained by subtracting out the value of the graphs at $q = 0$, i.e. the free energy. So let us give the result:

$$\frac{\Gamma}{(2\pi T)^2(N-1)} = \frac{1}{g^2 N}\int dz(\frac{\partial}{\partial z}q)^2 + \frac{1}{3}T^2\int dzq^2(1-q)^2$$
$$+ \frac{1}{16\pi^2}\frac{11}{3}\int dz(\frac{\partial}{\partial z}q)^2(\psi(q)+\psi(1-q)+\frac{1}{11})$$
$$+ \frac{1}{16\pi^2}2(1-\xi)\int dz(\frac{\partial}{\partial z}q)^2$$
$$+ \frac{g^2 N}{16\pi^2}T^2\int dz(q^2(1-q)^2 - \frac{2}{3}q(1-q))$$
$$- \frac{g^2 N}{16\pi^2}T^2\frac{4}{3}(1-\xi)\int dz(-q^2(1-q)^2 + \frac{1}{4}q(1-q))$$

$$(3.3)$$

The dependence on the variable q is, as explained above, modulo 1. The first two terms are the classical gradient term and the one-loop constant potential term. The one loop term (see fig.7) has minima at $\pm q = 0, 1, ...$ The form of this term shows that the $Z(3)$ symmetry is spontaneously broken, the minima being the vacua we introduced in the previous section.

Then follow the one loop gradient terms, where the logarithmic derivative of Euler's gammafunction is written as ψ. There appears a pole in dimension 4 in this term that we have absorbed in the coupling g in the traditional way: g picks up a logarithmic dependence in T, consistent with asymptotic freedom.

The last two terms are the two-loop constant potential terms[6]. Notable is the presence of gauge dependence; on top of that the simple vacuum structure of the lowest order result (periodic vacua at $\pm q=0,1,2,...$) is undone by the two-loop terms; they provide a linear term in q with a negative sign so that the periodic vacua are split into pairs $O(g^2)$ apart.

Much has been made of this "shift of the groundstate": as the q-variable transforms like the fourth component of the vector potential, it seems that any symmetry transformation *changing* that component of the vector potential would be spontaneously broken.

But one thing has not been taken into account in (3.3): to two loop order in the potential the one-loop renormalisation effects of the Polyakov loop should be relevant[5]!

Once this is done (see fig.5) we have a shift δq in the variable q:

$$q_r = q + \delta q = q + \frac{g^2 N}{16\pi^2}(3 - \xi)(q - \frac{1}{2}) \qquad (3.4)$$

The dependence of 3.3 on the renormalised q_r cures the gauge dependence problem, and at the same time restores the original vacuum structure at the integer values of q:

$$\frac{\Gamma}{(2\pi T)^2(N-1)} = \frac{1}{g^2 N} \int dz (\frac{\partial}{\partial z} q_r)^2 (1 + \frac{g^2}{16\pi^2} \frac{11}{3} N(\psi(q_r) + \psi(1 - q_r) - 1)$$

$$+ \frac{1}{3} T^2 (1 - 5 \frac{g^2 N}{16\pi^2}) \int dz q_r^2 (1 - q_r)^2 \qquad (3.5)$$

In fact, the two-loop constant potential is identical to the one-loop up to a constant!

This two fold miracle is brought about by the gauge independence of the effective action *as function of a gauge invariant variable* and by the underlying $Z(N)$ symmetry of the action.

It is now an easy matter to get the profile $q(z)$ and its free energy (that is, the surface tension), by minimising Γ. This can be done by completing the square of kinetic (K) and potential (V) terms in eqn(3.5), and subtracting the mixed term between the two. So the right hand side of eqn(3.5) becomes in this notation:

$$\int dz([(K^{\frac{1}{2}} - V^{\frac{1}{2}}]^2 + 2K^{\frac{1}{2}} V^{\frac{1}{2}}) \qquad (3.6)$$

Since the square root of the kinetic term is proportional to the derivative of q, the mixed term does not depend on what configuration $q(z)$ we take. The minimum of Γ is then obtained by setting the square to zero. We find from the remaining mixed term the surface tension:

$$\alpha = \frac{4\pi^2 T^3 (N-1)}{\sqrt{3}\sqrt{g^2 N}}(1 + \frac{g^2 N}{16\pi^2}\left\{11\int_0^1 dq(2q-1)log\frac{\Gamma(q)}{\Gamma(1-q)} - \frac{13}{3}\right\})$$

(3.7)

This is a remarkably simple result for a two loop computation. The leading term is of order $\frac{1}{g}$, not $\frac{1}{g^2}$ as is traditional in semi classical cases. This can be understood easily from eqn(3.6): the kinetic term is order $\frac{1}{g^2}$, the potential term of order 1, so the mixed term is order $\frac{1}{g}$.

The profile itself is obtained from solving the first order differential equation one gets from setting the square in eqn(3.6) zero. This leads to an analytic solution[2]. It is, for the reasons just mentioned above, a function of gz, so slowly varying. For SU(2) and SU(3) it is an easy matter to show that the path along the q-direction is indeed the minimising path. For general N we do not yet know.

Comparison to lattice results for SU(3) is shown in fig.2, and things work very well.

4. Adding the fermions

Up till now we worked in an academic universe. Adding in the quarks is destroying the original $Z(3)$ symmetry in the action. The reason is that quarks do feel the centergroup transformations so they will pick up a discontinuity under the twisted gauge transformations. On a less formal level, the presence of virtual quarks renders the expectation value of the Polyakov loop non-zero, no matter what the temperature is. So its modulus will be the exponentiated free energy of an infinitely heavy quark bound to a light quark. Again we will compute the free energy of the whole system as a function of the loop average as in eqn (3.1).

The propagator of the quark will contain the background field C through

$$(\partial\!\!\!/ + i\mathcal{C}\!\!\!/)^{-1}$$

while the quark couples *only* to the quantum field Q. So the presence of quarks renormalizes the Polyakov loop starting from order g^4, and eqn (3.4) is valid for the two loop case *with* fermions.

Let us give the contributions to the free energy coming from the one and two loop (see fig.6):

$$
\frac{\Gamma}{(2\pi T)^2(N-1)} = \frac{1}{16\pi^2}(\frac{-2}{3})n_f \int dz (\frac{\partial}{\partial z}q_r)^2 [\frac{1}{N^2}(\psi(r) + \psi(1-r) - 1)
$$
$$
+ \frac{N-1}{N^2}(\psi(s) + \psi(1-s) - 1)]
$$
$$
- n_f \frac{1}{3}T^2 \int dz[r^2(1-r)^2 + \frac{1}{(N-1)}s^2(1-s)^2]
$$
$$
n_f \frac{g^2}{16\pi^2}T^2 \int dz[(-\frac{5}{6} + \frac{1}{2N^2})q_r^2 + \frac{2}{3}q_r^3
$$
$$
+ (\frac{2}{3} - \frac{3}{N} + \frac{2}{N^2} + \frac{3}{N^3} - \frac{3}{N^4})q_r^4]
$$

$$(4.1)$$

The fermionic variables r and s are related to the renormalized q_r by: $r = \frac{q_r}{N} + \frac{1}{2}$, $s = -\frac{(N-1)}{N}q_r + \frac{1}{2}$.

From the graphs one sees that the effect is proportional to the n_f flavours, which we have taken massless. The two loop term for the fermions is down by a factor of N in the large N expansion, compared to the two loop gluon term in eqn (3.5).

The two loop term in 4.1 is only valid for small values of q_r, up to $q_r = \frac{3}{4}$. It would take too much space to give the full expression. In fig.7 the behaviour for all q_r is shown, for a large value of the coupling, in order to render the correction visible. The number of colours is three.

Note the local minima at non zero values of q. As we said before, this implies a spontaneous breaking of charge conjugation in these meta stable vacua, since it transforms q into $-q$.

In the two loop graph (fig.6) there is a linear term in q and also gauge dependent terms. Both disappear after taking the renormalisation effects of the Polyakov line into account (eqn (3.4). The linear term (C violating!) is, to the best of our knowledge, not forbidden by any general theorem[8]. It remains intriguing why the stable vacuum at $q = 0$ is C conserving, to at least two loop order.

Of course, one must know whether perturbation theory is not beset by infrared divergencies in higher orders, especially in the region where q_r is on the order of g^2. From our discussion below eqn(3.2) we see that in this region the infra red protection of our background field propagators gets weak. Since the infra red divergencies are *three* dimensional, the perturbative series may get upset by terms of order $\frac{1}{g^2}$. This is quite a non trivial question. Further comments are relegated to the last section.

5. The small bangs

Taking only the one loop effects of eqn(4.1) gives already an amusing effect. Let us go to very high temperatures and take 6 massless flavours. Then we see in fig.7 a rather shallow local minimum at a value of $q = q_{ms}$ near 1. This is a vestige of the old global minimum of pure gluon theory at $q = 1$, and C is spontaneously broken in this metastable minimum!

One can now ask the following question[4]: if just after the Big Bang parts of the universe got stuck in this metastable minimum, at what later time these bubbles of false vacuum will have decayed into the true vacuum of the theory?

To answer this question it is necessary to know the nucleation rate γ of bubbles of true vacuum inside the region of the metastable vacuum. This rate is taken per unit volume and unit time; it can be computed from our effective action 3.5 and 4.1. It is for dimensional reasons of the form:

$$\gamma \sim T^4 \exp -S(q_{ms}) \tag{5.1}$$

The action needed to tunnel from the metastable vacuum q_{ms} to the stable vacuum at $q = 0$ figures in the exponent in 5.1. It will be computed[4] in a simple minded thin wall approximation.

In this approximation the free energy for forming a small droplet of stable vacuum is dominated by the surface energy one has to pay for small radius, for large radius one gains energy due to the volume term. The critical radius is where one breaks even, and the corresponding free energy is in terms of our surface tension α and the vacuum density difference ϵ:

$$S(q_{ms}) = \frac{16\pi}{3} \frac{\alpha^3}{\epsilon^2} \tag{5.2}$$

Both α and ϵ follow from our calculation and can be read off from fig.7.

Consider the time t at which the nucleation rate γ obeys

$$\gamma t^4 \sim 1 \qquad (5.3)$$

to be the time where the nucleation has substantially taken place. Time is in Big Bang scenarios related to the temperature T by

$$T^2 t \sim M_{Planck} \qquad (5.4).$$

From eqns 5.1, 5.2, 5.4 and the known behaviour of $g(T)$ from the QCD beta function one can immediately estimate the temperature at which 5.3 is satisfied. The details of the calculation can be found in Kajantie et al.[4], and the outcome is a temperature in the Tev region. This bubble scenario may be a testing ground for C-violating effects, due to the metastable vacuum.

6. Outlook

One conclusion is that perturbation theory works wonderfully well in hot QCD, to two loop order. Our final answer, eqn(3.5), is gauge invariant, has stable $Z(3)$ symmetric minima and is infra red finite. The latter starts to become problematic in higher loops: it is an established fact that at $q = 0$ the free energy develops such singularities. But remember we are only interested in the difference of free energies: we normalize our result to be zero at $q = 0$ and it could well be that the infrared singularities cancel out in this difference! Of course the renormalisation of the Polyakov loop has to be taken into account in every order.

How to check whether this happens? One answer is to compute explicitly, but this would not lead very far.... However our gauge choice renders the infrared singularities Abelian in the $q \neq 0$ sector, as we indicated in section 3. So it may be possible to chase them in a systematic way.

Thus the statement would be that the infra red singularities of hot QCD are present only in the $q = 0$ sector. This contains quantities like the bulk free energy.

Once taken care of, they might leave us with an effective action, calculable to any order. To bring the C(*and* CP) violating properties into play deserves further attention[9]. For the moment the C-violating potential seems a curious- and hot-conspiracy,that some might like[10].

Acknowledgements

I'm indebted to Andreas Gocksch, Keijo Kajantie, John Taylor and Mike Teper for useful discussions. I did profit much from the insights, comments and caveats of Rob Pisarski, who brought the C-non-invariance of the meta stable vacua to my attention.

References

[1]   R.D.Pisarski, Nucl. Phys. B309, 476 (1988).

      J.C.Taylor, S.M.H.Wong, Nucl. Phys. B346 (1990), 115.

[2]   T.Bhattacharya, A.Gocksch, C.P.Korthals Altes and R.D.Pisarski, Phys. Rev. Lett., 66 (1991) 988.

      C.P.Korthals Altes, to be published in the Proceedings of the Hot Summer Daze workshop, BNL, August 1991.

      T.Bhattacharya, A.Gocksch, C.P.Korthals Altes and R.D.Pisarski, BNL preprint in preparation.

[3]   G.'tHooft, M.J.G.Veltman, in Proceedings of the Marseille conference 1972, parus au Centre de Physique Theorique au CNRS, Marseilles (1972), ed. C.P.Korthals Altes.

[4]   M.Ogilvie, R.Dixit, Washington University preprint, 1991.

      J.Ignatius, K.Kajantie, K.Rummukainen, Helsinki preprint, 1991.

[5]   V.M.Belyaev, Phys. Lett. B254, 153 (1991).

[6]   K.Enkvist, K.Kajantie, Zeitschr. Phys. C47, 291 (1990).

This work contains the two loop constant potential,with a different coefficient from ours in the gauge dependent term.

[7]   K.Kajantie,L.Karkainen, R.Rummukainen, Nucl. Phys. B333,100 (1990).

S.Huang, J.Potvin, C.Rebbi and S.Sanielovici, Phys. Rev. D42, 2864 (1990).

[8]   C.Vafa, E.Witten, Phys.Rev.Lett. 53,535 (1984).

[9]   C.P.Korthals Altes, R.D.Pisarski, in preparation.

[10]   R.D.Pisarski, conspiratory communication, to remain secret.

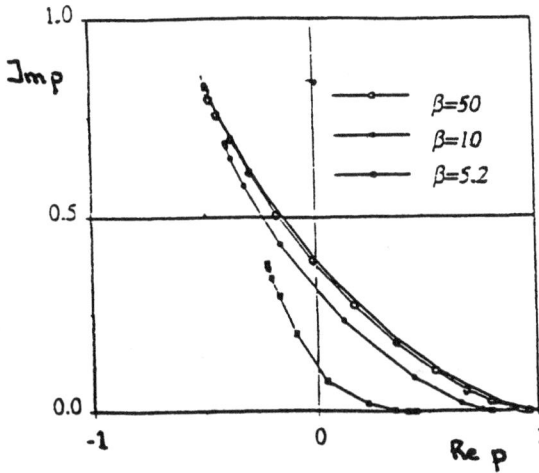

Fig. 1   Numerically determined trajectories (Kajantie et al. [7]) for various $T(\sim \beta)$, from $p = 1$ to $p = exp\ i\ \frac{2\pi}{3}$. For the highest T the trajectory follows the border of the allowed region for p, defined in section 2.

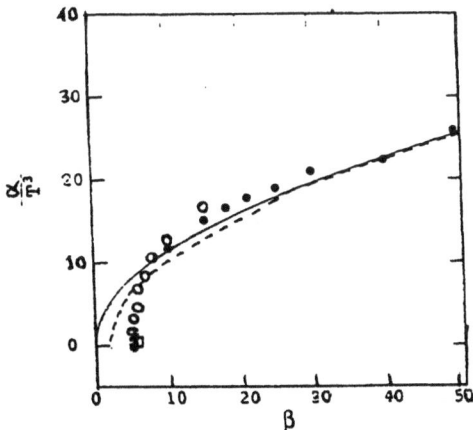

Fig. 2   Results from a numerical calculation (Kajantie et al. [7]) for the surface tension. Dots are data points ; continuous line is lowest order result(eqn. (3.7)), broken line the $O(g^2)$ correction.

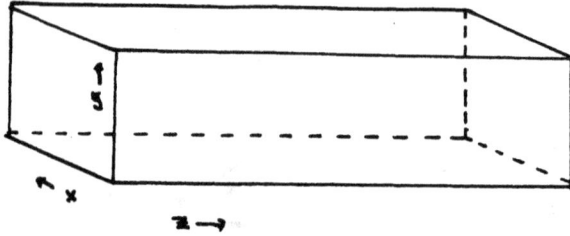

Fig. 3   Box of size $L_t^2 \times L$, $L \gg L_t$. In the two $x - y$ planes at the endpoints in the elongated z-direction we have fixed the Polyakov loops to the values 1 (at $z = 0$) and $exp\ i\ \frac{2\pi}{3}$ (at $z = L$). A domain wall develops in between, in the hot QCD phase.

Fig. 4   Diagrams in the gluonic sector, contributing to the effective action to two loop order (eqn. (3.3)), and symmetry factors. Continuous line is gluon propagator, broken line is ghost propagator.

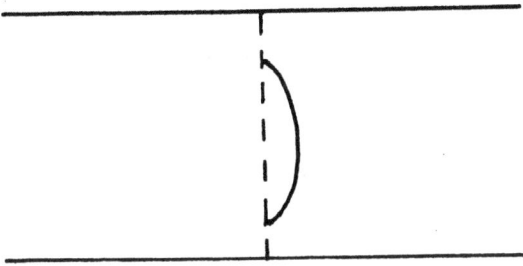

Fig. 5   The one loop renormalization of the Polyakov loop (eqn. (3.4)). Continuous line is gluon propagator.

Fig. 6   Diagrams for the fermion contribution to two loop order, contributing to the effective action (eqn. (4.1)). The dotted line is the fermion propagator, continuous line the gluon propagator.

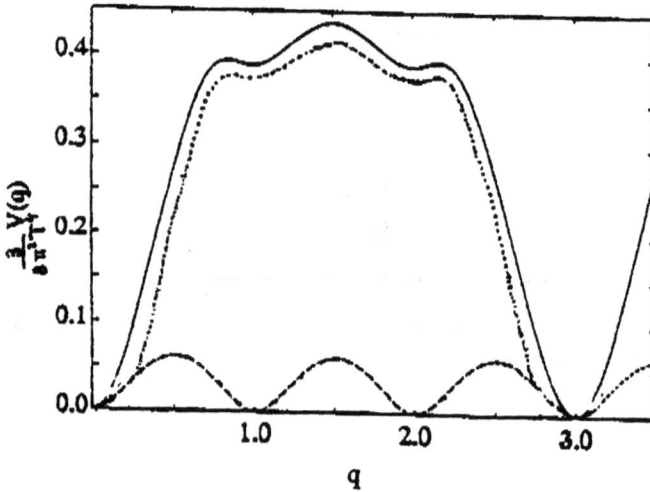

Fig. 7 The dimensionless potential $\frac{3}{8\pi^2 T^4} V$ as function of the renormalized parameter q, from eqns. (3.5) and (4.1).

Dashed line   : one-loop pure gluon result, N=3.
Continuous line   : full one-loop result including six massless quarks.
Dotted line   : the full two-loop result with six massless quarks,
                           for $\frac{g^2(T)}{4\pi} \approx 0.1$.

THE LAST TEN YEARS OF RADIATIVE CORRECTIONS

Giampiero Passarino

Randall Laboratory
University of Michigan
Ann Arbor, MI 48109-1120, USA
and

Dipartimento di Fisica Teorica
INFN, Sezione di Torino, Italy
Università di Torino
Torino, Italy[*]

ABSTRACT

Some of the old and new concepts in the field of radiative corrections for gauge theories are discussed in perspective and in the light of the recent high precision measurements at LEP.

---

[*] *Permanent address*

1. Introduction and Diagrammatica

The main purpose of this note will be to discuss in a selective way the foundation and the successive development of what could be termed the modern era of radiative corrections. Basically this covers the period from 1977 to the present day and nothing could give a better illustration of the progress in this field than the following comparison:

1979 The principal feature of this model* is the renormalizability and one of the fundamental ingredients is the Higgs particle ... Moreover, at some point the Higgs meson will play an essential role in the actual result, and perhaps we may get information on the Higgs meson even without actually producing it.*

1991 $m_H > 48\,\text{GeV}$, LEP experiments.

We have not yet found the Higgs but at least we moved from a pure speculative field to some well defined experimental fact: there is no light Higgs! At the same time we have found the Z^0 (why not W^0 is a real mystery) and the W, we know that there are only three light neutrinos, the top quark is still missing and there is a failure of new particle searches to find anything.

If we go back to 1978 and consider the relevant literature we will discover several examples of radiative corrections concerning static quantities[2] or QED processes.[3] A major step towards a systematic calculation of radiative corrections is given by the long term project developed by Martinus Veltman that can be synthesized in the following four points:

- develop a scheme that makes the calculation feasible
- collect formulae valid for scalars one loop diagrams[4]
- reduce all one loop integrals to these scalar expressions[1]
- develop a computer program that provides for numerical answers.[5]

This strategy is general enough to deal with all kind of electroweak processes at the one loop level. For instance an arbitrary scalar three point function will be given in terms of 12 dilogarithms while an arbitrary scalar four point function is a more complicated object. Obviously if one is interested in some particular case, say massless fermions or QED, then the result can be cast in a much simpler form. An example is given by the following diagram

$$\frac{1}{i\,\pi^2} \int d^n q \frac{1}{q^2((q+p_1)^2 + M^2)(q+p_1+p_2)^2} = -\frac{1}{Q^2}\left[Li_2\left(1 - \frac{Q^2}{M^2}\right) - \tfrac{1}{6}\pi^2\right]$$

for $p_i^2 = 0, p_1 + p_2 = Q$. However the original idea was to have a program which works

---

* The standard model of electroweak interactions.
* From ref. 1

without modifications and with the necessary precision for all kinds of processes, therefore $e^+e^- \rightarrow \mu^+\mu^-$ as well as $e^+e^- \rightarrow W^+W^-$ or e ven $W^+W^- \rightarrow W^+W^-$.

The feasibility of the procedure was demonstrated by three distinct calculations.[1,6] All of them were done in Utrecht, more or less in the same period, and to a large extent made possible by a massive use of SCHOONSCHIP, the program for algebraic manipulations written by M. Veltman.

In these calculations the ultraviolet infinities are represented by a parameter Δ and the value of the counter terms depends on this Δ. We emphasize that the renormalization was a numerical procedure. After renormalization the answer must be independent of Δ, which is a strong test on the internal and numerical consistency of the whole calculation. Nowadays this approach could be termed the Δ-scheme.

Among the various technical developments which took place afterward we would like to mention the so called helicity amplitude method. Instead of describing the general idea[7] let us consider a simple example which applies for massless fermions. A lmost all diagrams that contributes to $e^+(p_+)e^-(p_-) \rightarrow \bar{f}(q_+)f(q_-)$ are proportional to $\Gamma_e(\lambda) \cdot \Gamma_f(\sigma)$ where $\lambda, \sigma = \pm 1$ denote the helicities of the fermions and

$$\Gamma_e^\mu(\lambda) = \bar{v}_{-\lambda}(p_+)\,\gamma^\mu\left(V_e + A_e\,\gamma^5\right)u_\lambda(p_-)$$
$$\Gamma_f^\mu(\sigma) = \bar{u}_\sigma(q_-)\,\gamma^\mu\left(V_f + A_f\,\gamma^5\right)v_{-\sigma}(q_+) \tag{1.1}$$

where

$$u_\lambda(p) = \Pi_\lambda\, u(p), \qquad \Pi_\lambda = \tfrac{1}{2}\left(1 + \gamma^5\right) \tag{1.2}$$

etc. ... Therefore for the in-states we can write

$$\Gamma_e^\mu(\lambda) = \sum_{\beta = \pm 1} V_e^\beta\, \mathcal{P}(\lambda, \beta)\, \bar{v}(p_+)\gamma^\mu\Pi_\lambda u(p_-) \tag{1.3}$$

where we have introduced a projection operator $\mathcal{P}$

$$\mathcal{P}(\lambda, \lambda') = \tfrac{1}{2}\left(1 + \lambda\lambda'\right)$$

and also $V^\pm = V \pm A$. A similar expression holds for the out states and we can rewrite the diagram, for a particular choice of the fermion helicities, as

$$\Gamma_e(\lambda) \cdot \Gamma_f(\sigma) = \sum_{\alpha, \beta = \pm 1} V_e^\alpha V_f^\beta\, \mathcal{P}(\lambda, \alpha)\mathcal{P}(\sigma, \beta)\, S(\lambda, \sigma) \tag{1.4}$$
$$S(\lambda, \sigma) = \bar{u}_\sigma\gamma^\mu v_{-\sigma}\bar{v}_{-\lambda}\gamma^\mu u_\lambda$$

S can be given directly in terms of invariants by simply taking its trace

$$S(\lambda, -\lambda) = 4\left(p_+ \cdot q_+ p_- \cdot q_-\right)^{1/2}$$
$$S(\lambda, \lambda) = 4\left(p_+ \cdot q_- p_- \cdot q_+\right)^{1/2} \tag{1.5}$$

The spinors have disappeared and we can compute a set of complex numbers and from them derive the unpolarized cross section. It follows in particular that in order to compute vertex corrections to the process we only need two particular combinations of scalar three point form factors and they are $-2\,C_{24}+(C_{23}+C_{11})s+1$ for the case of two internal fermion lines and $6\,C_{24}-(C_{23}+C_{11}+C_0)s-1$ for two internal vector lines. Also two combinations of four point scalar form factors suffice in accounting for the boxes.

2. The ρ parameter

After the first period several calculations appeared in the literature (in Spires 581 documents are recorded with the two words *radiative corrections* in the title). Roughly we can divide them into two classes: radiative corrections to measurable quantities, as cross sections or asymmetries, and a search for virtual effects through radiative corrections to measured observables. Whenever we have a set of data points at some energy scale we can also try to put bounds on the new physics which is supposed to take place at a larger scale, if non-decoupling occurs. There are two rather important examples and we start by discussing them.

Nowadays the so called ρ parameter has become an essential ingredient in discussing electroweak physics but this has not always been recognized appropriately and therefore we find opportune to reestablish its original interpretation.[8] Generally speaking we have the following. A certain parameter p may appear in different places in the original Lagrangian and therefore it will receive different interpretations $P_1,\ldots,P_l$ and different radiative corrections but the requirement of gauge invariance is that when we take the ratios $r_{ij} = P_i/P_j$ and remove radiative corrections then $r_{ij} = 1, \forall i,j$. In the so called minimal standard model there is an important relation among the bare parameters of the Lagrangian

$$M^2 = M_0^2 c_\theta^2 \tag{2.1}$$

where $M(M_0)$ is the bare $W(Z^0)$ mass and c_θ is the cosine of the bare weak mixing angle. Given any set of four data points we can compute the same quantities in the tree approximation, extract M, M_0, s_θ and obtain

$$\rho_{exp} = \left(\frac{M}{M_0 c_\theta^2}\right)_{tree} \tag{2.2}$$

which is in general not one. In 1977, well before the advent of LEP, only low energy data were available and M, M_0, s_θ were extracted from α (Thomson scattering), G_μ (μ decay) and $\nu, \overline{\nu}$ neutral current cross sections. Today, as it will be discussed in the following sections, we can use instead α, G_μ and the vector boson masses. In this case we have in lowest order

$$M^2 = M_w^2, \qquad M_0^2 = M_z^2, \qquad s_\theta^2 = \frac{\pi\alpha}{2\,G_\mu M_w^2} \qquad (2.3)$$

The presence of new particles induces quantum corrections in these tree relations and since the new particles come in multiplets an eventual mass splitting generates further mass difference between the W and the Z^0. We can subtract at this point radiative corrections from the experimental data and obtain the bare parameters of the Lagrangian. In particular in the minimal standard model ρ_{bare} is finite as a consequence of isospin invariance. Therefore ρ_{bare} can be predicted and consequently we will be able to verify if the corrections satisfy, within the experimental errors, the expected relation

$$\rho_{bare} = \frac{M^2}{M_0^2 c_\theta^2} = 1 \qquad (2.4)$$

When applied to a new generation of fermions this argument gives a constraint on the well known Veltman factor, namely for a fermion doublet the corrections are proportional to

$$
\begin{aligned}
& G_\mu \left[m_1^2 + m_2^2 + 2\,\frac{m_1^2 m_2^2}{m_1^2 - m_2^2} \ln \frac{m_2^2}{m_1^2} \right] \\
& \sim \tfrac{4}{3}\,G_\mu \left(\overline{m}\delta\right)^2, && \delta \to 0 \qquad (2.5) \\
& \sim G_\mu\, m_1^2, && \delta \to \infty
\end{aligned}
$$

where m_1, m_2 are the masses of the fermions in the doublet with $2\overline{m}^2 = \left(m_1^2 + m_2^2\right)$ and $2\overline{m}^2\delta = m_1^2 - m_2^2$. This allows us to make a prediction for $\Delta m = m_1 - m_2$. Due to the large uncertainties in the neutrino cross sections the bound in 1977 was $\Delta m < 750 GeV$. This bound became better and better in the following years and already before the LEP results on the number of light neutrinos it was clear that only a fourth fermion generation could possibly be allowed.

There is another important reason to consider the corrections to the vector boson masses and to the ρ parameter. In the limit of a large Higgs mass the minimal standard model becomes the massive Yang Mills theory, at least up to one loop diagrams.[9] Therefore the vector boson masses are sensitive to this limit but at the one loop level the correction factor to the ρ parameter is only logarithmic.[10] This result is just the so called Veltman screening theorem which already in 1977 was proving that all measurements, including those in the 100 GeV region, are rather weakly dependent on the value of the Higgs boson mass. Indeed the theoretical predictions that we are now considering for quantities to be measured at LEP are always reported with an error arising from a variation on m_H between 50 GeV and 1 TeV. If m_H is above 1 TeV however perturbation theory breaks down.

Using the most recent data as input, namely $M_z = 91.174 \pm 0.021\,\text{GeV}$ from LEP and $M_w/M_z = 0.8794 \pm 0.0035$ from UA2+CDF, gives $\rho_{bare} = 1$ with an error margin of about 0.5% if the top quark has a mass around 110 GeV after which ρ_{bare} increases as $G_\mu m_t^2$. In particular there is a shift in ρ from $\rho_{exp} = 0.9971$ to $\rho_{bare} = 0.9991$ and

$$0.9952 < \rho_{bare} < 1.0029, \quad \text{at } 90\% \text{ C.L. for } \quad m_H = 100\,\text{GeV} \quad m_t = 110\,\text{GeV}$$

This statement is valid only within the minimal standard model but in principle other types of fermion or boson multiplets could contribute. Given the present experimental and phenomenological situation these contributions are of some interest only if they cancel against the top one. This turns out not to be the case for supersymmetric theories[11] but other examples in which this happens have been found, some of them quite recently.[12]

More generally if higher representations for the Higgs field are also present then ρ_{bare} is in general not finite and it must be fitted from a fourth measurement, just as the rest of the bare parameters g, M and s_θ.[13] In this case therefore four data points are needed for the renormalization procedure and a fifth measurement is needed in order to test the theory.

3. Renormalization Schemes

During the following years there has been quite some debate on the choice of a renormalization scheme as can be seen for instance in the Proceedings of the 89 Brighton Conference.[14] Indeed if we move beyond the tree approximation then the notion of a counter term must be introduced and in order to fix the parameters of the Lagrangian we need a set of conventions that should always be stated before the final numerical predictions. Already at the time when the Utrecht project started we had a well defined and unambiguous procedure. Given a Lagrangian $\mathcal{L}$

$$\mathcal{L} = \mathcal{L}(p_1, \ldots, p_n) \tag{3.1}$$

depending on n parameters we choose n data points d_i and compute the corresponding quantities corrected up to one loop. Thus we obtain a set of n fitting equations of the form

$$d_i^{exp} = d_i^{th}(p_1 \ldots, p_n, \Delta), \qquad (i = 1, \ldots, n) \tag{3.2}$$

These equations can be solved to first order in perturbation theory and the solution fixes the parameters. Obviously they contain ultraviolet divergencies which are represented in the actual calculation by a quantity Δ

$$\Delta = -\frac{2}{n-4} + \gamma - \ln \pi \tag{3.3}$$

At this point we can choose a scheme by prescribing precisely what Δ is and all the procedure is totally equivalent to a prescription for the counter terms. A good example is given by QED where we only have one parameter, the charge e. First we write the one loop $\gamma - \gamma$ transition as

$$S(p) = \frac{e^2}{16 \pi^2} \Pi(p^2) p^2 \tag{3.4}$$

Next we can use Thomson scattering between two charged particles $AB \to AB$ and consider the γ exchange diagram in the t channel

$$M(AB \to AB) = (2 \pi)^4 \, i \, \frac{Q_A Q_B}{t} \frac{e^2}{1 - \frac{e^2}{16 \pi^2} \Pi(-t)} + \cdots \tag{3.5}$$

where $Q_{A(B)}$ is the $A(B)$ charge in units of e. As it is well known the additional terms in eq. (3.5) do not contribute to the residue of the pole at $t = 0$ and we find the following relation between the data point α and the parameter e^2

$$\frac{e^2}{4 \pi} = \frac{\alpha}{1 + \frac{\alpha}{4 \pi} \Pi(0)} \sim \alpha \left[1 - \frac{\alpha}{4 \pi} \Pi(0) + \mathcal{O}\left(\alpha^2 \right) \right] \tag{3.6}$$

If we compute explicitly $\Pi(0)$ we find

$$\Pi(0) = -\frac{52}{9} \Delta + \frac{4}{3} \sum_l \ln m_l^2 + \frac{16}{9} \ln m_t^2 + \Pi_q(0) \tag{3.7}$$

where $\Pi_q(0)$ denotes the light quark contribution. The counter term can now be prescribed, for instance in the $\overline{MS}$ scheme[15] at any scale μ, for example at $\mu = M_z$. In this case we simply choose $\Delta = \ln M_z^2$ and obtain

$$\Pi(0)\Big|_{\overline{MS}} = \frac{4}{3} \sum_l \ln \frac{m_l^2}{M_z^2} + \frac{16}{9} \ln \frac{m_t^2}{M_z^2} - \frac{220}{27} + \Pi_q(0) - \mathrm{Re}\Pi_q \left(-M_z^2 \right) \tag{3.8}$$

where the last two terms can be computed via a dispersion relation. This formulation allows us to introduce the concept of $\overline{MS}$ parameters, in this case $e_{\overline{MS}}$. However we could as well use any value for Δ and discuss only measurable quantities which are Δ-independent by definition. This becomes especially relevant within the context of the standard model where there is no one-to-one correspondence between the parameters of the Lagrangian and some set of data points, namely eq. (3.2) form a coupled system. Typically there is no way in which $\sin \theta$ can be defined uniquely beyond the tree approximation. It is hard to understand why one should talk about $\sin \theta$ at all but if this is really needed the best thing is to refer to $\sin \theta_{\overline{MS}}$ with a specification of the processes considered. The dependence of the parameters on Δ will cancel when we compute the $(n + 1)$-th quantity: this is finally a prediction of the model.

It is somehow ironic to observe that in the calculation of $e^+e^- \to \mu^+\mu^-$ [1] another prescription was used. Although the correct renormalization procedure is fully described at the beginning of sect. 10 of ref. [1], to all effects we ended up by introducing the so called on shell scheme. This was only due to same exhaustion but few years later the full one loop calculation was presented for $e - \mu$ scattering, μ decay and $\nu - e$ scattering, again by Veltman and collaborators,[16] therefore completing the original program. What was done in [1], as a sort of approximate procedure, amounts to the following. Given the physical masses M_w and M_z they are immediately related to the bare masses by

$$M^2 = M_w^2 + \frac{g^2}{16\,\pi^2} \operatorname{Re}\Sigma_{WW}\left(-M_w^2\right)$$

$$M_0^2 = M_z^2 + \frac{g^2}{16\,\pi^2 c_\theta^2} \operatorname{Re}\Sigma_{ZZ}\left(-M_z^2\right)$$

$$(3.9)$$

after which the counter term for $\sin^2\theta$ can be easily found

$$\delta \sin^2\theta = \cos^2\theta \left(\frac{\delta M_0^2}{M_0^2} - \frac{\delta M^2}{M^2}\right)$$

$$(3.10)$$

with $\delta M^2 = M^2 - M_w^2$ etc... and $\cos^2\theta = M_w^2/M_z^2$ by definition.

4. Renormalization from low energy data

The advent of LEP and SLC sets a boundary between renormalization from low energy data and renormalization from high energy data (the Z^0 mass). In this first period one could use α, G_μ and the cross section from $\nu - e$ or ν−quark scattering and make predictions, notably the vector boson masses. We have already discussed the Thomson scattering in QED and for the standard model it is enough to use the full expression for the photon vacuum polarization. As far as μ decay is concerned we write the expression for the muon lifetime

$$\tau_\mu^{-1} = \frac{G_\mu^2 m_\mu^5}{96\,\pi^3}\left[1 + \frac{\alpha}{2\,\pi}\left(\tfrac{25}{4} - \pi^2\right)\right]$$

$$(4.1)$$

In lowest order $G_\mu = g^2/8M^2$ whereas at the one loop level

$$G_\mu = \frac{g^2}{8\,M^2}\left[1 + \frac{\alpha}{4\,\pi}\delta_G + \frac{g^2}{8\,M^2}\frac{\Sigma_{WW}(0)}{2\,\pi^2}\right]$$

$$(4.2)$$

where δ_G takes into account the ultraviolet and infrared finite contributions from vertices and boxes[16]. dwhereas $\Sigma_{WW}(0)$ is the self energy correction to the W line at zero momentum. This equation can be solved for g^2/M^2 while α gives $g^2 s_\theta^2$. Finally we consider the ratio of the two neutral current interaction processes

$$R = \frac{\sigma\left(\overline{\nu}_\mu\, e^-\right)}{\sigma\left(\nu_\mu\, e^-\right)} = \frac{\xi^2 - \xi + 1}{\xi^2 + \xi + 1} \tag{4.3}$$

where ξ is the data point. The theoretical prediction will be $R = R_{lo} + R_{1l}$ and at zero momentum transfer R_{lo} is a function of s_θ^2 alone. Thus a solution for s_θ^2 is given by

$$s_\theta^2 = \frac{1 - \xi}{4} - 4\,\frac{R_{1l}}{(1 - \xi)\,R'_{lo}} \tag{4.4}$$

Alternatively we could extract s_θ^2 from neutrino-nucleon scattering experiments, namely from $R_{\nu(\overline{\nu})}$ where

$$R_{\nu(\overline{\nu})} = \frac{\sigma_{NC}^{\nu(\overline{\nu})}}{\sigma_{CC}^{\nu(\overline{\nu})}} = \left(\frac{M}{M_0 c_\theta}\right)^4 \left[\tfrac{1}{2} - s_\theta^2 + \tfrac{5}{9}\left(1 + r_{\nu(\overline{\nu})}\right) s_\theta^4\right] \tag{4.5}$$

where $r_\nu = 1/r_{\overline{\nu}} = \sigma_{CC}^{\overline{\nu}}/\sigma_{CC}^{\nu} \approx 0.4$ and we have shown the result of the lowest order calculation.

As a general statement we can say that most radiative corrections are actually small, less than 1%, with two notable exceptions. First we have the running of the parameters, typically $\alpha(0) \to \alpha(M_z^2)$ where the large logarithms present in the renormalized photon self energy must be summed to all orders. Secondly there are radiative corrections to the vector boson masses due to light fermions, approximately 1 GeV from the leptonic sector and 2 GeV from the hadronic sector, and to the unknown top quark mass. The consequence is a shift in the vector boson masses as predicted from low energy data which was discovered independently from Veltman and Consoli and collaborators.[17] For a large top quark mass this amounts to

$$\frac{\delta M_W^2}{M_W^2} \sim \text{const}, \qquad \frac{\delta M_Z^2}{M_Z^2} \sim -\frac{m_t^2}{M_Z^2} \qquad \text{for} \quad m_t \to \infty \tag{4.6}$$

The total shift in the vector boson masses with respect to the lowest order value is a nice example of the predictive power of the underlying renormalizable theory. To show how it works we consider the corrections to the Z^0 mass induced by fermions. In this case eq. (4.4) can be simplified and its solution gives

$$s_\theta^2 = s_0^2 + \frac{\alpha}{4\pi}\left[\Pi_{3\gamma}^f(0) - s_0^2\,\Pi_{\gamma\gamma}^f(0)\right], \qquad s_0^2 = \frac{1 - \xi}{4} \tag{4.7}$$

where $\Pi_{\gamma\gamma}^f$ is given in eq. (3.7) and the term in parenthesis is the coefficient of p^2 in the fermionic part of the $Z^0 - \gamma$ transition. The Z^0 mass, defined as the zero of the real part of the inverse Z^0 propagator, can now be computed: the lowest order value is

$$M_z^2\big|_{LO} = \frac{8\,\pi\alpha}{G_\mu\,(3 - 2\xi - \xi^2)} \qquad (4.8a)$$

while the one loop corrected value is a solution of

$$\left(1 + \frac{\alpha}{4\,\pi}\,P\right) M_z^2 = M_z^2\big|_{LO} + \frac{\alpha}{4\,\pi}\,Q \qquad (4.8b)$$

where

$$\begin{aligned}
s_0^2 c_0^2 P &= \left(c_0^2 - s_0^2\right) \Pi_{3\gamma}^f(0) + s_0^4\,\Pi_{\gamma\gamma}^f(0) \\
s_0^2 c_0^2 Q &= \Sigma_{WW}^f(0) - \mathrm{Re}\Sigma_{ZZ}^f\left(-M_z^2\right)
\end{aligned} \qquad (4.9)$$

Predictions for the vector boson masses from low energy data, including all radiative corrections, were given in[18] and the agreement with the successive UA1 and UA2 data turned out to be impressive. It should be noticed that besides vector boson masses also all low energy processes receive a contribution due to the fermion families that amounts to the replacements $M^2 \to M^2 - S_{WW}(0)$ or $M_0^2 \to M_0^2 - S_{ZZ}(0)$. It is interesting to note that once we define $\sin\theta_{\overline{MS}}$ from low energy data we find no term quadratic in the quark masses. The fermionic contribution is given in this case by

$$\sin^2\theta_{\overline{MS}}(\nu e) = s_0^2 + \frac{\alpha}{6\,\pi} \sum_f N_{cf} Q_f \left(I_{3f} - 2\,Q_f s_0^2\right) \ln\frac{m_f^2}{M_z^2} \qquad (4.10)$$

Similarly we can define a $\sin^2\theta_{\overline{MS}}$ from the leptonic forward-backward asymme try or the left-right asymmetry at the Z^0 peak; there would again be no contribut ion proportional to the top mass squared but the Z^0 mass is another story.

5. Renormalization at the Z^0 peak

Not all the renormalization schemes have ρ as a free parameter, indeed the most used scheme before the advent of LEP was the so called on shell scheme,[19] where $\sin^2\theta = 1 - M_w^2/M_z^2$ to all orders of perturbation theory and therefore $\rho = 1$ by construction. This is perhaps not the right place to analyze the conceptual differences between the two choices since, after all, defining a scheme is not the same as making a prediction and moreover if you choose to work with $\rho = 1$ the corrections will show up in some other place. Even after the comeback of the $\overline{MS}$ scheme[20] it has remained a common practice to present results in terms of a prediction for $\sin^2\theta$; $\sin^2\theta_W$ from the W mass or $\sin^2\overline\theta$ from the leptonic Z^0 partial width and the leptonic forward-backward asymmetry or whatever.

Of course, there is nothing wrong in reporting the data in this form if only everyone would agree on some procedure. However what happens is that some experiments properly subtract vertex corrections in the width and some others don't. Also it is not clear if everyone agrees on the procedure for extracting the Z^0 couplings from A^l_{FB} at the Z^0 peak where a term coming from the interference of the Z^0 exchange diagram with the imaginary part of the photon vacuum polarization should be subtracted. Basically we face a situation where the experimental data are not only corrected for the acceptance of the apparatus and so on, but they are also dressed with some radiative corrections (QED and also some of the pure weak corrections) which differ from experiment to experiment. In this conditions is not clear if we should average the $\sin^2\bar\theta$ obtained from the four LEP experiments. Apart from that it would seem appropriate if the results of the experiments were to be expressed in terms of $\sin^2\theta_{\overline{MS}}$ and the ρ parameter.

After all there is no definition needed for $\sin^2\theta$ which we prefer to consider as one of the bare parameters of the Lagrangian. Seen it in this way there is very little motivation for predicting a $\sin^2\theta$ and we should always talk in terms of measurable quantities, M_w from UA2 and CDF and $\Gamma_l, A^l_{FB}, \ldots$ from LEP. It took nearly a decade to go back to the $\overline{MS}$ scheme let's hope that in another decade $\sin^2\theta$ will disappear.

In comparing the first calculations for $e^+e^- \to \overline{f}f$ with the most recent ones[21] we have to consider the level of accuracy reached by the experiments. To give an idea we consider the LEP average for the leptonic width, $\Gamma_l = 83.31 \pm 0.40\,\mathrm{MeV}$, which corresponds to an error of about 0.5%. This means that the theoretical predictions had to move in the permille region where some of the higher order effects must be properly taken into account. Of course we are still missing the full two loop calculation but the leading corrections are under control. At this point different schemes disagree on the treatment of the sub-leading corrections and a comparison between different calculations gives an estimate of the theoretical error. A useful quantity to consider when we compare two different calculations, $\mathcal{C}_1$ and $\mathcal{C}_2$, of the same observable $\mathcal{O}$ is

$$e_t(\mathcal{O}) = \frac{|\mathcal{C}_1 - \mathcal{C}_2|}{\Delta\mathcal{O}}$$

where $\Delta\mathcal{O}$ is the reported experimental error. For instance we take $\mathcal{O} = \Gamma_Z$, which is now measured with an error of $9\,\mathrm{MeV}$, and compare our calculation[24] and the one of Hollik[24] (for $\alpha_s = 0$). Over a large range of values for m_t and m_H it is found that $\max e_t(\Gamma_Z) < 0.24$.

In treating radiative corrections around the Z^0 peak we encounter the problem of separating pure weak effects from the large QED corrections. It is well known that in order to describe the Z^0 line shape or the peak asymmetries we must treat properly the radiation from the initial states and for that an $\mathcal{O}(\alpha)$ calculation is not enough to reproduce correctly the parameters of the resonance. A major achievement has been represented by the introduction of the structure function approach[22] which can deal with multi-photon emission. Once the QED corrections are subtracted we are left with the *interesting physics* which we analyze in the following by strictly sticking to the minimal standard model.

The main ingredient for discussing one loop radiative corrections is represented by the vector boson transitions. They can be cast into the following form

$$S_{\gamma\gamma} = \frac{g^2 s_\theta^2}{16\,\pi^2}\,\Pi_{\gamma\gamma}\,p^2, \qquad S_{WW} = \frac{g^2}{16\,\pi^2}\,\Sigma_{WW}$$

$$S_{ZZ} = \frac{g^2}{16\,\pi^2 c_\theta^2}\,\Sigma_{ZZ} = \frac{g^2}{16\,\pi^2 c_\theta^2}\left[\Sigma_{33} - 2\,s_\theta^2 \Sigma_{3\gamma} + s_\theta^4 \Pi_{\gamma\gamma} p^2\right] \qquad (5.1)$$

$$S_{Z\gamma} = \frac{g^2 s_\theta}{16\,\pi^2 c_\theta}\,\Sigma_{Z\gamma} = \frac{g^2 s_\theta}{16\,\pi^2 c_\theta}\left[\Sigma_{3\gamma} - s_\theta^2 \Pi_{\gamma\gamma} p^2\right]$$

where the explicit expression for the various Σ's can be found in.[23] Moreover there are in the literature several examples of how to implement the relation $\Sigma_{Z\gamma}(0) = 0$ even in the presence of bosonic corrections; the same recipe makes the vector boson-fermion-antifermion vertices ultraviolet finite.[24] Equipped with the self energies our primary task will be to introduce a set of three fitting equations for g, M and s_θ: these will be solved in some approximation and we will get the bare parameters in the $\overline{MS}$ scheme. After that we can make a prediction which obviously will depend on the input data, in the sense that we have to specify the processes considered for the fitting equations. The variations in the corrections for different high precision experiments are minimal.

First we use the fine structure constant as already done in the previous sections

$$\frac{1}{g^2 s_\theta^2} = \frac{1}{4\,\pi\alpha} + \frac{1}{16\,\pi^2}\,\Pi_{\gamma\gamma}(0) \qquad (5.2)$$

where now $\Pi_{\gamma\gamma}(0)$ contains both fermionic and bosonic terms. Second from μ decay we get

$$G_\mu = \frac{g^2}{8\,M^2}\left[1 + \frac{\alpha}{4\pi}\,\delta_G + \frac{g^2}{8\,M^2}\,\frac{\Sigma_{WW}(0)}{2\,\pi^2}\right] \qquad (5.3)$$

Even if not strictly necessary it is however possible to obtain a definition of the Fermi coupling constant which is independent from the particular process. The correct procedure consists in identifying the gauge dependent part of δ_G[25]

$$\delta_G^I = \delta_G - \delta_G^{NI}, \qquad \Sigma_{WW}^I(0) = \Sigma_{WW}(0) + s_\theta^2 M^2 \delta_G^{NI} \qquad (5.4)$$

after which we introduce the universal Fermi coupling constant G

$$G = \frac{G_\mu}{1 + \frac{\alpha}{4\pi}\delta_G} \qquad (5.5)$$

and iterate the self energy term in eq.(5.3) to get

$$8\frac{M^2}{g^2} = \frac{1}{G} + \frac{1}{2\pi^2}\Sigma^I_{WW}(0) \tag{5.6}$$

The superscript I will be however neglected in the following. As the third data point we have the value of the Z^0 mass as extracted from th e line shape whose average is $M_z = 91.174 \pm 0.021\,\text{GeV}$. We define M_z as the zero of the real part of the inverse Z^0 propagator. Any other definition woul d do since M_z is an input to the procedure of fitting the bare parameters. A calculation $\mathcal{O}(g^2)$ gives

$$\frac{M^2}{c_\theta^2} = M_z^2 + \frac{g^2}{16\pi^2 c_\theta^2}\,\text{Re}\Sigma_{ZZ}\left(-M_z^2\right) \tag{5.7}$$

however we can keep track of large higher order effects by considering the Z^0 dressed propagator written in a form which includes two loop reducible diagrams

$$\Delta_{ZZ} = \frac{1}{(2\pi)^4 i}\left[\Delta_Z - \left(\frac{g^2 s_\theta}{16\pi^2 c_\theta}\right)^2 \frac{\Sigma^2_{Z\gamma}}{\Delta_\gamma p^2}\right]^{-1}$$

$$\Delta_\gamma = 1 - \frac{g^2 s_\theta^2}{16\pi^2}\Pi_{\gamma\gamma}, \qquad \Delta_Z = p^2 + \frac{M^2}{c_\theta^2} - \frac{g^2}{16\pi^2 c_\theta^2}\Sigma_{ZZ} \tag{5.8}$$

Actually we are only interested in those higher order corrections which are numerically dominant and form a gauge invariant subset of the full $\mathcal{O}(g^4)$ expression. Consequently eq. (5.7) becomes

$$\frac{M^2}{c_\theta^2} = M_z^2 + \frac{g^2}{16\pi^2 c_\theta^2}\,\text{Re}\Sigma_{ZZ}\left(-M_z^2\right) - \frac{g^4 s_\theta^2}{256\pi^4 c_\theta^2}\frac{1}{M_z^2}\,\text{Re}\left[\frac{\Sigma^2_{Z\gamma}\left(-M_z^2\right)}{1 - \frac{g^2 s_\theta^2}{16\pi^2}\Pi_{\gamma\gamma}\left(-M_z^2\right)}\right]_f \tag{5.9}$$

where the subscript f stands for fermionic contribution only. Before discussing the effects due to the Z^0 transitions we observe that eq. (5.2) alone allows us to write the corrections to the γ - exchange in a process $e^+e^- \to \bar{f}f$ (self energy only)

$$M_{\gamma\gamma} = -(2\pi)^4 i \frac{4\pi Q_f \hat{\alpha}(s)}{s}\left[1 + i\frac{\hat{\alpha}(s)}{4\pi}\text{Im}\Pi^f_{\gamma\gamma}(-s)\right]$$

$$\times \left\{1 + \frac{\hat{\alpha}(s)}{4\pi}\left[\Pi^b_{\gamma\gamma}(-s) - \Pi^b_{\gamma\gamma}(0)\right]\right\}\gamma^\mu \otimes \gamma^\mu \tag{5.10}$$

where b stands for bosonic and we have introduced $\hat{\alpha}$, the running α

$$\hat{\alpha}(s) = \frac{\alpha}{1 - \frac{\alpha}{4\pi}\Pi^f}, \qquad \Pi^f = \text{Re}\Pi^f_{\gamma\gamma}(-s) - \Pi^f_{\gamma\gamma}(0) \tag{5.11}$$

therefore summing all large logarithms present in Π^f while the bosonic part is always expanded to first order to guarantee the gauge invariance of the final result that we obtain by including vertices and boxes. As a next step we need a solution for s_θ. By combining in the appropriate way the three fitting equations we get

$$\left[1 + \frac{\alpha}{4\pi} \Pi_{\gamma\gamma}(0)\right] s_\theta^2 c_\theta^2 = \frac{\pi\alpha}{2\,GM_z^2} \left\{ 1 + \frac{G}{2\pi^2} \left[\Sigma_{WW}(0) - \text{Re}\Sigma_{ZZ}\left(-M_z^2\right)\right] \right.$$

$$\left. + \frac{\alpha}{4\pi} \frac{G}{2\pi^2} \text{Re} S_{ho}^f \right\} \qquad (5.12)$$

$$S_{ho}^f = \frac{1}{M_z^2} \frac{\Sigma_{Z\gamma}^2\left(-M_z^2\right)}{1 - \frac{\alpha}{4\pi}\left[\Pi_{\gamma\gamma}\left(-M_z^2\right) - \Pi_{\gamma\gamma}(0)\right]}\Bigg|_f$$

Since Σ_{ZZ} and $\Sigma_{Z\gamma}$ contain factors of s_θ^2 this becomes a quadratic equation in s_θ^2, with radiative corrections included up to a certain order in perturbation theory, that must be solved consistently. There is a simple lowest order solution

$$s_0^2 = \tfrac{1}{2} \left[1 - \sqrt{1 - \frac{2\pi\alpha}{GM_z^2}}\, \right] \qquad (5.13)$$

Strictly speaking we usually solve an implicit equation $f(a, \lambda)$, where a is some parameter and λ is a coupling constant, by expanding a around a_0, solution of $f(a_0, 0) = 0$. However we could as well expand a around $\bar{a}$, solution of some equation $h(\bar{a}, \lambda) = 0$, as long as $\bar{a}$ is gauge invariant. In this way we can include in $\bar{a}$ the bulk of some large effect arising in higher orders in λ. Given $\hat{\alpha} \equiv \hat{\alpha}\left(-M_z^2\right)$ we expand s_θ^2 around a $\hat{s}^2$ defined by

$$\hat{s}^2 = \tfrac{1}{2} \left[1 - \sqrt{1 - \frac{2\pi\hat{\alpha}}{GM_z^2}}\, \right] \qquad (5.14)$$

We also put

$$s_\theta^2 = \hat{s}^2 + \frac{\alpha}{4\pi} \delta s_1 + \left(\frac{\alpha}{4\pi}\right)^2 \delta s_2 + \mathcal{O}(\alpha^3) \qquad (5.15)$$

where in $\mathcal{O}(\alpha^2)$ only the fermionic parts are retained. It is straightforward to find

$$\delta s_1 = \frac{\Sigma}{\hat{c}^2 - \hat{s}^2} - \frac{1}{M_z^2} \text{Re}\Sigma_{Z\gamma}\left(-M_z^2\right)$$

$$\delta s_2 = \frac{\left(\Sigma^f\right)^2}{\left(\hat{c}^2 - \hat{s}^2\right)^3} - \Pi_{\gamma\gamma}^f(0)\, \delta s_1^f \qquad (5.16)$$

where the Σ is the following combination of vector boson self energies

$$M_z^2 \Sigma = \Sigma_{WW}(0) - \text{Re}\Sigma_{33}\left(-M_z^2\right) + \text{Re}\Sigma_{3\gamma}\left(-M_z^2\right) + \hat{c}^2\hat{s}^2 M_z^2 \Pi^b. \qquad (5.17)$$

Eq. (5.15) defines again an $\overline{MS}$ parameter, $\sin^2\theta = \sin^2\theta_{\overline{MS}}\left(M_z^2\right)$. The previous procedure deserves some comment. Our goal was a solution for s_θ^2 (in the $\overline{MS}$

scheme) which is corrected to $\mathcal{O}(\alpha)$ and includes the leading terms in the top quark mass up to $\mathcal{O}(\alpha^2)$. By explicitly computing the relevant two point functions we see that in the limit $m_t \to \infty$ the Σ in eq. (5.17) is nothing else than the familiar $\Delta\rho$ and therefore

$$\Sigma \sim -\frac{3}{4}\frac{m_t^2}{M_z^2} \qquad \text{for} \qquad m_t \to \infty \tag{5.18}$$

Since $\Sigma^2 \sim m_t^4$ we must recover the same leading behavior from the two loop irreducible diagrams. Despite the fact that a complete analysis of the vector boson self energies at the two loop level and with arbitrary external momentum is not yet available we have an explicit formula valid at $p^2 = 0$ which effectively gives the leading behavior in m_t.[26]

$$\Sigma^{1loop} \sim -\frac{3}{4}\frac{m_t^2}{M_z^2}, \qquad \left(\Sigma^{2loop}\right)^{irr} \sim -\frac{\alpha}{4\pi}\frac{19 - 2\pi^2}{\hat{s}^2}\frac{3\,m_t^4}{16\,M_w^2\,M_z^2} \tag{5.19}$$

Therefore we replace Σ everywhere by $\left(\Sigma^{1loop} + \Sigma^{2loop}\right)^{irr}$ and truncate the expansion to $\mathcal{O}(\alpha^2)$ by retaining terms proportional to m_t^4 only. Having a solution for s_θ^2 we can transform it into a solution for g^2, always in the $\overline{MS}$ scheme, by using Thomson scattering

$$\frac{4\pi\alpha}{g^2} = \left[1 + \frac{\alpha}{4\pi}\Pi_{\gamma\gamma}(0)\right]s_\theta^2 \tag{5.20}$$

It follows that

$$\frac{4\pi\alpha}{g^2} = \hat{s}^2 + \frac{\alpha}{4\pi}\delta g_1 + \left(\frac{\alpha}{4\pi}\right)^2\delta g_2 + \mathcal{O}(\alpha^3)$$

$$\delta g_1 = \frac{\Sigma}{\hat{c}^2 - \hat{s}^2} - \hat{s}^2\Pi - \frac{1}{M_z^2}\mathrm{Re}\Sigma_{3\gamma}\left(-M_z^2\right) \tag{5.21}$$

$$\delta g_2 = \frac{3}{16}\frac{1}{\hat{c}^2 - \hat{s}^2}\left[\frac{3}{(\hat{c}^2 - \hat{s}^2)^2} - \frac{19 - 2\pi^2}{\hat{c}^2\hat{s}^2}\right]\left(\frac{m_t^2}{M_z^2}\right)^2$$

with $\Pi = \mathrm{Re}\Pi_{\gamma\gamma}\left(-M_z^2\right) - \Pi_{\gamma\gamma}(0)$. In this way we obtained the parameters of the Lagrangian, three in the case of the standard model, and they can be inserted into the expression of the radiative corrections for any other quantity. Examples of the correct procedure as well as few numerical results are discussed in the next section.

6. Electroweak physics around the scale of the vector bosons

According to the general strategy described before we are now ready to make predictions for testing the minimal standard model. The main issue has not been to define s_θ but rather to obtain the building blocks which render any calculation feasible. Having solved for g, M and s_θ in terms of α, G_μ and M_z it is almost immediate to derive a finite expression for the W mass which will be a certain function of m_t and to some extent of m_H, and the result can be compared with the mass measurements from the colliders. The W mass is again defined as the zero of

$$M_W^2 + \frac{g^2}{16\,\pi^2}\,\mathrm{Re}\Sigma_{WW}\left(-M_W^2\right) = M^2 \tag{6.1}$$

The two coupling constants $\hat{\alpha}$ and G are related through the relation

$$\frac{GM_z^2}{2\,\pi^2} = \frac{\hat{\alpha}}{4\,\pi\hat{s}^2\hat{c}^2} \tag{6.2}$$

thus using M^2/g^2 from the second fitting equation we obtain

$$M_z^2\hat{c}^2 + \frac{\hat{\alpha}}{4\,\pi\hat{s}^2}\left[\Sigma_{WW}(0) - \mathrm{Re}\Sigma_{WW}\left(-M_w^2\right)\right] - M_w^2\,\frac{4\,\pi\hat{\alpha}}{\hat{s}^2g^2} = 0 \tag{6.3}$$

Starting from eq.(5.21) we can show that up to the desired order

$$\frac{4\,\pi\hat{\alpha}}{g^2} = \hat{s}^2 + \frac{\hat{\alpha}}{4\,\pi}\left(\frac{\Sigma^f}{\hat{c}^2 - \hat{s}^2} + \Sigma^R\right) + \left(\frac{\hat{\alpha}}{4\,\pi}\right)^2\,\frac{\left(\Sigma^f\right)^2}{\left(\hat{c}^2 - \hat{s}^2\right)^3} \tag{6.4}$$

where $\Sigma = \Sigma^f + \Sigma^R$ and $\Sigma^f \equiv \Sigma^f_{irr}$. Σ^R is isospin conserving while the contributions proportional to Σ^f break isospin and are potentially large and gauge invariant but due to the structure of the corrections they can be absorbed into the ρ parameter, namely we define

$$\rho^{-1} = 1 + \frac{GM_z^2}{2\,\pi^2}\,\Sigma^f. \tag{6.5}$$

Eq. (6.3) simplifies if we also redefine the *lowest order* s_θ^2 as a $\hat{s}_\rho^2$, solution of

$$2\,\rho\hat{s}_\rho^2\hat{c}_\rho^2 = \frac{\pi\hat{\alpha}}{GM_z^2} \tag{6.6}$$

Indeed in this case we can show that

$$\frac{4\,\pi\hat{\alpha}}{g^2} = \hat{s}_\rho^2 + \frac{\hat{\alpha}}{4\,\pi}\,\Sigma^R \tag{6.7}$$

In conclusion the W mass will be given by the sum of two terms one of which reduces to the lowest order relation in the limit $\rho \to 1$ and properly includes all potentially large corrections while the second is usually small

$$M_w^2 = M_z^2 \rho \hat{c}_\rho^2 + \frac{\hat{\alpha}}{4\pi \hat{s}_\rho^2} \left[\Sigma_{WW}(0) - \mathrm{Re}\Sigma_{WW}\left(-M_w^2\right) - M_w^2 \Sigma^R \right]$$

$$M_w^2 \Big|_{LO} = \tfrac{1}{2} M_z^2 \left[1 + \sqrt{1 - \frac{2\pi\alpha}{GM_z^2}} \right]$$

(6.8)

This relation is the exact counterpart of the mass shift relation obtained in the low energy renormalization scheme. In particular from there we find that in the limit $m_t \to \infty$ the W mass stays constant while the Z^0 mass decreases. Here M_z is an input parameter and M_w becomes larger for large m_t. Numerical predictions for $1 - M_w^2/M_z^2$ are given in Table 1. These predictions should be compared with the average from UA2 and CDF which is 0.2265 ± 0.006. The typical error on M_w coming from variations in M_z and α_{had} is about 30 MeV. We can compare the result of this calculation for M_w with other results present in the literature. For instance the agreement with Degrassi et al.[20] is within 10 MeV for $m_H = 100$ GeV and even better for higher values of m_H.

Table 1

Predictions for $s_W^2 = 1 - M_W^2/M_Z^2$ *as a function of* m_t *and* m_H, *for* $M_Z = 91.174\,GeV$

m_t/m_H (GeV)	$1 - M_W^2/M_Z^2$		
	100	500	1000
100	0.2309	0.2328	0.2338
130	0.2276	0.2295	0.2305
160	0.2238	0.2258	0.2268
190	0.2196	0.2216	0.2226

Another interesting quantity is the partial width $Z^0 \to \bar{f}f$. Again the most relevant contribution is given by the self energy corrections which include in this case both the $Z^0 - Z^0$ and the $Z^0 - \gamma$ transitions. Since the vertex corrections are finite we use the solution found for s_θ and derive for the amplitude

$$\frac{1}{(2\pi)^4 i} M\left(Z^0 \to \bar{f}f\right) = i\,(2G)^{1/2}\,M_z \left(1 + \frac{GM_z^2}{2\pi^2} \Pi_Z\right)^{-1/2}$$

$$\times \gamma^\mu \left[I_{3f} - 2Q_f \left(\hat{s}_\rho^2 + \frac{\alpha}{4\pi} \frac{\Sigma^R}{\hat{c}_\rho^2 - \hat{s}_\rho^2} \right) + I_{3f}\gamma^5 \right] + \text{vertices}$$

(6.9)

where the square root gives the Z^0 wave function renormalization factor

$$\Pi_Z = \Sigma^f + \Pi_Z^R$$

$$M_z^2 \Pi_Z^R = \left(2\hat{s}_\rho^2 - 1\right) \mathrm{Re}\Sigma_{3\gamma}\left(-M_z^2\right) + M_z^2 \, \mathrm{Re}\left[\hat{s}_\rho^4 \Pi_{\gamma\gamma}\left(-M_z^2\right) - \Sigma'_{ZZ}\left(-M_z^2\right)\right]$$

(6.10)

Further we can use

$$\left[1 + \frac{GM_z^2}{2\pi^2}\Pi_Z\right]^{-1} = \rho\left[1 - \frac{GM_z^2}{2\pi^2}\rho\,\Pi_Z^R\right]$$

(6.11)

so that once more the leading fermionic corrections are absorbed into the ρ parameter. Vertex corrections can be reduced to a vector coupling and an axial coupling. Including all contributions the amplitude now becomes

$$\frac{1}{(2\pi)^4 i} M\left(Z^0 \to \overline{f}f\right) = i\,(2\,G\rho)^{1/2}\, M_z\,\gamma^\mu\left[\left(I_{3f} - 2\,Q_f\hat{s}_\rho^2 + I_{3f}\gamma^5\right)\right.$$

$$\left(1 - \frac{GM_z^2}{4\pi^2}\rho\,\Pi_Z^R\right).$$

$$\left.\left(-\frac{GM_z^2}{\pi^2}\rho\left(Q_f\frac{\hat{s}_\rho^2\hat{c}_\rho^2}{\hat{c}_\rho^2 - \hat{s}_\rho^2}\right)\Sigma^R - V - A\gamma^5\right)\right]$$

(6.12)

The structure of this amplitude is particularly simple. Apart from the vertices, which must always be included because of gauge invariance, we see a set of universal corrections which can be absorbed into a sort of improved lowest order approximation. Vertex corrections are written in terms of three point scalar form factors where we can use the massless limit for the fermions, unless $f = b$. From the amplitude the partial width follows by squaring and adding both QED and QCD corrections. For instance consider $Z^0 \to \overline{b}b$ for $M_z = 91.174 \pm 0.021\,\mathrm{GeV}$, $\alpha_s = 0.12^{+0.01}_{-0.02}$ and $m_t = 130\,\mathrm{GeV}$: then

$$m_H = 100\,\mathrm{GeV}, \qquad \Gamma_b = 376.4\,\mathrm{MeV}$$

$$\pm \overbrace{0.3}^{M_z} \pm \overbrace{0.2}^{\alpha_{had}} + \overbrace{1.1 - 2.2}^{\alpha_s} \quad \mathrm{MeV}$$

$$m_H = 1000\,\mathrm{GeV}, \qquad \Gamma_b = 375.1\,\mathrm{MeV}$$

where the errors are shown separately. We also give in Table 2(3) the different partial widths for few values of m_t and $m_H = 100(1000)\,\mathrm{GeV}$.

Table 2

Z^0 partial widths in MeV as a function of m_t, for $M_Z = 91.174$ GeV, $m_H = 100$ GeV and $\alpha_S = 0.12$

	Z^0 partial widths (MeV)						
m_t (GeV)	$\Gamma_{e^+e^-}$	$\Gamma_{\bar{u}u}$	$\Gamma_{\bar{d}d}$	$\Gamma_{\bar{b}b}$	$\Gamma_{\bar{\nu}\nu}$	Γ_{had}	Γ_{tot}
100	83.52	297.4	381.3	377.2	166.5	1735	2485
150	83.92	299.7	383.0	375.8	167.2	1741	2494
200	84.48	302.6	385.6	374.2	168.1	1751	2508

Table 3

Z^0 partial widths in MeV as a function of m_t, for $M_Z = 91.174$ GeV, $m_H = 1000$ GeV and $\alpha_S = 0.12$

	Z^0 partial widths (MeV)						
m_t (GeV)	$\Gamma_{e^+e^-}$	$\Gamma_{\bar{u}u}$	$\Gamma_{\bar{d}d}$	$\Gamma_{\bar{b}b}$	$\Gamma_{\bar{\nu}\nu}$	Γ_{had}	Γ_{tot}
100	83.31	296.3	380.1	376.0	166.2	1729	2477
150	83.71	298.6	381.7	374.5	166.9	1735	2487
200	84.27	301.5	384.4	373.0	167.8	1745	2501

7. The top quark

It is not a surprise that the bulk of large radiative corrections due to a heavy top quark can be given in terms of the ρ parameter. From this point of view very little has changed from Veltman paper on large mass differences in 1977. Of course our perspective has changed in the mean time because of the high precision measurements: the data are now much better and we know that only three generations can be accommodated. In addition to that the measured values of Γ_b and o f A_{FB}^b indicate beyond any doubt that the b quark cannot be a singlet. Therefore the problem is nowadays confined to just one parameter, namely the value of the top quark mass. Seen it in this way we can say that the study of the ρ parameter has represented the archetype of all the tests of the standard model.

While ρ in 1977 was defined at $p^2 = 0$ we can now use high energy data and the radiative corrections to ρ requires specification of the processes considered. We may for example use α, G_μ and M_z to compute $\sin^2 \theta_{\overline{MS}}^{(1)}$ and α, G_μ and M_W to compute $\sin^2 \theta_{\overline{MS}}^{(2)}$. Thus

$$\frac{\cos^2 \theta_{\overline{MS}}^{(1)}}{\cos^2 \theta_{\overline{MS}}^{(2)}} = \rho_{exp} \left(1 + \Delta\rho\right) = 1 \qquad (7.1)$$

and $\rho_{exp} = 0.9971$. But we could also use g_V^l and g_A^l to compute a $\sin^2 \theta_{\overline{MS}}^{(3)}$ and a different $\Delta\rho$ will follow by replacing $2 \to 3$ in the previous equation and by using $\rho_{exp} = 0.9966$.

Since we are missing an experimental evidence for the top, with a lower bound set by CDF at 89 GeV on m_t, radiative corrections to measurable quantities are still the best place where to look for indirect evidence. We can combine all data and express the result as a probability that they are consistent with a certain range of values for m_t. Usually the LEP data are combined with the W mass measurement from CDF and UA2. Naturally one has to include QCD corrections and a value for α_s must be chosen which adds some uncertainty to the final answer. Here the currently accepted interpretation is that the LEP results show an impressive confirmation of QCD even if recently some criticism has been raised on this orthodoxy.[27] However, in the ratios $r_q = \Gamma_{\bar{q}q}/\Gamma_{had}$ there is less dependence on α_s, for instance $r_b(r_d)$ varies of about 0.1%(0.2%) when α_s goes from 0.10 to 0.13. We sketch briefly the procedure and refer to the literature[28] for a complete treatment(the most updated fits can be found in ref. 29) . Given $M_z = 91.174$ Ge V, $\alpha_s = 0.12$ and

$$\Gamma_l = 83.31 \pm 0.40 \,\text{Mev}, \qquad \Gamma_h = 1740 \pm 9 \,\text{MeV}, \qquad \Gamma_Z = 2485 \pm 9 \,\text{MeV}$$

$$1 - \frac{M_w^2}{M_z^2} = 0.2265 \pm 0.006$$

a fit to m_t gives (90% C.L.)

$$m_H = 100 \,\text{GeV}, \qquad m_t = 110^{+41}_{-28} \,\text{GeV}$$
$$m_H = 500 \,\text{GeV}, \qquad m_t = 132^{+36}_{-40} \,\text{GeV}$$

Strictly speaking these upper limits on m_t are only valid within the context of the minimal standard model, i.e. the Weinberg model with three generations of fermions and a complex Higgs isodoublet. If we allow for a deviation from minimality the bounds can be evaded and some higher value for m_t can, in principle, be consistent with the present data.[30] Non minimal standard models have the bare vector boson masses as totally unrelated quantities and one should always specify the corresponding mechanism which will generate further radiative corrections, eventually large if the new scalar multiplets have large mass splittings. A precise measurement of the Z^0 partial width into b quarks may help in understanding both m_t and the eventual presence of new physics, but only from a qualitative point of view. Indeed this quantity, due to the structure of the vertex corrections, is the only place where we have a dependence on m_t away from self energy corrections. $\Gamma_{b\bar{b}}$ is therefore a genuine indicator of large m_t effects.[31] A model independent analysis is possible[30] but we cannot disentangle simultaneously all the effects competing in the ρ parameter. As we pointed out already eq. (5.15) defines $\sin^2 \theta_{\overline{MS}} (M_z^2)$ but especially in a model independent analysis it makes more sense to talk in terms of g_V and g_A, the vector and axial couplings of the Z^0 to fermions. The analysis of ref. 30, where ρ is an arbitrary number fixed by an additional data, gives for the b quark couplings the results shown in Table 4.

Table 4 Z^0 *couplings to the b quark for* $M_Z = 91.174$ *GeV in the model independent analysis of ref. [30]. The common error is* 0.0018 *for* g_V^2 *and* 0.0012 *for* g_A^2.

	b quark couplings to Z^0	
m_t (GeV)	g_V^2	g_A^2
100	0.1188	0.2489
150	0.1174	0.2469
200	0.1156	0.2443
250	0.1133	0.2411

On more general grounds however if the top quark is not around the corner we have to explain why the minimal standard model with an hypothetical and relatively light top can explain so well the data. This question becomes even more puzzling in the light of the fact that absolutely no signal for new physics has come from LEP.

8. The Higgs mass

Despite some problem in the correct interpretation of the data we have now a general consensus on the validity of the standard model since

$$\kappa = 1 - \frac{\text{th}}{\text{exp}} \approx 1\%$$

However we are not yet able to produce a satisfactory answer to the longstanding problem of understanding the structure of the Higgs sector. The only sound result available is represented by the so called Veltman screening theorem: as far as low energy observable quantities are concerned, the eventual Higgs absence is practically unnoticed. We can try to understand the validity of the theorem after the LEP results. A combined fit to m_t and m_H shows a relatively low value for m_H, however already the 68% C.L. curve in the $m_t - m_H$ plane goes well beyond the 1 TeV region.

Motivated, in some way, by this screening different authors in different times have tried to put upper bounds on m_H. To understand the physical meaning to be attached to these bounds we should remember that a heavy Higgs translates into a breakdown of perturbation theory. As far as we know upper bounds on m_H are always a consequence of applying perturbation theory, which is doomed to fail if m_H is large.

After the pioneering papers of Veltman and Lee et al.[32] we know that at some energy scale the longitudinal vector bosons become strongly interacting if indeed the Higgs is so heavy that we may assume its absence. The idea that vector bosons are strongly interacting with themselves is certainly older than the standard model[33] and led to the hypothesis that in all partial waves the amplitudes are always small in order to protect radiative effects from affecting the low energy picture. The main question regarding the Higgs system is therefore where do we expect the onset of a strongly interacting standard model? In Born approximation for instance the $j = 0$ partial wave behaves as

$$T_B^0 \sim Gs, \qquad M_W^2 \ll s \ll m_H^2$$

with a critical energy $s_c = 8\pi G^{-1}$. The problem of computing one loop radiative corrections to WW scattering in all possible two-particle channels is a very difficult one and therefore for some time it was just impossible to study the modification of s_c due to the first order radiative corrections. A major breakthrough is represented by the so called equivalence theorem[34] which allows us to compute the scattering by using the effective scalar theory. This has been done[35] but we feel that only qualitative answers can be given. One loop correction are in general large because perturbation theory tends to break down for a large Higgs mass and in several cases they lower the critical energy, however to predict any definite number is not very convenient.

First not everyone agrees on the formulation of the unitarity constraints beyond the tree level. Second we are always analyzing perturbative unitarity, i.e. two-particle unitarity in a truncated perturbative expansion. Failure of the resulting bounds simply implies that at the corresponding energy higher orders become of the same size of the Born term. Finally we are unable to predict the scale where resonances possibly arise as a function of the scale where perturbation theory breaks down in its first few orders. Moreover we know from hadron physics that it is difficult to make predictions until a great deal of data exists. At the moment we have none even if for few months aro und 1984 the CERN anomalous events raised some hope that we could have seen the begi nning of a strongly interacting Higgs sector.[36]

9. Conclusions

In concluding this brief analysis on the last decade of radiative corrections one should point out that we have also been witness to a spectacular technological process. The possibility of having access to more and more powerful computer resources has contributed to a sizeable decrease in the time interval needed for such calculations. Also, as it appears, we are moving towards a scenario where the outcome of a calculation results from the combined efforts of several people instead of coming from isolated individuals.

There are few steps in the future of radiative corrections which we can predict offhand. With the second phase of LEP we will enter the era of vector boson pair production and starting from the pioneering calculation of ref. 6 we count already four separate calculations of $e^+e^- \to W^+W^-$.[37]

The successive and interesting step will bring us into the region where the Higgs boson should manifest itself. Perhaps we can rephrase the Higgs problem by saying that independently from its presence the next generation of accelerators hopefully will allow us to study the WW scattering via W fusion. In order to understand the signal however we need to compute all process which contribute to $pp \to VVX(V = W, Z^0)$ and not only WW scattering. Several groups are already at work.

Certainly we have made a lot of progress in understanding the technical subtleties of radiative corrections but admittedly the situation is far from satisfactory. In a sense we have now a model which after radiative corrections agrees very well with the data if we assume certain values for two fundamental ingredients as m_t and m_H but the top quark and the Higgs boson are still missing and both are essential for the mathematical consistence of the theory. In addition to that as soon as we move m_t or m_H above some threshold all kind of new and non perturbative phenomena will in principle appear but no evidence for that seems to be contained in the present experimental data.

The chief questions which already emerged a long time ago are still unanswered: how many generations? (today, why three generations?) Is the Higgs sector of the standard model an effective description of a more fundamental structure?

REFERENCES

1. G. Passarino and M. Veltman, *Nucl. Phys.* **B160** (1979) 151.

2. W. A. Bardeen, R. Gastmans and B. Lautrup, *Nucl. Phys.* **B46** (1972) 319.

3. P. Van Nieuwenhuizen, *Nucl. Phys.* **B28** (1971) 381;
   F. Berends, R. Gaemers and R. Gastmans, *Nucl. Phys.* **B57** (1973) 381; **B63** (1973) 381,
   B67 (1974) 541.

4. G. 't Hooft and M. Veltman, *Nucl. Phys.* **B153** (1979) 365.

5. M. Veltman, Formf, a CDC program for the computation of one loop form factors.

6. M. Consoli, *Nucl. Phys.* **B160** (1979) 208;
   M. Lemoine and M. Veltman, *Nucl. Phys.* **B164** (1980) 445.

7. Kalcul Coll., F. Berends et al., in Erice 83, Proceedings Electroweak Effects at High Energy, and references therein;
   M. Caffo and E. Remiddi, *Hel. Phys. Acta* **55** (1982) 339;
   G. Passarino, *Nucl. Phys.* **B237** (1984) 249.

8. D. Ross and M. Veltman, *Nucl. Phys.* **B95** (1975) 135;
   M. Veltman, *Nucl. Phys.* *B123* (1977) 89.

9. J.J. van der Bij and M. Veltman, *Nucl. Phys.* *B231* (1984) 205.

10. M. Veltman, *Acta Phys. Pol.* **B8** (1977) 475.

11. R. Barbieri and L. Maiani, *Nucl. Phys.* **B224** (1983) 32.

12. M. Einhorn, D.R.T. Jones and M. Veltman *Nucl. Phys.* **B191** (1981) 146;
    A. Denner, R.J. Guth and J.H. Kühn, *Phys. Lett.* **240B** (1990) 438;
    S. Bertolini and A. Sirlin, MPI-PAE-PTH-74-90.

13. B.W. Lynn and E. Nardi, CERN preprint, CERN-TH 5876/90, S.I.S.S.A. 121 EP;
    G. Passarino, Torino preprint DFTT/G-90-3, to appear in *Nucl. Phys.* B.

14. *Proceedings of the Workshop on Radiative Corrections: Results and Perspectives*, eds. N. Dombey and F. Boudjema, 1990 Plenum Press.

15. G. 't Hooft, *Nucl. Phys.* **B61** (1973) 455;
    W.A. Bardeen, A.J. Buras, D.W. Duke and T. Muta, *Phys. Rev.* **D18** (1978) 3998.

16. M. Green and M. Veltman, *Nucl. Phys.* **B169** (1980) 90.

17. M. Veltman, *Phys. Lett.* **91B** (1980) 95;
    F. Antonelli et al., *Nucl. Phys.* **B183** (1981) 195, *Phys. Lett.* **91B** (1980) 90.

18. W. Marciano, *Phys. Rev.* **D20** (1979) 274;
    A. Sirlin, *Phys. Rev.* D22 (1980) 971;
    W. Marciano and A. Sirlin, *Phys. Rev.* *D22* (1980) 2695

19. A. Sirlin, *Phys. Rev.* **D29** (1984) 89.

20. G. Passarino and M. Veltman, *Phys. Lett.* **237B** (1990) 537;
    G. Degrassi, S. Fanchiotti and A. Sirlin, *Nucl. Phys.* **B351** (1991) 49.

21. M. Consoli and W. Hollik, in *Physics at LEP1*, G. Altarelli, R. Kleiss and C. Verzegnassi eds., CERN-89-08 (1989) p.7;
    G. Degrassi and A. Sirlin, MPI-PAE-PTH-48-90;
    D.C. Kennedy and B. W. Lynn, *Nucl. Phys.* **B322** (1989) 1.

22. E.A. Kuraev and V.S. Fadin, *Sov. J. Nucl. Phys.* **41** (1985) 41,466;
    O. Nicrosini and L. Trentadue, *Phys. Lett.* **196B** (1987) 551;
    O. Nicrosini and L. Trentadue, in *Radiative Corrections for e^+e^- Collisions*, J.H. Kühn ed., Springer and Verlag 1989;
    F. Berends, W.L. van Nerven and G.J.H. Burgers, *Nucl. Phys.* B297 (1988) 429;
    F. Berends et al., in *Physics at LEP1*, G. Altarelli, R. Kleiss and C. Verzegnassi eds., CERN-89-08 (1989) p.89.

23. G. Passarino, in *Radiative Corrections for e^+e^- Collisions*, J.H. Kühn ed., Springer and Verlag 1989, p.179.

24. W. Hollik, *Fortschr. Phys.* **38** (1990) 165;
    D.C. Kennedy and B.W. Lynn, ref. 21.;
    G. Passarino, Torino preprint DFTT/G-90-3, to appear in *Nucl. Phys.* B.

25. M. Kuroda, G. Moultaka and D. Schildknecht, *Nucl. Phys.* **B350** (1991) 25.

26. J.J. van der Bij and F. Hoogeven, *Nucl. Phys.* **B283** (1987) 41977.

27. M. Consoli, talk given at the XXVIth Rencontres de Moriond, Electroweak Interactions and Unified Theories, Les Arcs, Savoie, March 10-17, 1991.

28. V. Barger, J.L. Hewett and T.G. Rizzo, *Phys. Rev. Lett.* **65** (1990) 1313;
    J. Ellis and G.L. Fogli, *Phys. Lett.* **249B** (1990) 543;
    G. Passarino, *Phys. Lett.* **B255** (1991) 127;
    G. Altarelli and R. Barbieri, *Phys. Lett.* **253B** (1991) 161;
    M.E. Peskin and T. Takeuchi, *Phys. Rev. Lett.* **65** (1990) 964;
    D.C. Kennedy and P. Langacker, *Phys. Rev. Lett.* **65** (1990) 2967;
    F. Dydak, Results from LEP and SLC, Rapporteur talk at the 25th InternationalConference on High Energy Physics, Singapore, 2-8 August 1990, CERN-PPE/91-14;
    R. Clare, talk given at the XXVIth Rencontres de Moriond, Electroweak Interactions and Unified Theories, Les Arcs, Savoie, March 10-17, 1991
    P. Ratoff, talk given at the XXVIth Rencontres de Moriond, Electroweak Interactions and Unified Theories, Les Arcs, Savoie, March 10-17, 1991.

29. D. Schaile, talk given at the Ringberg Workshop on High Precision vs. High Energy in e^+e^- Collisions, April 15-19, 1991.

30. G. Passarino, talk given at the Ringberg Workshop on High Precision vs. High Energy in e^+e^- Collisions, April 15-19, 1991.

31. A. Djouadi, G. Girardi, C. Verzegnassi, W. Hollik and F. Renard, *Nucl. Phys.* **B349** (1991) 48.

32. M. Veltman, *Acta Phys. Pol.* **B8** (1977) 475; *Phys. Lett.* **70B** (1977) 254:
    B. Lee, C. Quigg and R. Thacker, *Phys. Rev. Lett.* **38** (1977) 205;
    Phys. Rev. **D16** (1977) 1519.

33. T. Appelquist and J.D. Bjorken, *Phys. Rev.* **D4** (1971) 3726.

34. J.M. Cornwall, D.N. Levin and G. Tiktopoulos, *Phys. Rev.* **D10** (1974) 1145;
    H. Veltman, *Phys. Rev.* **D41** (1990) 2294.

35. G. Passarino, *Phys. Lett.* **156B** (1985) 231; *Nucl. Phys.* **B343** (1990) 31;
    L. Durand, J.M. Johnson and J.L. Lopez, *Phys. Rev. Lett.* **64** (1990) 1215;
    M. Veltman and F. Ynduráin, *Nucl. Phys.* **B325** (1989) 1.

36. M.Veltman, *Phys. Lett.* **139B** (1984) 307;
    M.J. Duncan and M. Veltman, *Phys. Lett.* **139B** (1984) 310;
    G. Passarino, *Phys. Lett.* **152B** (1985) 271.

37. R. Philippe, *Phys. Rev.* **D26** (1982) 1588;
    M. Böhm et al., *Nucl. Phys.* **B304** (1988) 463;
    J. Fleischer, F. Jegerlehner and M. Zralek, *Z. Phys.* **C42** (1989) 409.

DFUB 91-02
Revised 11 April 1991

The analytic value of the atomic three electron correlation integral with Slater wave functions

Ettore Remiddi

Dipartimento di Fisica, Università di Bologna, I-40126 Bologna, Italy
INFN, Sezione di Bologna, I-40126 Bologna, Italy

Abstract.
The three electron atomic correlation integral with Slater Type wave functions is evaluated in closed analytic form. The result is expressed in terms of rational functions, logarithms and dilogarithms of simple arguments, so that its precise and fast numerical evaluation is straightforward.

To Tini Veltman, for his 60th birthday.

1. It is the purpose of this paper to provide with a closed analytic expression for the atomic three electron correlation integral

$$Z(w_1, w_2, w_3) \equiv \int d\vec{r}_1 d\vec{r}_2 d\vec{r}_3 e^{-w_1 r_1} e^{-w_2 r_2} e^{-w_3 r_3} |\vec{r}_1 - \vec{r}_2| \frac{1}{|\vec{r}_3 - \vec{r}_1|} |\vec{r}_2 - \vec{r}_3| . \qquad (1)$$

The integral naturally arises in the study of atoms with three electrons or more when using Hylleraas wave functions to account for two-electron correlation effects. To the author knowledge, the integral eq.(1) has been evaluated so far only by means of approximated numerical techniques, by expanding for instance one of the $|\vec{r}_i - \vec{r}_j|$ factors in Legendre polynomials, performing in closed form the integration of the resulting terms and then summing a suitable number of terms of the so obtained infinite series.

The closed analytic formula obtained in this paper involves, besides rational fractions and logarithms, a few dilogarithmic functions of simple arguments. The basic properties of the dilogarithm are recalled in Appendix A for the benefit of the unfamiliar reader; let us just stress here that the dilogarithm of argument x has the same analytic properties in x as the logarithm of argument $(1 - x)$ and, for practical purposes, its accurate numerical evaluation presents the same problems as the evaluation of the logarithm. The dilogarithm is often encountered in the calculation of radiative corrections in QED [1]; to the author pleasure, it turned out that the techniques developped for the computational problems arising there can be used, with obvious extensions, also in the analytic evaluation of the atomic integral eq.(1).

The result looks (perhaps is) somewhat cumbersome; but it is in fact astonishingly simple when compared to the large amount of algebra which was needed to obtain it, suggesting the existence of some underlying (and yet unknown) structure. To process the algebra, the use of an algebra manipulating program was mandatory. The author relied, in all the steps of the calculation, on the program SCHOONSCHIP by M. Veltman [2], which provided with the needed flexibility and computing power.

The plan of the paper is as follows. In Section 2, which contains the essential part of the calculation, an auxiliary "fundamental" integral is introduced and evaluated by means of the "differentiate and integrate" algorithm which is the bulk of the approach. In Section 3 the result is extended to a number of related integrals, including that of eq.(1). Section 4 contains the conclusions, while Appendix A recalls definition and properties of the dilogarithm and Appendix B the derivation of some of the formulae used in the text.

2. To start with, let us introduce the auxiliary integral

$$A(w_1, w_2, w_3; u_1, u_2, u_3) \equiv \int d\vec{r}_1 d\vec{r}_2 d\vec{r}_3 \; \frac{e^{-w_1 r_1}}{r_1} \frac{e^{-w_2 r_2}}{r_2} \frac{e^{-w_3 r_3}}{r_3}$$
$$\frac{e^{-u_3 |\vec{r}_1 - \vec{r}_2|}}{|\vec{r}_1 - \vec{r}_2|} \frac{e^{-u_2 |\vec{r}_3 - \vec{r}_1|}}{|\vec{r}_3 - \vec{r}_1|} \frac{e^{-u_1 |\vec{r}_2 - \vec{r}_3|}}{|\vec{r}_2 - \vec{r}_3|} ; \qquad (2)$$

it is obvious that the integral (1) and a wide family of related integrals with the same exponentials and different powers of the factors r_i, $|\vec{r}_i - \vec{r}_j|$ can be obtained from it by differentiating with respect to the variables w_i, u_i and then putting $u_i = 0$. In this work, we will limit ourselves to provide with closed analytic formulae for eq.(2) and its first u_i-derivatives only at $u_i = 0$ but for arbitrary values of the w_i, so that all the integrals with non-negative powers of the r_i's can also be obtained by differentiation. Their is some hope that the $u_i \neq 0$ case, which is of interest for simple molecules, can also be worked out with similar techniques, but that generalization has not yet been attempted.

As a first step, we use the Fourier representation

$$\frac{e^{-wr}}{r} = \int \frac{d\vec{p}}{(2\pi)^3} \frac{4\pi}{p^2 + w^2} e^{i\vec{p}\vec{r}}$$

for the six exponentials appearing in eq.(2), integrate on all the $d\vec{r}_i$ so obtaining three Dirac δ-functions, and then integrate the δ-functions in three of the momenta; eq.(2) becomes

$$A(w_1, w_2, w_3; u_1, u_2, u_3) = \int \frac{d\vec{p}_3}{(2\pi)^3} \frac{d\vec{p}_2}{(2\pi)^3} \frac{d\vec{p}_1}{(2\pi)^3} \frac{4\pi}{p_3^2 + u_3^2} \frac{4\pi}{p_2^2 + u_2^2} \frac{4\pi}{p_1^2 + u_1^2}$$
$$\frac{4\pi}{(\vec{p}_1 - \vec{p}_2)^2 + w_3^2} \frac{4\pi}{(\vec{p}_2 - \vec{p}_3)^2 + w_1^2} \frac{4\pi}{(\vec{p}_3 - \vec{p}_1)^2 + w_2^2} . \tag{3}$$

To perform the angular integrations, define

$$z_1 \equiv \frac{p_2^2 + p_3^2 + w_1^2}{2p_2 p_3} , \qquad z_2 \equiv \frac{p_3^2 + p_1^2 + w_2^2}{2p_3 p_1} , \qquad z_3 \equiv \frac{p_1^2 + p_2^2 + w_3^2}{2p_1 p_2} , \tag{4}$$

so that

$$\frac{1}{(\vec{p}_2 - \vec{p}_3)^2 + w_1^2} = -\frac{1}{2p_2 p_3} \frac{1}{\hat{p}_2 \hat{p}_3 - z_1} ,$$

where $\hat{p}_i$ is the unit vector in the direction of $\vec{p}_i$, and similarly for the other denominators; by introducing polar coordinates through $d\vec{p} = p^2 dp d\Omega(\hat{p})$ eq.(3) can be written as

$$A(w_1, w_2, w_3; u_1, u_2, u_3) =$$
$$\frac{1}{32\pi^3} \int_0^\infty \frac{dp_3^2}{p_3^2 + u_3^2} \int_0^\infty \frac{dp_2^2}{p_2^2 + u_2^2} \int_0^\infty \frac{dp_1^2}{p_1^2 + u_1^2} B(z_1, z_2, z_3) , \tag{5}$$

where

$$B(z_1, z_2, z_3) \equiv -\frac{1}{p_1 p_2 p_3} \int \frac{d\Omega(\hat{p}_1) d\Omega(\hat{p}_2) d\Omega(\hat{p}_3)}{(\hat{p}_1 \hat{p}_2 - z_3)(\hat{p}_1 \hat{p}_3 - z_2)(\hat{p}_2 \hat{p}_3 - z_1)} . \tag{6}$$

The analytic integration on the spherical angle $d\Omega(\hat{p}_3)$ can be performed by means of the formula

$$\int \frac{d\Omega(\hat{p}_3)}{(\hat{p}_1 \hat{p}_3 - z_2)(\hat{p}_2 \hat{p}_3 - z_1)} = \frac{2\pi}{\sqrt{\delta(z_1, z_2, z)}} \ln \left| \frac{z_1 z_2 - z + \sqrt{\delta(z_1, z_2, z)}}{z_1 z_2 - z - \sqrt{\delta(z_1, z_2, z)}} \right| , \tag{7}$$

where $z = \hat{p}_1\hat{p}_2$ is the cosine of the angle formed by $\hat{p}_1, \hat{p}_2$ and

$$\delta(z_1, z_2, z) \equiv z_1^2 + z_2^2 + z^2 - 2z_1z_2z - 1 . \tag{8}$$

Note that $\delta(z_1, z_2, z)$ is a second order polynomial in z, a property which will play an essential role in the following; as a side remark, it is easy to verify that $\delta(z_1, z_2, z) > 0$ for $w_1, w_2 > 0$ and $|z| \leq 1$.

In terms of z one has $d\Omega(\hat{p}_2) = dz d\phi_2$; when z and eq.(7) are used the integrand of eq.(6) is seen to be independent of ϕ_2 and $\hat{p}_1$, so that

$$B(z_1, z_2, z_3) = - \frac{16\pi^3}{p_1 p_2 p_3} \int_{-1}^{1} \frac{dz}{z - z_3} \frac{1}{\sqrt{\delta(z_1, z_2, z)}} \ln \left| \frac{z_1 z_2 - z + \sqrt{\delta(z_1, z_2, z)}}{z_1 z_2 - z - \sqrt{\delta(z_1, z_2, z)}} \right| . \tag{9}$$

At this point one might try the "brute force" analytic integration in z of eq.(9); formulae for doing that exist in the litterature, but the result is a combination of dilogarithmic functions of complicated arguments, which provide with no hint for the subsequent integrations on the p_i's. Our method consists instead in postponing any explicit analytic integration for a while, rewriting the required integral in a form which will be found more convenient later. Rather than integrating explicitly eq.(9), therefore, we introduce the function

$$C(z_1, z_2, z_3) \equiv - \int_{-1}^{1} \frac{dz}{z - z_3} \frac{\sqrt{-\delta(z_1, z_2, z_3)}}{\sqrt{\delta(z_1, z_2, z)}} \ln \left| \frac{z_1 z_2 - z + \sqrt{\delta(z_1, z_2, z)}}{z_1 z_2 - z - \sqrt{\delta(z_1, z_2, z)}} \right| ; \tag{10}$$

$B(z_1, z_2, z_3)$ is positive definite, as $z_3 > 1 \geq z$; $\delta(z_1, z_2, z_3)$, on the contrary, has no definite sign, and the choice of the minus sign in front of it in eq.(10) is suggested only by irrelevant aesthetical considerations.

Let us also introduce

$$\Delta(p_1^2, p_2^2, p_3^2) \equiv - 4p_1^2 p_2^2 p_3^2 \, \delta(z_1, z_2, z_3), \tag{11}$$

i.e., on account of eq.s(4)

$$\Delta(p_1^2, p_2^2, p_3^2) = w_1^2 p_1^4 + w_2^2 p_2^4 + w_3^2 p_3^4 + w_1^2 w_2^2 w_3^2 - (p_1^2 p_2^2 - w_3^2 p_3^2)(w_1^2 + w_2^2 - w_3^2)$$
$$- (p_2^2 p_3^2 - w_1^2 p_1^2)(w_2^2 + w_3^2 - w_1^2) - (p_3^2 p_1^2 - w_2^2 p_2^2)(w_3^2 + w_1^2 - w_2^2) . \tag{12}$$

It is to be noted, again, that $\Delta(p_1^2, p_2^2, p_3^2)$ is a quadratic form on each of the three variables p_i^2 (as well as on the w_i^2, not explicitly written in the arguments of Δ for simplicity). With the newly introduced symbols, eq.(9) becomes

$$B(z_1, z_2, z_3) = \frac{32\pi^3}{\sqrt{\Delta(p_1^2, p_2^2, p_3^2)}} C(z_1, z_2, z_3) ; \tag{13}$$

correspondingly, eq.(5) reads

$$A(w_1, w_2, w_3; u_1, u_2, u_3) =$$
$$\int_0^\infty \frac{dp_3^2}{p_3^2 + u_3^2} \int_0^\infty \frac{dp_2^2}{p_2^2 + u_2^2} \int_0^\infty \frac{dp_1^2}{p_1^2 + u_1^2} \frac{1}{\sqrt{\Delta(p_1^2, p_2^2, p_3^2)}} C(z_1, z_2, z_3) . \tag{14}$$

We further introduce the functions

$$T(w_1, w_2, w_3, p_2^2, p_3^2) \equiv \int_0^\infty \frac{dp_1^2}{p_1^2 + u_1^2} \frac{\sqrt{\Delta(-u_1^2, p_2^2, p_3^2)}}{\sqrt{\Delta(p_1^2, p_2^2, p_3^2)}} C(z_1, z_2, z_3) , \qquad (15)$$

$$Q(w_1, w_2, w_3, p_3^2) \equiv \int_0^\infty \frac{dp_2^2}{p_2^2 + u_2^2} \frac{\sqrt{\Delta(-u_1^2, -u_2^2, p_3^2)}}{\sqrt{\Delta(-u_1^2, p_2^2, p_3^2)}} T(w_1, w_2, w_3, p_2^2, p_3^2) , \qquad (16)$$

$$P(w_1, w_2, w_3, u_1, u_2, u_3) \equiv \int_0^\infty \frac{dp_3^2}{p_3^2 + u_3^2} \frac{\sqrt{\Delta(-u_1^2, -u_2^2, -u_3^2)}}{\sqrt{\Delta(-u_1^2, -u_2^2, p_3^2)}} Q(w_1, w_2, w_3, p_3^2) , \qquad (17)$$

where, for the sake of brevity only, the variables u_i are not explicitly written among the arguments of some of the above functions. Eq.s(2),(3),(14-17) then amount to

$$A(w_1, w_2, w_3; u_1, u_2, u_3) = \frac{1}{\sqrt{\Delta(-u_1^2, -u_2^2, -u_3^2)}} P(w_1, w_2, w_3; u_1, u_2, u_3) . \qquad (18)$$

We will now show that the above way of rewriting the integrals is indeed of help for obtaining a convenient expression for the derivatives with respect to the variables w_i of the function $P(w_1, w_2, w_3; u_1, u_2, u_3)$. Quite in general, let

$$S(a, x) \equiv s_0 + s_1(x - b) + \frac{1}{2}s_2(x - b)^2 \qquad (19)$$

be a second order polynomial in the variable x, with the coefficients depending on some unspecified parameter a, so that $s_i = s_i(a)$, $i = 0, 1, 2$ and consider the integral

$$K(a) \equiv \int_{x_1}^{x_2} \frac{dx}{x - b} \frac{\sqrt{S(a, b)}}{\sqrt{S(a, x)}} H(a, x) , \qquad (20)$$

where the otherwise unspecified function $H(a, x)$ depends also on a and x, while x_1, x_2 and b are independent of a. Thanks to the presence of the factor $\sqrt{S(a, b)}$ in the numerator of eq.(20), one finds the following formula for the a-derivative of $K(a)$:

$$\frac{\partial}{\partial a}K(a) = \int_{x_1}^{x_2} \frac{dx}{x - b} \frac{\sqrt{S(a, b)}}{\sqrt{S(a, x)}} \frac{\partial}{\partial a}H(a, x) + \frac{1}{s_1^2 - 2s_0 s_2} \frac{1}{\sqrt{S(a, b)}} \int_{x_1}^{x_2} \frac{dx}{\sqrt{S(a, x)}} \cdot$$
$$\left\{ -\frac{1}{2}\left[\frac{\partial s_0}{\partial a}s_1 s_2 - 2s_0\frac{\partial s_1}{\partial a}s_2 + s_0 s_1\frac{\partial s_2}{\partial a}\right](x - b) + \left[\frac{\partial s_0}{\partial a}(s_0 s_2 - s_1^2) + s_0\frac{\partial s_1}{\partial a}s_1 - s_0^2\frac{\partial s_2}{\partial a}\right]\right\}$$
$$\left[\delta(x - x_2) - \delta(x - x_1) - \frac{\partial}{\partial x}\right] H(a, x) .$$
$$(21)$$

Its derivation, which is elementary, is reported in Appendix B for convenience of the reader: its usefulness relies in the fact that it expresses the a-derivative of $K(a)$ in terms of quantities which can be evaluated without carrying out explicitly the original integral, namely the end-points of the function $H(a,x)$, given by to the two Dirac δ-functions $\delta(x - x_i)$ in eq.(21), and an integral involving the x-derivative of $H(a,x)$.

Eq.(21) can be used for obtaining the w_i-derivatives of $P(w_1, w_2, w_3; u_1, u_2, u_3)$ eq.(17), with p_3^2, $(p_3^2 + u_3^2)$ and $Q(w_1, w_2, w_3, p_3^2)$ in the role of x, $(x-a)$ and of the unspecified function $H(a,x)$, while $\Delta(-u_1^2, -u_2^2, p_3^2)$ is the second order polynomial in p_3^2 corresponding to $S(a,x)$. A closer inspection to the definition of $Q(w_1, w_2, w_3, p_3^2)$ shows that the end point values actually vanish, so that the required w_i-derivatives of $P(w_1, w_2, w_3, u_1, u_2, u_3)$ are expressed as the integral on p_3^2 of a combination of rational functions of p_3^2 times the corresponding w_i- and p_3^2-derivatives of $Q(w_1, w_2, w_3, p_3^2)$.

Eq.(21) can be used again for evaluating the derivatives of $Q(w_1, w_2, w_3, p_3^2)$ because $\Delta(-u_1^2, p_2^2, p_3^2)$, which appears in the r.h.s. of eq.(16), is also a second order polynomial in p_2^2; as in the previous case the end-point contributions are found to vanish and the required w_i- and p_3^2-derivatives of $Q(w_1, w_2, w_3, p_3^2)$ are expressed in terms of the various derivatives of $T(w_1, w_2, w_3, p_2^2, p_3^2)$.

The process can be iterated once more, so that all the w_i- and p_i^2-derivatives of $C(z_1, z_2, z_3)$ are eventually needed. As $C(z_1, z_2, z_3)$ depends on the w_i and the p_i^2 only through the three variables z_i, eq.(4), it is in fact sufficient to evaluate its three z_i-derivatives. The derivatives with respect to z_1, z_2 can also be worked out by means of formula (21), because $\delta(z_1, z_2, z)$, as already observed, is a second order polynomial in z (the change of sign in the argument of the square root in the numerator of eq.(10) is an overall constant factor which does not affect the applicability of the formula). The case of the w_3-derivative is slightly different - is easiest derivation is perhaps through eq.(37) which will be introduced below - but the result is similar and exhibits explicitly the expected symmetry of $C(z_1, z_2, z_3)$ for the exchange of the arguments.

When carrying out the above procedure, the factor $(s_1^2 - 2s_0 s_2)$ appearing in the denominator of eq.(21) takes the value $4(z_1^2 - 1)(z_2^2 - 1)$, while the "unspecified function" in the r.h.s. of the definition of $C(z_1, z_2, z_3)$, eq.(10), is in fact the explicitly known logarithm of eq.(10). Its derivative, a fraction, contains, among the others, terms in $1/\sqrt{\delta(z_1, z_2, z)}$, which get multiplied by the same square root factor appearing in eq.(21) to generate the denominator $1/\delta(z_1, z_2, z)$. After some fully straightforward albeit lenghty algebra that denominator is found to disappear; the z-integration is then elementary and the explicit analytic values of the required z_i-derivative are rather simple. One finds for instance

$$\frac{\partial C(z_1, z_2, z_3)}{\partial z_1} =$$

$$\frac{2}{\sqrt{\delta(z_1, z_2, z_3)}} \left[\frac{z_1 z_2 - z_3}{z_1^2 - 1} \ln\left(\frac{z_2 + 1}{z_2 - 1}\right) + \frac{z_1 z_3 - z_2}{z_1^2 - 1} \ln\left(\frac{z_3 + 1}{z_3 - 1}\right) - \ln\left(\frac{z_1 + 1}{z_1 - 1}\right) \right]; \qquad (22)$$

due to the already recalled symmetry of $C(z_1, z_2, z_3)$ in its arguments it is not necessary to write down explicitly the derivatives with respect to z_2 and z_3.

Once the derivatives of $C(z_1, z_2, z_3)$ are evaluated, one can proceed bacwards to evaluate the derivatives of $T(w_1, w_2, w_3, p_2^2, p_3^2)$, which were seen to be an integral on p_1^2 of the derivatives of $C(z_1, z_2, z_3)$ times suitable rational factors. At this stage, to simplify the calculation, we put $u_1 = 0$ in the denominator $(p_1^2 + u_1^2)$ of eq.(15). When eq.s(4),(12) are used for eliminating the z_i's, everything is expressed in terms of the p_i^2's and w_i's, and the denominator $1/\Delta(p_1^2, p_2^2, p_3^2)$ appears, in the same way as $1/\delta(z_1, z_2, z)$ appeared in the previous case. After a really long algebraic manipulation, that denominator is also found to disappear (such a result is always expected in this kind of calculations, although a satisfactory formal proof of this fact is missing; the elimination of the denominator provides in practice with one of the most important guideline in the organization of the whole calculation). One is eventually left with a relatively simple expression, say about one hundred terms, or less, for each of the derivatives. That expression has in general the form of a ratio of polynomials in the integration variable p_1^2 as well as in the other variables, times the three logarithms appearing in eq.(22). More explicitly, one finds that an essential role is plaid by the three polynomials, of second order in the arguments p^2, q^2, w^2,

$$R_2(p_1^2, p_2^2, -w_3^2), \qquad R_2(p_2^2, p_3^2, -w_1^2), \qquad R_2(p_3^2, p_1^2, -w_2^2), \tag{23}$$

where

$$R_2(p^2, q^2, -w^2) \equiv p^4 + q^4 + w^4 - 2p^2q^2 + 2w^2p^2 + 2w^2q^2 . \tag{24}$$

Remarkably, the actual value of the factor $(s_1^2 - 2s_0s_2)$ appearing in eq.(21) is in this case $R_2(w_1^2, w_2^2, w_3^2)R_2(p_2^2, p_3^2, -w_1^2)$.

All the p_1^2-integrals consist of an algebraic factor, whose possible denominators are $1/p_1^2$, which corresponds to $1/(p_1^2 + u_1^2)$ of eq.(15) at $u_1 = 0$, $1/R_2(p_1^2, p_2^2, -w_3^2)$ and $1/R_2(p_3^2, p_1^2, -w_2^2)$, times one of the logarithms of eq.(22). A closer inspection shows that in general any p-integral involving a factor $1/R_2(p^2, q^2, -w^2)$ can be conveniently written in terms of the four basic combinations

$$dp^2 \frac{wq}{R_2(p^2, q^2, -w^2)} , \qquad dp^2 \frac{(p^2 - q^2 + w^2)}{R_2(p^2, q^2, -w^2)} ,$$
$$dp \frac{w(p^2 + q^2 + w^2)}{R_2(p^2, q^2, -w^2)} , \qquad dp \frac{q(p^2 - q^2 - w^2)}{R_2(p^2, q^2, -w^2)} . \tag{25}$$

When that is done, one obtains a limited number of p_1^2-integrals such as, for instance,

$$\int_0^\infty dp_1^2 \frac{p_3 w_2}{R_2(p_3^2, p_1^2, -w_2^2)} \ln\left(\frac{z_3 + 1}{z_3 - 1}\right) ,$$
$$\int_0^\infty dp_1^2 \frac{p_1^2 - p_2^2 + w_3^2}{R_2(p_3^2, p_1^2, -w_2^2)} \ln\left(\frac{z_2 + 1}{z_2 - 1}\right) . \tag{26}$$

Terms in $\ln\left(\frac{z_1 + 1}{z_1 - 1}\right)$ also exist; as this logarithm does not depend on p_1, those terms can be integrated at once; the actually occurring integrals are

$$\int_0^\infty \frac{dp\ p^2}{R_2(p^2, q^2, -w^2)} = \frac{\pi}{4w} ,$$
$$\int_0^\infty \frac{dp}{R_2(p^2, q^2, -w^2)} = \frac{\pi}{4w} \frac{1}{q^2 + w^2} . \tag{27}$$

For the continuation of the calculation it is not necessary to evaluate explicitly the other p_1^2-integrals, but it is in fact convenient to keep them in the form of eq.(26), giving them *ad hoc* names, or just "protecting" them with suitable brackets in the subsequent steps. To summarize, each of the 50 to 100 terms occurring in the expressions of the derivatives of $T(w_1, w_2, w_3, p_2^2, p_3^2)$ with respect to any of its five arguments is therefore the product of one of the above p_1^2-integrals times a rational fraction in the five variables p_2^2, p_3^2 and w_i's, times the overall factor $1/\sqrt{\Delta(0, p_2^2, p_3^2)}$, generated when using eq.(21) for the derivatives of $T(w_1, w_2, w_3, p_2^2, p_3^2)$, eq.(15).

With the so obtained expression for the derivatives of $T(w_1, w_2, w_3, p_2^2, p_3^2)$ we can use again eq.(21) to obtain the derivatives of $Q(w_1, w_2, w_3, p_3^2)$, eq.(16), at $u_2 = 0$; the pattern is the same, the denominator $1/\Delta(0, p_2^2, p_3^2)$ is generated but actually found to disappear after some algebra, all the p_2-integrals involving $R_2(p_2^2, p_3^2, -w_1^2)$ can be written in one of the four forms of eq.(25), the explicit integration in p_2 is neither necessary nor convenient, an overall denominator $1/\sqrt{\Delta(0, 0, p_3^2)}$ appears.

With one more iteration of the algorithm one obtains the three w_i-derivatives of $P(w_1, w_2, w_3, 0, 0, 0)$. The denominator $1/\Delta(0, 0, p_3^2)$ is generated but found to disappear, while everything is multiplied by the simple overall factor $1/\sqrt{\Delta(0, 0, 0)} = 1/(w_1 w_2 w_3)$. The derivatives consist of a limited number (a couple of dozens) of integrals like

$$\int_0^\infty \frac{w_1 \, dp_3}{p_3^2 + w_1^2} \int_0^\infty dp_2 \frac{p_3(p_2^2 - p_3^2 - w_1^2)}{R_2(p_2^2, p_3^2, -w_1^2)} \int_0^\infty dp_1 \frac{p_1^2 - p_2^2 + w_3^2}{R_2(p_1^2, p_2^2, -w_3^2)} \ln\left(\frac{z_2 + 1}{z_2 - 1}\right) = \frac{1}{2} \pi^3 \ln\left(\frac{w_1 + w_2 + w_3}{3w_1 + w_2 + w_3}\right), \tag{28}$$

$$\int_0^\infty dp_3^2 \left(\frac{1}{p_3^2 + w_1^2} - \frac{1}{p_3^2}\right) \int_0^\infty \frac{dp_2^2 p_3 w_1}{R_2(p_2^2, p_3^2, -w_1^2)} \int_0^\infty dp_1^2 \frac{p_1^2 - p_3^2 + w_2^2}{R_2(p_3^2, p_1^2, -w_2^2)} \ln\left(\frac{z_3 + 1}{z_3 - 1}\right) = \pi^3 \ln\left(\frac{w_1 + w_2 + w_3}{3w_1 + w_2 + w_3}\right). \tag{29}$$

All the appearing triple integrals are in general equal to a factor π^3 times a logarithm whose arguments are linear combinations of the w_i's with simple integer coefficients, such as for instance $(w_1 + 3w_2 + w_3)$, $(2w_1 + w_3)$, $(w_2 + w_3)$ etc. To establish the above results, one can differentiate with respect to one of arguments w_i the integral in p_3, by using formulae which are the extension of eq.(21) to the present case (two of them are reported in Appendix B), so obtaining end-point values and derivatives with respect to w_i and p_3 of the p_3-integrand, which is an integral in p_2; by repeated use of the same formulae one can propagate the derivatives through the subsequent p_2 and p_1 integrations, until only the derivatives of the logarithm are needed. In so doing one finds that the p_1, p_2 and p_3 integrations are elementary (almost ironically, only the two integrals of eq.(27) occur), and the required w_i-derivative of the triple integral on p_3, p_2, p_1 is found to be equal to π^3 times simple rational denominators in the w_i. The the r.h.s. of eq.s(28),(29) can then be obtained by quadrature. The otherwise arbitrary additive constant of the quadrature is fixed by checking that l.h.s. and r.h.s. coincide for some special set of values of the w_i's; the l.h.s. and r.h.s. and eq.s(28),(29), for instance, both vanish at $w_2 = \infty$.

Collecting results, one finally obtains

$$\frac{\partial}{\partial w_1} P(w_1, w_2, w_3, 0, 0, 0) = 32\pi^3 \left\{ \frac{1}{w_1 + w_2 - w_3} \ln\left(\frac{w_1 + w_2}{w_3}\right) \right.$$

$$- \frac{1}{w_1 - w_2 + w_3} \ln\left(\frac{w_2 + w_3}{w_1}\right) + \frac{1}{w_1 - w_2 - w_3} \ln\left(\frac{w_3 + w_1}{w_2}\right) \tag{30}$$

$$\left. - \frac{1}{w_1 + w_2 + w_3} \left[\ln\left(\frac{w_1 + w_2}{w_3}\right) + \ln\left(\frac{w_2 + w_3}{w_1}\right) + \ln\left(\frac{w_3 + w_1}{w_2}\right) \right] \right\} ,$$

and similar formulae for the other derivatives, which are not written down explicitly due to the symmetry for the exchange of the w_i's.

From inspection, one sees that one can easily evaluate the integral eq.(2), at $u_i=0$ and in the limit $w_3 \gg w_1, w_2$, by performing the change of variable $\vec{r}_3 \to \vec{r}$, $\vec{r} = w_3 \vec{r}_3$, and then approximating $|\vec{r}_1 - \vec{r}_3| \simeq r_1$, $|\vec{r}_2 - \vec{r}_3| \simeq r_2$; in that limit the integration is elementary, giving the result

$$A(w_1, w_2, w_3, 0, 0, 0) \simeq \frac{64\pi^3}{w_1 w_2 w_3^2} \left\{ w_1 \ln\left(\frac{w_1 + w_2}{w_1}\right) + w_2 \ln\left(\frac{w_1 + w_2}{w_2}\right) \right\} , \quad w_3 \gg w_1, w_2 . \tag{31}$$

We can at last integrate eq.(30) by quadrature in w_1, fixing the otherwise undetermined additive constant by comparison with eq.(31) at large w_3. The result is

$$P(w_1, w_2, w_3, 0, 0, 0) = 32\pi^3 D(w_1, w_2, w_3), \tag{32}$$

where

$$D(w_1, w_2, w_3) \equiv$$

$$\ln\left(\frac{w_1 + w_2}{w_3}\right) \ln\left(\frac{w_1 + w_2 + w_3}{w_1 + w_2}\right) - \mathrm{Li}_2\left(-\frac{w_3}{w_1 + w_2}\right) - \mathrm{Li}_2\left(1 - \frac{w_3}{w_1 + w_2}\right)$$

$$+ \ln\left(\frac{w_2 + w_3}{w_1}\right) \ln\left(\frac{w_1 + w_2 + w_3}{w_1 + w_3}\right) - \mathrm{Li}_2\left(-\frac{w_2}{w_1 + w_3}\right) - \mathrm{Li}_2\left(1 - \frac{w_2}{w_1 + w_3}\right) \tag{33}$$

$$+ \ln\left(\frac{w_3 + w_1}{w_2}\right) \ln\left(\frac{w_1 + w_2 + w_3}{w_2 + w_3}\right) - \mathrm{Li}_2\left(-\frac{w_1}{w_2 + w_3}\right) - \mathrm{Li}_2\left(1 - \frac{w_1}{w_2 + w_3}\right) ;$$

the function $\mathrm{Li}_2(x)$ appearing in eq.(33) is the Euler dilogarithm, whose definition and main properties are recalled in Appendix A for convenience of the reader; it suffices to repeat once more here that it can be numerically evaluated as fast and accurately as the logarithm.

According to eq.s(2),(12),(18) one can also write

$$A(w_1, w_2, w_3, 0, 0, 0) = \frac{1}{w_1 w_2 w_3} P(w_1, w_2, w_3, 0, 0, 0) ,$$

i.e.

$$\int d\vec{r}_1 d\vec{r}_2 d\vec{r}_3 \frac{e^{-w_1 r_1}}{r_1} \frac{e^{-w_2 r_2}}{r_2} \frac{e^{-w_3 r_3}}{r_3} \frac{1}{|\vec{r}_2 - \vec{r}_3|} \frac{1}{|\vec{r}_3 - \vec{r}_1|} \frac{1}{|\vec{r}_1 - \vec{r}_2|} =$$

$$\frac{32\pi^3}{w_1 w_2 w_3} D(w_1, w_2, w_3) . \tag{34}$$

3. In order to procede from eq.(34) towards the integral eq.(1), let us differentiate twice eq.(2) with respect to u_1 and then put $u_i = 0$ for all the i; in so doing we obtain

$$\frac{\partial^2}{\partial u_1^2} A(w_1, w_2, w_3, u_1, u_2, u_3)\bigg|_{u_i=0} =$$
$$\int d\vec{r}_1 d\vec{r}_2 d\vec{r}_3 \frac{e^{-w_1 r_1}}{r_1} \frac{e^{-w_2 r_2}}{r_2} \frac{e^{-w_3 r_3}}{r_3} |\vec{r}_2 - \vec{r}_3| \frac{1}{|\vec{r}_3 - \vec{r}_1|} \frac{1}{|\vec{r}_1 - \vec{r}_2|} . \tag{35}$$

According to eq.(14), for finite u_i one has also

$$\frac{\partial^2}{\partial u_1^2} A(w_1, w_2, w_3, u_1, u_2, u_3) =$$
$$\int_0^\infty \frac{dp_3^2}{p_3^2 + u_3^2} \int_0^\infty \frac{dp_2^2}{p_2^2 + u_2^2} \int_0^\infty dp_1^2 \frac{6u_1^2 - 2p_1^2}{(p_1^2 + u_1^2)^3} \frac{1}{\sqrt{\Delta(p_1^2, p_2^2, p_3^2)}} C(z_1, z_2, z_3) . \tag{36}$$

To evaluate such an integral, let us observe quite in general that if $S(x)$ is, in the notation of eq.s(19),(20) (but dropping the dependence on the parameters a, which play no role here), a second order polynomial in x, the following "integration by parts" formula holds

$$\int_{x_1}^{x_2} \frac{dx}{(x-b)^n} \frac{1}{\sqrt{S(x)}} H(x) =$$
$$-\frac{1}{n-1} \frac{1}{S(b)} \int_{x_1}^{x_2} dx \left\{ \frac{1}{2\sqrt{S(x)}} \left[\frac{2n-3}{(x-b)^{n-1}} s_1 + \frac{n-2}{(x-b)^{n-2}} s_2 \right] \right. \tag{37}$$
$$\left. + \frac{\sqrt{S(x)}}{(x-b)^{n-1}} \left[\delta(x - x_2) - \delta(x - x_1) - \frac{\partial}{\partial x} \right] \right\} H(x) .$$

Eq.(37) can be used to express the p_1^2 integral of eq.(36) in terms of the integral with a single inverse power of $(p_1^2 + u_1^2)$, essentially equivalent to eq.(15), plus terms involving the derivatives of $C(z_1, z_2, z_3)$, which are explicitly known and are given in eq.(22), plus end-point values, which are easy to get (in the concerned case they actually vanish). After the "integration by parts" the $u_1 \to 0$ limit is trivial. Putting also $u_2 = u_3 = 0$ one immediately identifies a term proportional to $D(w_1, w_2, w_3)$, plus a number of terms which, after some by now standard algebra, can be brought in the form of the integrals already encountered in the w_i-derivatives of $D(w_1, w_2, w_3)$, such as those listed in eq.s(28),(29). The result reads

$$\int d\vec{r}_1 d\vec{r}_2 d\vec{r}_3 \frac{e^{-w_1 r_1}}{r_1} \frac{e^{-w_2 r_2}}{r_2} \frac{e^{-w_3 r_3}}{r_3} \frac{1}{|\vec{r}_1 - \vec{r}_2|} \frac{1}{|\vec{r}_3 - \vec{r}_1|} |\vec{r}_2 - \vec{r}_3| = \frac{64\pi^3}{w_2^2 w_3^2} .$$
$$\left\{ \frac{w_2^2 + w_3^2 - w_1^2}{2w_1 w_2 w_3} D(w_1, w_2, w_3) - \left(\frac{1}{w_1 + w_2} + \frac{1}{w_1 + w_3} \right) \right. \tag{38}$$
$$\left. + \frac{w_1 - w_2}{w_1 w_2} \ln\left(\frac{w_1 + w_2}{w_3}\right) + \frac{w_1 - w_3}{w_1 w_3} \ln\left(\frac{w_2 + w_3}{w_1}\right) + \frac{w_2 + w_3}{w_2 w_3} \ln\left(\frac{w_3 + w_1}{w_2}\right) \right\} .$$

The whole procedure can be repeated for the variable u_3, which multiplies $|\vec{r}_1 - \vec{r}_2|$ in the exponential of eq.(2), to obtain the formula

$$\frac{\partial^2}{\partial u_3^2}\frac{\partial^2}{\partial u_1^2}A(w_1, w_2, w_3, u_1, u_2, u_3)\Big|_{u_i=0} = \int_0^\infty dp_3^2\frac{6u_3^2 - 2p_3^2}{(p_3^2 + u_3^2)^3}$$
$$\int_0^\infty \frac{dp_2^2}{p_2^2 + u_2^2}\int_0^\infty dp_1^2\frac{6u_1^2 - 2p_1^2}{(p_1^2 + u_1^2)^3}\frac{1}{\sqrt{\Delta(p_1^2, p_2^2, p_3^2)}}C(z_1, z_2, z_3)\ . \tag{39}$$

After integrating "by parts" in u_1 and u_3 the $u_i = 0$ is trivial and the result reads

$$\int d\vec{r}_1 d\vec{r}_2 d\vec{r}_3\ \frac{e^{-w_1 r_1}}{r_1}\frac{e^{-w_2 r_2}}{r_2}\frac{e^{-w_3 r_3}}{r_3}|\vec{r}_1 - \vec{r}_2|\frac{1}{|\vec{r}_3 - \vec{r}_1|}|\vec{r}_2 - \vec{r}_3| = \frac{64\pi^3}{w_1^3 w_2^5 w_3^3}\cdot$$
$$\left\{\frac{1}{2}\left[w_2^2(2w_1^2 + w_2^2 + 2w_3^2) - 3(w_1^2 - w_3^2)^2\right]D(w_1, w_2, w_3)\right.$$
$$+ w_1\left[(w_3 - w_2)w_2^2 + 3w_3(w_1^2 + w_3^2) + 3w_2(w_1^2 - w_3^2)\right]\ln\left(\frac{w_2 + w_3}{w_1}\right)$$
$$+ w_3\left[(w_1 - w_2)w_2^2 + 3w_1(w_1^2 + w_3^2) - 3w_2(w_1^2 - w_3^2)\right]\ln\left(\frac{w_1 + w_2}{w_3}\right) \tag{40}$$
$$+ w_2(w_1 + w_3)\left[3(w_1 - w_3)^2 - w_2^2\right]\ln\left(\frac{w_3 + w_1}{w_2}\right)$$
$$+ 2w_1 w_3\left[w_2^3\frac{w_1^2 + 3w_1 w_3 + w_3^2}{(w_1 + w_3)^3} - w_2\frac{2w_1^2 + w_1 w_3 + 2w_3^2}{w_1 + w_3}\right.$$
$$\left.\left.+ \frac{w_1^3}{w_1 + w_2} + \frac{w_3^3}{w_2 + w_3} - w_1^2 + w_2^2 - w_3^2\right]\right\}\ .$$

We can now turn to the evaluation of the integral eq.(1), which is the main purpose of this paper. We have obviously

$$Z(w_1, w_2, w_3) = -\frac{\partial}{\partial w_1}\frac{\partial}{\partial w_2}\frac{\partial}{\partial w_3}\frac{\partial^2}{\partial u_3^2}\frac{\partial^2}{\partial u_1^2}A(w_1, w_2, w_3, u_1, u_2, u_3)\Big|_{u_i=0}\ . \tag{41}$$

As eq.s(39),(40) are exact in all the w_i's, for obtaining the value of the desired integral in closed analytic form it is sufficient to differentiate eq.(40) three times with respect to w_1, w_2 and w_3, and then change the overall sign. The derivatives of $D(w_1, w_2, w_3)$ are already known, see eq.(30), so that the actual differentiation of eq.(40) is completely straightforward, especially when an algebraic program is used; as matter of fact it was much easier to perform the derivatives than properly retype the obtained terms to exhibit the obvious properties of symmetry for the exchange of w_1 with w_3 and of regularity at

$w_3 = w_1 + w_2$ etc. The result can be written as

$$\int d\vec{r}_1 d\vec{r}_2 d\vec{r}_3 \ e^{-w_1 r_1} \ e^{-w_2 r_2} \ e^{-w_3 r_3} \ |\vec{r}_1 - \vec{r}_2| \frac{1}{|\vec{r}_3 - \vec{r}_1|} |\vec{r}_2 - \vec{r}_3| = \frac{64\pi^3}{w_1^4 w_2^6 w_3^4} \cdot$$

$$\left\{ \frac{3}{2} \left[3w_2^4 + 6(w_1^2 + w_3^2)w_2^2 + 5(3w_1^4 + 2w_1^2 w_3^2 + 3w_3^4) \right] D(w_1, w_2, w_3) \right.$$

$$+ 8w_1^2 w_2^3 w_3^2 \left[\frac{\ln_2(\frac{w_1 + w_2}{w_3})}{(w_1 + w_2 - w_3)^3} - \frac{\ln_2(\frac{w_2 + w_3}{w_1})}{(w_1 - w_2 - w_3)^3} \right.$$

$$\left. - \frac{\ln_2(\frac{w_1 + w_3}{w_2})}{(w_1 - w_2 + w_3)^3} - \frac{\mathrm{Sln}(w_1, w_2, w_3)}{(w_1 + w_2 + w_3)^3} \right]$$

$$+ 12w_1 w_2^2 w_3(w_1^2 + w_3^2) \left[\frac{\ln_1(\frac{w_1 + w_2}{w_3})}{(w_1 + w_2 - w_3)^2} + \frac{\ln_1(\frac{w_2 + w_3}{w_1})}{(w_1 - w_2 - w_3)^2} \right.$$

$$\left. - \frac{\ln_1(\frac{w_1 + w_3}{w_2})}{(w_1 - w_2 + w_3)^2} + \frac{\mathrm{Sln}(w_1, w_2, w_3)}{(w_1 + w_2 + w_3)^2} \right]$$

$$+ 12w_2 \left[(w_1^2 + w_3^2)(w_1 + w_3)^2 - 3w_1^2 w_3^2 \right] \left[\frac{\ln(\frac{w_1 + w_2}{w_3})}{w_1 + w_2 - w_3} - \frac{\ln(\frac{w_2 + w_3}{w_1})}{w_1 - w_2 - w_3} \right]$$

$$- 12w_2 \left[(w_1^2 + w_3^2)(w_1 - w_3)^2 - 3w_1^2 w_3^2 \right] \left[\frac{\ln(\frac{w_1 + w_3}{w_2})}{w_1 - w_2 + w_3} + \frac{\mathrm{Sln}(w_1, w_2, w_3)}{w_1 + w_2 + w_3} \right] \quad (42)$$

$$+ 3w_3 \left[3(w_1 - w_2)w_2^2 - (9w_1^2 + 7w_3^2)w_2 - 15(w_1^2 + w_3^2)w_1 \right] \ln\left(\frac{w_1 + w_2}{w_3} \right)$$

$$+ 3w_1 \left[3(w_3 - w_2)w_2^2 - (7w_1^2 + 9w_3^2)w_2 - 15(w_1^2 + w_3^2)w_3 \right] \ln\left(\frac{w_2 + w_3}{w_1} \right)$$

$$- 3w_2 \left[3(w_1 + w_3)w_2^2 + 7w_1^3 + 9w_1^2 w_3 + 9w_1 w_3^2 + 7w_3^3 \right] \ln\left(\frac{w_3 + w_1}{w_2} \right)$$

$$- 2(2w_1^2 - 9w_1 w_3 + 2w_3^2)w_2^2 + 80w_1 w_3(w_1^2 + w_3^2)$$

$$+ 2(8w_1^3 + 13w_1^2 w_3 + 13w_1 w_3^2 + 8w_3^3)w_2$$

$$+ 8\frac{w_1 w_2^2 w_3}{(w_1 + w_2 + w_3)^2}(w_1^2 + w_1 w_3 + w_3^2)$$

$$- 4\frac{w_2}{w_1 + w_2 + w_3}(4w_1^4 + 3w_1^3 w_3 + 5w_1^2 w_3^2 + 3w_1 w_3^3 + 4w_3^4)$$

$$+ 48\frac{w_1^3 w_2^3 w_3^3}{(w_1 + w_3)^5} + 24\frac{w_1^2 w_2 w_3^2}{(w_1 + w_3)^3}(w_2^2 + 2w_1 w_3) + 8\frac{w_1 w_2 w_3}{w_1 + w_3}(3w_2^2 - 2w_1 w_3)$$

$$- 8w_1 w_3 \left[\frac{w_1^5}{(w_1 + w_2)^3} + \frac{w_3^5}{(w_2 + w_3)^3} \right] + 48w_1 w_3 \left[\frac{w_1^4}{(w_1 + w_2)^2} + \frac{w_3^4}{(w_2 + w_3)^2} \right]$$

$$\left. - 120w_1 w_3 \left[\frac{w_1^3}{w_1 + w_2} + \frac{w_3^3}{w_2 + w_3} \right] \right\} .$$

For ease of typing, we have introduced in eq.(40) the quantities

$$\text{Sln}(x, y, z) \equiv \ln\left(\frac{x+y}{z}\right) + \ln\left(\frac{x+z}{y}\right) + \ln\left(\frac{y+z}{x}\right),$$

$$\ln_1(x) \equiv \ln(x) - (x-1),$$

$$\ln_2(x) \equiv \ln(x) - (x-1) + \frac{1}{2}(x-1)^2.$$

Eq.(40) looks, and perhaps is, somewhat cumbersome, but in fact its structure is remarkably simple. It involves only the dilogarithmic combination $D(w_1, w_2, w_3)$ defined in eq.(33), the three logarithms $\ln\left(\frac{w_1+w_2}{w_3}\right)$, $\ln\left(\frac{w_1+w_3}{w_2}\right)$, $\ln\left(\frac{w_2+w_3}{w_1}\right)$, and simple rational functions of the w_i's; further, it is ready for the actual numerical evaluation, being in particular regular (as expected, of course) at, say, $w_1 + w_2 = w_3$, because the numerators of all the terms with $1/(w_1 + w_2 - w_3)^n$ vanish as $(w_1 + w_2 - w_3)^n$ (the same is true for the other two denominators without definite sign).

It is clear from the derivation why eq.(42) has a size larger then eq.s(40),(38) and the really compact eq.(34); it is also clear that similar formulae can be obtained, when needed, for all the related three electron correlation integrals with higher positive powers of the r_i and $|\vec{r}_i - \vec{r}_j|$ in the numerator, so that the algorithm presented in this paper is by no means restricted to $s-$wave functions only.

4. The use of eq.(42) as well as of the related formulae for the three electron correlation integrals with higher positive powers of the r_i's and $|\vec{r}_i - \vec{r}_j|$ in the numerator is expected to speed up considerably the compuational time required for the proper accounting of two electron correlation effects in atoms, so making high precision calculations easier. It is being investigated whether and how the techniques introduced here can be of use for the analytic evaluation of atomic integrals with many electron correlation effects or of two electron correlation effects in *ab initio* calculations of simple molecules.

Acknowledgements. The author wants to thank G. Bendazzoli and F. Bernardi for bringing his interest on this problem, S. Turrini who kindly provided with direct numerical checks of the key formulae and M. Veltman for the assistance in the use of the algebra manipulating program SCHOONSCHIP.

Note added in Proof.

After the completion of this work, the author discovered the existence of Ref.[4], where the Analytic evaluation of three-electron integrals was also worked out; the result of Ref.[4] is in fact more general, as it provides with a formula valid for any value of the parameters u_i, not just at the point $u_i = 0$ (notation of this paper). In the common region of applicability, the results are in perfect numerical agreement. In the terminology of Ref.[4], at the Auxiliary Reference Point (ARP) $w_1 = w_2 = w_3 = 1$, $u_1 = u_2 = u_3 = 0$, the numerical values of eq.s(34),(38),(40) are $4.382\ 174\ 441\ 144 \times 10^2$, $1.204\ 780\ 633\ 933 \times 10^3$ and $8.504\ 405\ 304\ 091 \times 10^3$, coinciding with the entries (000000, 000200, 000220) of Table III of Ref.[4], while at the same ARP for eq.(42) we find $9.155\ 447\ 160\ 887 \times 10^4 = 2^{10}\pi^3 \times 2.883\ 566\ 595\ 319$, to be compared with 2.883 566 595, as quoted in Ref.[5].

The approach of Ref.[4] and of the present paper is the same, *i.e.* Fourier transform for the auxiliary integral eq.(2), called generating function in Ref.[4]; Ref.[4] however differs strongly in the technique followed for performing the quadruple definite integral corresponding to eq.(14); rather than using the "differentiate and integrate" algorithm of this paper, the first integration corresponding to eq.(9) is performed by "brute force" by means of eq.(4.34) of Ref.[4]. The result is (see the remarks after eq.(9) of this paper) a combination of dilogarithmic functions of complicated arguments, whose subsequent integration on the three momenta is then ingeniously carried out in Ref.[4] by contour integration. As a consequence of the original "brute force" integration, the key result, eq.(2.1) of Ref.[4] and following formulae, is obtained in a form somewhat discouraging the reader, as it contains many dilogarithms of complicated and complex (*i.e.* not real) arguments, exhibiting a variety of spurious singularities which cancel out in the final result, and whose actual numerical evaluation requires among the other subtleties a careful preliminary branch tracking. That contrasts with the plainness of eq.s(33) and following of this paper, where the results are expressed as a combination of real functions of real variables, whose numerical evaluation is immediate, in the whole range of variability of the arguments w_i.

Without underestimating the complications inherent in the much more general $u_i \neq 0$ case, it is likely that also eq.(2.1) of Ref.[4], after a major rewriting effort, can be recast in the form of an expression, much simpler and of much greater practical use, in which the compensating singularities do not appear at all. Once expressed in that way, the result could receive the acknowledgement that it deserves and produce the expected impact in high precision correlation calculations.

Appendix A. The Euler dilogarithm $\text{Li}_2(y)$, sometimes also called Spence function, appears naturally when integrating a logarithm multiplied by a rational expression, in the same way as logarithms appear when integrating rational expressions. It is defined by

$$\text{Li}_2(y) \equiv -\int_0^1 \frac{dt}{t} \ln(1 - yt) = -\int_0^y \frac{dt}{t} \ln(1 - t) \ . \qquad (A.1)$$

It is real for real $y \leq 1$ and develops an immaginary part for real $y > 1$. From Eq.(A.1) one has the power series expansion

$$\text{Li}_2(y) = \sum_{m=1}^{\infty} \frac{y^m}{m^2} \ , \qquad (A.2)$$

which converges for $y \leq 1$, $\text{Li}_2(0) = 0$ and the values at $y = 1$

$$\text{Li}_2(1) = -\int_0^1 \frac{dt}{t} \ln(1 - t) = \zeta(2) \ , \qquad (A.3)$$

where $\zeta(p) \equiv \sum_{m=1}^{\infty} \frac{1}{m^p}$ is the Riemann ζ-function of argument p.

From the very definition, one has also

$$\frac{d}{dy}\text{Li}_2(y) = -\frac{1}{y}\ln(1 - y) \ . \qquad (A.4)$$

By elementary use of the above equations one can easily obtain a number of relations between dilogarithms of related arguments. One has for instance

$$\frac{1}{2}\text{Li}_2(y^2) = \text{Li}_2(y) + \text{Li}_2(-y) \ ; \qquad (A.5)$$

the relation holds at $y = 0$, while the derivatives of the l.h.s and of the r.h.s. are equal for any y; therefore eq.(A.5) is true for any value of y. From it at $y = -1$ one obtains

$$\int_0^1 \frac{dt}{t} \ln(1 + t) = -\text{Li}_2(-1) = \frac{1}{2}\zeta(2). \qquad (A.6)$$

By the same technique one can also obtain identities between dilogarithms whose arguments are related by the transformations $y \to 1/y$, $y \to (1 - y)$ and combinations thereof; all the identities can best be established by checking their validity at some convenient particular value of y and then differentiating with respect to y, so obtaining an identity between logarithms, whose validity is trivial to be ascertained. One finds, for real $y > 0$

$$\text{Li}_2(y) = -\text{Li}_2(1 - y) - \ln y \ln(1 - y) + \zeta(2) \qquad (A.7)$$

$$\text{Li}_2(-y) = -\text{Li}_2\left(-\frac{1}{y}\right) - \frac{1}{2}\ln^2 y - \zeta(2) ; \qquad (A.8)$$

by analytic continuation of eq.(A.8) to $y = -1$, as $\ln^2(-1) = -\pi^2$ one derives the known relation $\zeta(2) = \pi^2/6$.

With the above formulae, it is quite easy to evaluate numerically $\text{Li}_2(y)$; for small y one can use directly the expansion (A.2); more systematically, one can use the proper combination of eq.s (A.7),(A.8), to express the required value of $\text{Li}_2(y)$ in terms of a dilogarithm whose argument x is in the range $-1 < x < \frac{1}{2}$; in that interval the dilogarithm is analytic and can can be evaluated by a quickly convergent power expansion (see Ref.[3] for an implementation of the method).

Appendix B. We sketch here a proof of eq.(21) of the text. When differentiating $K(a)$ eq.(20) with respect to a, one obtains in the r.h.s. three terms, corresponding to
i) the derivative of the function $H(a, x)$;
ii) the derivative of $\sqrt{S(a, b)}$ in the numerator;
iii) the derivative of $\sqrt{S(a, x)}$ in the denominator.
For the third term, with the definition of $S(a, x)$ eq.(19) one has obviously

$$\frac{\partial}{\partial a}\frac{1}{\sqrt{S(a, x)}} = -\frac{1}{2}\frac{1}{\sqrt{S(a, x)}}\frac{1}{S(a, x)}\left[\frac{\partial s_0}{\partial a} + \frac{\partial s_1}{\partial a}(x - b) + \frac{1}{2}\frac{\partial s_0}{\partial a}(x - b)^2\right] .$$

Only in the term with $\partial s_0/\partial a$ the denominator $1/(x - b)$ appearing in eq.(20) is still present; to process it one can use the algebraic identity

$$\frac{1}{S(a, x)}\frac{1}{x - b} = \frac{1}{S(a, b)}\left[\frac{1}{x - b} - \frac{s_1 + \frac{1}{2}s_2(x - b)}{S(a, x)}\right],$$

and the first term in the r.h.s. is found to cancel out exactly with the contribution ii) above. One can then integrate by parts the remaining terms $\dfrac{dx}{\left(\sqrt{S(a, x)}\right)^3}, \dfrac{dx(x - b)}{\left(\sqrt{S(a, x)}\right)^3},$
and then collect results, till eq.(21) is obtained.

By a similar approach one can deal with the integrals involving the factors of eq.(25) in the text. As the analog of eq.(21) one finds for instance, for any function $F(p, q, w)$,

$$\frac{\partial}{\partial q}\int_{p_1}^{p_2} dp^2 \frac{wq}{R_2(p^2, q^2, -w^2)} F(p, q, w) = \int_{p_1}^{p_2} dp^2 \frac{wq}{R_2(p^2, q^2, -w^2)}\frac{\partial F(p, q, w)}{\partial q}$$

$$- \int_{p_1}^{p_2} dp^2 \frac{wq}{R_2(p^2, q^2, -w^2)}\frac{p^2 + q^2 + w^2}{2q^2}\left[\delta(p - p_2) - \delta(p - p_1) - \frac{\partial}{\partial p}\right] F(p, q, w) , \qquad (B.1)$$

$$\frac{\partial}{\partial w}\int_{p_1}^{p_2} dp^2 \frac{wq}{R_2(p^2, q^2, -w^2)} F(p, q, w) = \int_{p_1}^{p_2} dp^2 \frac{wq}{R_2(p^2, q^2, -w^2)}\frac{\partial F(p, q, w)}{\partial w}$$

$$- \int_{p_1}^{p_2} dp^2 \frac{wq}{R_2(p^2, q^2, -w^2)}\frac{p^2 - q^2 - w^2}{2w^2}\left[\delta(p - p_2) - \delta(p - p_1) - \frac{\partial}{\partial p}\right] F(p, q, w) . \qquad (B.2)$$

Similar formulae hold for all the other terms listed in eq.(25).

References.
[1] M.J. Levine, E. Remiddi and R. Roskies, in *Quantum Electrodynamics*, p.162-321, T. Kinoshita Ed., *Advanced series "Directions in High Energy Physics"*, Vol. **7**, Pub. World Scientific, Singapore 1990, and references therein to previous works .

[2] H. Strubbe, *Compt. Phys. Commun.* **8**,1 (1974); **18**, 1 (1979);
M. Veltman, *SCHOONSCHIP, A Program For Symbol Handling*, unpublished.

[3] K.S.Kölbig, J.A.Mignaco and E.Remiddi, *BIT* **10**, 38 (1970).

[4] D.M.Fromm and R.N.Hill, *Phys.Rev.* **A36**, 1013 (1987).

[5] W.Kutzelnigg and W.Klopper, *J.Chem.Phys.* **94**, 1985 (1991).

Example of a quantum field theory based on a nonlinear Lie algebra

K. Schoutens[1]

Institute for Theoretical Physics
State University of New York at Stony Brook
Stony Brook, NY 11794-3840

A. Sevrin[2]

Department of Physics, University of California
and
Theoretical Physics Group, Lawrence Berkeley Laboratory
Berkeley, California 94720

and

P. van Nieuwenhuizen[1,3]

Theory Division, CERN
CH-1211 Geneva 23
Switzerland

---

[1]Work supported in part by grant NSF-91-08054
[2]This work was supported in part by the Director, Office of Energy Research, Office of High Energy and Nuclear Physics, Division of High Energy Physics of the U.S. Department of Energy under Contract DE-AC03-76SF00098 and in part by the National Science Foundation under grant PHY90-21139.
[3]On leave from ITP, SUNY at Stony Brook

1 Introduction.

In this contribution to Tini Veltman's Festschrift we shall give a paedagogical account of our work [1]-[6] on a new class of gauge theories called W gravities. They contain higher spin gauge fields, but the usual no-go theorems for interacting field theories with spins exceeding two do not apply since these theories are in two dimensions. It is, of course, well known that ghost-free interacting massless spin 2 fields ('the metric') are gauge fields, and correspond to the geometrical notion of general coordinate transformations in general relativity, but it is yet unknown what extension of these ideas is introduced by the presence of massless higher spin gauge fields. A parallel with supergravity may be drawn: there the presence of massless spin 3/2 fields (gravitinos) corresponds to local fermi-bose symmetries of which these gravitinos are the gauge fields. Their geometrical meaning becomes only clear if one introduces superspace (with bosonic and fermionic coordinates): they correspond to local transformations of the fermionic coordinates. For W gravity one might speculate on a kind of W-superspace with extra bosonic coordinates.[4]

A reason for being interested in W gravities is that they are, like ordinary $d = 2$ gravity, 'integrable'. The integrability of these theories makes it possible to find exact answers, to all order in the number of loops, for various quantities in the theory such as the effective action. With such exact, all-order-in-perturbation-theory results one might start the study of nonperturbative properties. There one expects a connection with the so-called matrix models but at this moment no precise relations are known.

In this contribution we focus on (classical and quantum) W_3 gravity, which is the simplest version of W gravity and contains, besides the spin 2 gravity fields, gauge fields of spin 3. The main part of our review will be to discuss how the effective action

---

[4]Incidentally, whether he likes it or not, Tini Veltman contributed indirectly to the discovery of supergravity. In 1974 he and Gerard 't Hooft had computed the 1-loop divergences of pure gravity (which was finite), and gravity coupled to scalars, (which was not finite) using their new covariant quantization methods. Their work was extended by various people, mostly Deser and one of us. We calculated the one-loop divergences of all kinds of different matter systems. None was finite. In this connection, Tini suggested in the fall of 1975 to consider the coupling of spin 3/2 fields to gravity. This theory became supergravity. It was first constructed as the classical gauge theory of supersymmetry, but later it was indeed found that at the quantum level it was 1-loop finite. Now that recent results ([7]) confirm earlier work ([8]) that gravity is not 2-loop finite, many people believe that supergravity is not 3-loop finite, and that one should go to string theory for a consistent quantum theory of gravity. However, a proof that supergravity is not 3-loop finite is lacking.

of this theory can be obtained from 'the $c \to \infty$ limit of the induced action' (which we explicitly found) by a renormalization of the spin 2 and spin 3 gauge fields and the $Sl(3, \mathbf{R})$ level k (with Z factors which are infinite series in c^{-1}).

Pure gravity in $d = 2$ is based on a linear algebra, the Virasoro algebra, which is given by

$$[L_m, L_n] = (m - n)L_{m+n}, \quad m, n \in \mathbf{Z}. \tag{1.1}$$

In contrast, W_3 gravity is based on a nonlinear algebra. The precise form will be given in the next section, but let us mention that its classical (Poisson bracket) version is of the form

$$[T_A, T_B] = f_{AB}{}^C T_C + V_{AB}{}^{CD} T_C T_D. \tag{1.2}$$

The work on W_3 gravity can therefore be considered as work on a prototype model for nonlinear gauge theories. Further models based on nonlinear algebras might be searched for and studied.

In [9], we have extended our work on non-linear gauge theories to four dimensions, where we have found that the kind of gauge theories we consider necessarily contain one real scalar field for each gauge field, and we have studied whether these scalars could act as Higgs scalars. [Scalar fields arise in general in gauge field theories based on nonlinear algebras. For example, consider the algebra given in (1.2). It yields, upon 'gauging', gauge fields $h_\mu{}^A$ corresponding to the generators T_A. However, the presence of the TT term in the abstract algebra leads, in addition, to scalar fields t_A in the coadjoint representation of the gauge group. As such these scalar fields are an integral part of the gauge multiplet. In W_3 gravity, the scalar fields are $t_{++}(x^+, x^-)$ and $w_{+++}(x^+, x^-)$ and look like currents. We have found [5] that they play the role of auxiliary fields needed to close the gauge algebra.] We intend to follow up on that work in the future, and shall not discuss it here as we already need a lot of clemency from the side of the 'jubilaris', who, of course, has probed the Higgs problem deeper than anybody else over the years.[5]

---

[5]In fact, as mentioned by Cabibbo, it is remarkable that one physicist (i) made such an important theoretical discovery as quantization of nonabelian gauge theories, (ii) then applied this theoretical framework to realistic models, doing much explicit numerical work useful for experiments, and (iii) developed his own software programs ('Schoonschip') and P.C.'s. With such a set of accomplishments

Let us now stop enumerating the reasons why it is interesting to study W gravities, as the relevance of a subject is inversely proportional to the number of such reasons. Instead, we start our discussion of W_3 gravity. We shall first, in sections 2 and 3, discuss the W_3 algebra and various classical theories of W_3 gravity. In subsequent sections we shall then develop in some detail the quantum theory of W_3 gravity in the chiral gauge.

2 The W_3 algebra, subalgebras and classical limits.

The quantum W_3 algebra [10] contains the Virasoro algebra and further spin 3 generators W_m,

$$
\begin{aligned}
{[L_m, L_n]} &= \frac{c}{12} m(m^2 - 1)\delta_{m+n,0} + (m-n)L_{m+n} \\
{[L_m, W_n]} &= (2m-n)W_{m+n},
\end{aligned}
\tag{2.1}
$$

which satisfy a quadratically nonlinear quantum algebra

$$
\begin{aligned}
{[W_m, W_n]} =\ & \frac{c}{360} m(m^2 - 1)(m^2 - 4)\delta_{m+n,0} \\
& + (m-n)\left\{ \frac{1}{15}(m+n+3)(m+n+2) - \frac{1}{6}(m+2)(n+2) \right\} L_{m+n} \\
& + \beta(m-n)\Lambda_{m+n}.
\end{aligned}
\tag{2.2}
$$

In order that the Jacobi identities be satisfied, the central charge c in the $[W, W]$ commutators must be the same as in the Virasoro algebra, and further

$$
\beta = \frac{16}{22 + 5c}.
\tag{2.3}
$$

The objects Λ_m are nonlinear in L_m

$$
\Lambda_m = \sum_n : L_{m-n}L_n : -\frac{3}{10}(m+3)(m+2)L_m,
\tag{2.4}
$$

---

he is entitled to his criticism of modern speculative theoretical developments. However, in the absence of obvious, fundamental and solvable problems in the standard model, it seems to us that quantum gravity stands out as the challenge of the next century.

where the normal ordered product[6] is defined as

$$: L_p L_q := L_p L_q \qquad \text{if} \quad p \leq -2$$
$$: L_p L_q := L_q L_p \qquad \text{if} \quad p > -2 . \tag{2.5}$$

They satisfy

$$[L_m, \beta \Lambda_n] = (3m - n)\beta \Lambda_{m+n} + \frac{8}{15}(m^3 - m)L_{m+n} \tag{2.6}$$

and, were it not for the last term, they would be primary fields of dimension 4.

In terms of operator fields

$$L(z) = \sum L_m z^{-m-2}, \quad W(z) = \sum W_m z^{-m-3}$$
$$\Lambda(z) = (TT)(z) - \frac{3}{10}T''(z) = \sum \Lambda_m z^{-m-4}, \tag{2.7}$$

one can write down operator product expansions (OPE's), which are equivalent to (2.1), (2.2)

$$L(z)L(w) = \frac{c/2}{(z-w)^4} + \frac{2T(w)}{(z-w)^2} + \frac{T'(w)}{z-w} + \cdots$$

$$L(z)W(w) = \frac{3W(w)}{(z-w)^2} + \frac{W'(w)}{z-w} + \cdots$$

$$W(z)W(w) = \frac{c/3}{(z-w)^6} + \frac{2T(w)}{(z-w)^4} + \frac{T'(w)}{(z-w)^3}$$
$$+ \frac{1}{(z-w)^2}\left[2\beta\Lambda(w) + \frac{3}{10}T''(w)\right]$$
$$+ \frac{1}{z-w}\left[\beta\Lambda'(w) + \frac{1}{15}T'''(w)\right] + \cdots \tag{2.8}$$

We now observe the following.

(i) The subset of generators $w = \{L_{\pm 1}, L_0, W_{\pm 2}, W_{\pm 1}, W_0\}$ does not form a subalgebra if $\beta \neq 0$.

---

[6]Different authors use different conventions for normal ordering, and consequently obtain slightly different forms of the algebra. Our normal ordering conventions follow from (6.10) and the mode expansions as in (2.7).

(ii) The half-infinite sets $\{L_m, m \geq -1\}$ and $\{W_n, n \geq -2\}$ do not form a subalgebra either.

(iii) In the limit $c \to \pm\infty$, the subset w in (i) yields a closed, linear algebra, which is $SU(2,1)$.

(iv) In the non-chiral case, with left-handed and right-handed generators L_n, W_n and $\overline{L}_n, \overline{W}_n$, and in the limit $c \to \pm\infty$, the set of generators $\left\{ L_0 - \overline{L}_0, L_{-1}, \overline{L}_{-1}, W_0 - \overline{W}_0, W_{-2}, \overline{W}_{-2}, W_{-1}, \overline{W}_{-1} \right\}$ forms a subalgebra which one might call the W_3 Poincaré algebra.

(v) The *classical* (*i.e.*, with Poisson brackets) version of the algebra without central charge reads

$$
\begin{aligned}
[L_m, L_n] &= (m-n)L_{m+n} \\
[L_m, W_n] &= (2m-n)W_{m+n} \\
[W_m, W_n] &= (m-n)\left(\sum_k L_{m+n-k}L_k \right).
\end{aligned}
\tag{2.9}
$$

It is this algebra which is reproduced if one evaluates the commutator algebra of local symmetries under which the classical action of chiral W_3 gravity coupled to scalar matter fields is invariant [11]. The (LL) term is then related to the appearance field-dependent structure functions in the commutator algebra. A classical W_3 algebra *with* central extension exists and was discussed in [12].

(vi) One can also take linear combintions of W_m and L_m and rescale such that if c is a function of $\hbar$, all commutators vanish for $\hbar \to 0$ [13].

While the algebra (v) is related to the purely classical formulation, it has been found that the algebras (iii) and (iv) are related to (quantum) induced W_3 gravity [5] and topological W_3 gravity [14].

3   The classical formulations.

At the classical level, one can consider covariant formulations of W gravity theories, or consider these theories in special gauges such as the light-cone gauge or the chiral gauge (which is the light-cone gauge with only one chirality present).

In our approach to covariant W_3 gravity [2, 3], the gauge multiplet contains four vielbein fields $e_\mu{}^+$, $e_\mu{}^-$ and four W-vielbein fields $b_\mu{}^{++}$, $b_\mu{}^{--}$. The local gauge invariances are: general coordinate, Weyl and local Lorentz symmetries, together with their W analogues. By fixing some of the algebraic symmetries, one obtains the light-cone theory, and by fixing additional symmetries one obtains the chiral theory.

In the chiral theory there are only two local symmetries left: ϵ symmetry and λ symmetry, with parameters ϵ_+ and λ_{++}. They arise as particular linear combinations of all local symmetries present in the covariant formulation, which are chosen such that the chiral gauge is preserved.

Let us now discuss the coupling of the W_3 gauge fields to scalar matter fields ϕ^i, $i = 1, 2, \ldots, N$. We should first explain the observation made by Hull [11], that a free action for N scalar fields admits a rigid W_3 symmetry. The chiral gauge transformations of the scalar fields read

$$\delta_\epsilon \phi^i = \epsilon_+ \, \partial_- \phi^i \,, \quad \delta_\lambda \phi^i = \lambda_{++} \, d^{ijk} \, \partial_- \phi^j \partial_- \phi^k \,, \tag{3.1}$$

where d^{ijk} is a symmetric 3-index tensor. [We will denote the complex coordinates of $d = 2$ euclidean space-time by z, $\bar{z}$ and write ∂_- and ∂_+ for ∂_z and $\partial_{\bar{z}}$, respectively.] One can now promote these symmetries to local gauge invariances by introducing gauge fields h_{++} and b_{+++} in the standard way. It turns out [11] that the scalar field action with only the minimal coupling to these gauge fields,

$$S_{\text{ch}} = \frac{1}{\pi} \int d^2 z \, \left[-\frac{1}{2}\partial_+\phi^i\partial_-\phi^i + \frac{1}{2}h_{++}\partial_+\phi^i\partial_+\phi^i + \frac{1}{3}b_{+++}d^{ijk}\partial_+\phi^i\partial_+\phi^j\partial_+\phi^k \right] \,, \tag{3.2}$$

is gauge-invariant, provided we choose the transformation rules of h_{++} and b_{+++} appropriately and we have the identity

$$d^{k(ij} \, d^{l)mk} = \delta^{(ij} \, \delta^{l)m} \,. \tag{3.3}$$

In a next step, we consider the light-cone formulation, with both chiralities present. We introduce gauge fields $h_{\pm\pm}$ and $b_{\pm\pm\pm}$ corresponding to the local symmtries $\epsilon_\pm$ and $\lambda_{\pm\pm}$. In [1] we found that the light-cone gauge action is non-polynomial in the spin-3 gauge fields. This difficulty can be circumvented by introducing auxiliary fields $F_+{}^i$ and $F_-{}^i$, which will later play the role of so-called *nested covariant derivatives*. With these variables, the gauge-invariant action takes the following form [1]

$$S_{lc} = \frac{-1}{\pi} \int d^2 z\, e \left[-\frac{1}{2} \nabla_+ \phi^i \nabla_- \phi^i - F_+{}^i F_-{}^i + F_+{}^i \left(\nabla_- \phi^i - \frac{1}{3} b_{---} d^{ijk} F_+{}^j F_+{}^k \right) \right.$$
$$\left. + F_-{}^i \left(\nabla_+ \phi^i - \frac{1}{3} b_{---} d^{ijk} F_-{}^j F_-{}^k \right) \right], \tag{3.4}$$

where $e = (1 - h_{++} h_{--})^{-1}$ and $\nabla_\pm = \partial_\pm - h_{\pm\pm} \partial_\mp$.

The field equation of $F_-{}^i$ is algebraic and leads to

$$F_+{}^i = \nabla_+ \phi^i - b_{+++} d^{ijk} F_-{}^j F_-{}^k \tag{3.5}$$

(with a similar result for $F_-{}^i$). When solving $F_\pm^i$ by iteration one obtains a generalization of a covariant derivative, which is infinitely nonlinear and is appropriately called a nested covariant derivative.

A variant of this light-cone formulation was later obtained in [15] where the spin-2 and spin-3 fields are treated symmetrically, leading to the action

$$S'_{lc} = \frac{-1}{\pi} \int d^2 z \left[-\frac{1}{2} \partial_+ \phi^i \partial_- \phi^i - F_+{}^i F_-{}^i \right.$$
$$\left. + \left\{ F_+{}^i \left(\partial_- \phi^i - \frac{1}{2} h_{--} F_+{}^i - \frac{1}{3} b_{---} d^{ijk} F_+{}^j F_+{}^k \right) + (+ \leftrightarrow -) \right\} \right]. \tag{3.6}$$

Upon elimination of $F_\pm{}^i$, the results from (3.4) and (3.6) should be equivalent up to redefinitions; this has been checked to some orders in fields [15].

A covariant formulation of W_3 gravity[7] can now be obtained by completing the gravitational covariantizations (giving for example $\nabla_\pm \varphi^i = e_\pm{}^\mu \partial_\mu \varphi^i$) and by incorporating all four spin-3 fields $b_+{}^{\pm\pm} \equiv e_+{}^\mu b_\mu{}^{\pm\pm}$ and $b_-{}^{\pm\pm}$. The $F_-{}^i$-field equations should now contain an extra term with $b_+{}^{++}$ according to

$$F_+{}^i - \nabla_+ \phi^i + \left(b_+{}^{++} F_+{}^j F_+{}^k + b_+{}^{--} F_-{}^j F_-{}^k \right) d^{ijk} = 0, \tag{3.7}$$

---

[7] This covariant formulation was first derived in [2, 3], where we employed a gauge procedure based on the algebra (2.9). This procedure naturally explains the origin of all eight local symmetries and the (complicated) form of the transformation rules of the fields.

and similarly for $F_+{}^i$. However, these equations cannot directly be written as $\delta S/\delta F_\pm{}^i = 0$, since that would violate the integrability conditions for these field equations. Instead, one may obtain the covariant action by putting

$$\frac{\delta S}{\delta F_\pm{}^i} = A_{ij}{}^{(\pm)} \left(F_\mp{}^j - \dots \right) , \tag{3.8}$$

where the $A_{ij}{}^{(\pm)}$ are integrating factors and the dots are as in (3.7). One can then determine the $A_{ij}{}^{(\pm)}$ and, eventually, the action. The latter reads [2]

$$\begin{aligned}
S_{\text{cov}} = \frac{-1}{\pi} \int d^2 z \; e \Big[&-\frac{1}{2}\nabla_+\phi^i\,\nabla_-\phi^i - F_+{}^i F_-{}^i \\
&+ \left\{ F_+{}^i \left(\nabla_-\phi^i - \frac{1}{3}b_-{}^{++}F_+{}^j F_+{}^k d^{ijk} \right) + (+ \leftrightarrow -) \right\} \\
&+ \left\{ -b_+{}^{++} d^{ijk} F_+{}^i F_+{}^j \left(F_-{}^k - \nabla_-\phi^k + \frac{1}{2}b_-{}^{\pm\pm} d^{klm} F_\pm{}^\ell F_\pm{}^m \right) + (+ \leftrightarrow -) \right\} \Big]
\end{aligned} \tag{3.9}$$

(in the last term we sum over the two combinations of indices). Some classical field equations of particular interest are

$$T_{++}(F) \equiv -\frac{1}{2}F_+{}^i F_+{}^i = 0, \qquad W_{+++}(F) \equiv -\frac{1}{3}d^{ijk} F_+{}^i F_+{}^j F_+{}^k = 0$$

$$T_{--}(F) = 0, \qquad W_{---}(F) = 0. \tag{3.10}$$

Let us remark that it is not yet clear that this particular 'covariant' formulation of W_3 gravity will turn out to be the most convenient for studying, for example, global characteristics of W_3 gravity. One also might consider a 'metric formulation', in terms of rank-2 and 3 completely symmetric tensors. The distinction, which is largely irrelevant at the classical level, might have some important consequences at the quantum level.

Apart from W_3 gravity, one can consider gravity theories associated to more general W algebras such as W_4, W_5, etc. and also w_∞ and W_∞ [15, 16, 17]. The latter theories contain spins ranging up to infinity. One may therefore consider the spin label as a Fourier index, in which case the w_∞ (W_∞) theories might correspond to some, as yet unknown, gauge theory in two complex (or four real) dimensions, perhaps describing self-dual gravitational instantons. The w_∞ algebra reads

$$\left[t_m{}^j, t_n{}^k \right] = [(j-1)n - (k-1)m] \, t_{m+n}^{j+k-2} \tag{3.11}$$

with $j, k \geq 2$ and $m \geq -j + 1, n \geq -k + 1$, and is clearly a linear algebra. Various aspects of W_∞ are discussed in [18].

Let us finally mention that W_3 supergravities exist, both for the chiral theory [11] and for the light-cone theory [19]. A covariant W supergravity has not yet been found.

4 Quantum aspects.

In the analysis of quantum W_3 gravity one can distinguish two approaches: the *critical* approach and the *noncritical* approach.

In the *critical* approach, one chooses a matter system and tries to construct an anomaly-free coupling to the W_3 gravity degrees of freedom. In particular, one can choose a matter system consisting of scalar fields, and study the theory in a Lagrange formulation. Upon quantization, one would then add the usual Faddeev-Popov ghosts to the action, and try to cancel all gauge anomalies by adding suitable counter terms to the quantum action. These extra terms will necessarily include so-called background couplings, which contain second or higher derivatives of the scalar fields ϕ^i. This can be seen as follows.

If one evaluates the 1 and 2 loop anomalies in the presence of ghosts [20, 21, 22, 23], one finds that all individual anomalies (including the so-called matter dependent anomalies proportional to $\partial_-\phi\partial_-^3\phi$) either vanish or can be cancelled by adding a certain matter independent counterterm to the quantum action. However, the simultaneous cancellation of all anomalies is only possible if, in addition, one adds certain matter dependent counterterms to the quantum action [24, 22, 25]. The easiest way to obtain the full, anomaly-free quantum action is by using the form of the quantum BRST charge for W_3, which has been known since 1987 [26]. Such a nilpotent BRST charge only exists if the matter system involved has exact W_3 symmetry at the quantum level. This then makes clear the origin of the background couplings, since it is known [27] that the realization of W_3 currents in terms of scalar fields necessarily (for $c \neq 2$) involves the introduction of background charges. The complete results for the anomaly-free coupling were first presented in [28].

In the *noncritical* approach (pioneered for ordinary gravity by Polyakov [29], KPZ [30] and Al.B. Zamolodchikov [31]) one does not try to cancel the anomalies, but, on the contrary, one keeps them in the theory, and uses the fact that they make the spin 2 and spin 3 gauge fields propagating at the quantum level. In the critical approach the effective action vanishes (it is proportional to $c - 100$, just like the effective action of gravity is proportional to $c - 26$ and in order that the anomalies cancel, c must be 100 in W_3 gravity), but in the noncritical approach the effective action is nonvanishing.

In the noncritical approach, one can distinguish between the induced action (obtained by integrating only over the matter system), and the effective action (obtained by integrating also over the gauge fields). One does not add ghosts, nor gauge-fixing terms, for the anomalous gauge symmetries. So far, only the chiral approach has been studied. The induced action is defined by

$$e^{\Gamma_{\text{ind}}[h, b; c]} = \langle e^{-\frac{1}{\pi} \int d^2 z \, (hT + bW)} \rangle, \qquad (4.1)$$

where T and W are abstract currents, satisfying the operator product expansions (OPE) of the W_3 algebra in (2.8). The simplest example would be to use two scalar fields, since

$$T = -\frac{1}{2}\partial_- \phi^i \partial_- \phi^i, \quad W = \frac{-i}{6} d^{ijk} \partial_- \phi^i \partial_- \phi^j \partial_- \phi^k, \qquad (4.2)$$

with $d^{000} = 1$, $d^{011} = -1$, form an exact W_3 algebra with $c = 2$. (Note that the field b in (4.1) has been rescaled by a factor of $i/2$ as compared to the field b in (3.2).) For a general number n of scalar fields, W_3 currents with adjustable central charge, can be constructed by using background charge couplings [27].

The OPE's for the W_3 algebra were given in section 2. Using that $\langle T \rangle = \langle W \rangle = 0$, one can compute $\Gamma_{\text{ind}}[h, b; c]$ order-by-order in perturbation theory. We shall write down two exact Ward identities (anomalous current conservation laws) for $\Gamma_{\text{ind}}[h, b; c]$, which are nonlocal and nonlinear. The first nonlocalities appear at the 3-loop level; we will compute them explicitly. We will see that these terms mark the onset of a $1/c$ expansion: whereas up to three loops all terms in $\Gamma_{\text{ind}}[h, b; c]$ are proportional to c, one finds that from the 3-loop level on there are also terms proportional to c^0, c^{-1}, etc . In the limit $c \to \pm\infty$, one obtains a local (but still nonlinear) Ward identity for the induced action.

The effective action is obtained by taking the Legendre transform of the generating functional $W[t, w; c]$ for connected graphs

$$\Gamma_{\text{eff}}[h, b; c] = W[t, w; c] - \frac{1}{\pi} \int d^2z \, (ht + bw), \tag{4.3}$$

with

$$\pi \frac{\delta W}{\delta t} = h, \quad \pi \frac{\delta W}{\delta w} = b. \tag{4.4}$$

The latter is defined by

$$e^{W[t, w; c]} = \int dh \, db \, e^{\Gamma_{\text{ind}}[h, b; c] + \frac{1}{\pi} \int d^2z(ht + bw)}. \tag{4.5}$$

The main results obtained in [5, 6], and which we shall discuss below, can be summarized as follows.

(i) $\Gamma_{\text{ind}}[h, b; c \to \pm\infty]$ can be obtained in closed form by 'reduction' (*i.e.*, imposing constraints, which can actually be solved) of the induced action for an $Sl(3, \mathbf{R})$ gauge theory coupled to matter.

(ii) The Ward identity for $\Gamma_{\text{eff}}[h, b; c]$ becomes local, since the nonlocalities of $\Gamma_{\text{ind}}[h, b; c]$ (which start at level c^o) are canceled by nonlocalities coming from h and b loops.

(iii) $\Gamma_{\text{eff}}[h, b; c]$ is obtained by (finitely) renormalizing $\Gamma_{\text{ind}}[h, b; c \to \pm\infty]$,

$$\Gamma_{\text{eff}}[h, b; c] = Z_c \, \Gamma_{\text{ind}}[Z_h h, Z_b b, c \to \pm\infty]. \tag{4.6}$$

(iv) By choosing suitable variables other than h, b, the effective action *itself* (and not just the defining Ward identity) becomes local. [For pure gravity, this was already found to be the case by Polyakov [29], who introduced a new variable f, related to h_{++} by $h_{++} = \partial_+ f/(\partial_- f)$. So our new variables extend Polyakov's f variable to the case of W_3 gravity.]

We finally comment on the tentative interpretation of scalar field theories coupled to W_3 gravity as W_3 strings [32, 33, 34]. Depending on the formulation, these would be critical or non-critical W_3 strings, respectively. However, the critical W_3 strings have the unusual (as compared to bosonic or supersymmetric strings) feature of the background charges that we discussed above. An interesting possibility would be that the two scalar fields that acquire background charges can be interpreted as remnants of W_3 gauge field degrees of freedom, which could lead to an equivalence with the formulation as non-critical W_3 strings. (This point of view is clearly supported by the results of [33].) This possibility, and other aspects of W_3 strings, are presently under study.

In the remaining sections we will present in some detail our derivations leading to the results (i) to (iv). Some of these derivations have not appeared elsewhere.

5 The Ward identities for the induced action.

To obtain the Ward identities for the induced action of chiral W_3 gravity, which we shall from now on denote by $\Gamma_{\text{ind}}^{W_3}[h, b]$ instead of $\Gamma_{\text{ind}}[h, b; c]$,

$$e^{\Gamma_{\text{ind}}^{W_3}[h, b]} = \langle e^{S_{\text{int}}} \rangle, \qquad S_{\text{int}} = -\frac{1}{\pi} \int d^2 z \, (hT + bW), \qquad (5.1)$$

we begin by varying h under ϵ symmetry as $\delta h = \partial_+\epsilon$, and b under λ symmetry as $\delta b = \partial_+\lambda$. Then we cancel any β-independent variation proportional to T or W by adding suitable extra terms in δh and δb, respectively. The left-over is then the anomaly of the induced action under *these* δh and δb variation rules, and the Ward identity is obtained by removing the gauge parameter from the equation $\delta h \frac{\delta}{\delta h}\Gamma_{\text{ind}}^{W_3} + \delta b \frac{\delta}{\delta b}\Gamma_{\text{ind}}^{W_3} =$ anomaly.

We begin with the easier case, the ϵ symmetry. Under $\delta h = \partial_+\epsilon$, the induced action varies as follows

$$\left(\delta\Gamma_{\text{ind}}^{W_3}\right) \exp \Gamma_{\text{ind}}^{W_3} = \langle \left(\frac{-1}{\pi} \int \partial_+\epsilon \, T \, d^2 z\right) e^{S_{\text{int}}} \rangle. \qquad (5.2)$$

The vacuum expectation value of a product of $T(z)$ with other operators $A_1(z_1) \ldots A_n(z_n)$ can be written, according to BPZ [35], as multiple contractions.[8] For example

$$\langle T(z)A(z_1)..A(Z_2) \rangle = \langle \underline{T(z)A(z_1)}A(z_2) \rangle + \langle A(z_1)\underline{T(z)A(z_2)} \rangle. \qquad (5.3)$$

---

[8]On-shell the component $T(z, \bar{z}) = T_{--}(z, \bar{z})$ does not depend on $\bar{z}$. In the OPE's only the on-shell part contributes and we shall write $T(z)$.

[This property can be derived by considering the behaviour under $z \to z + \epsilon(z)$ of the correlator $\langle A_1(z_1) \ldots A_n(z_n) \rangle$.]

It follows that

$$\left(\delta\Gamma^{W_3}_{\mathrm{ind}} \right) \exp \Gamma^{W_3}_{\mathrm{ind}} = \langle \left(-\frac{1}{\pi} \int (\partial_+ \epsilon)\, T d^2 z_1 \right) \left(-\frac{1}{\pi} \int (hT + bW) d^2 z_2 \right) e^{S_{\mathrm{int}}} \rangle, \qquad (5.4)$$

where the hooks indicate that one should take the singular terms in the TT and TW OPE's. From the TT OPE we obtain

$$\langle \frac{1}{\pi^2} \int \partial_+ \epsilon(z_1) h(z_2) \left[\frac{c/2}{(z_1 - z_2)^4} + \frac{2 T(z_2)}{(z_1 - z_2)^2} + \frac{T'(z_2)}{(z_1 - z_2)} \right] d^2 z_1 d^2 z_2\, e^{S_{\mathrm{int}}} \rangle. \qquad (5.5)$$

We now use the following identity for distributions

$$\frac{1}{(z_1 - z_2)^n} = \pi \frac{(-1)^{n+1}}{(n-1)!} \frac{\partial_{z_1}{}^{n-1}}{\partial_{\bar{z}_1}} \delta^2(z_1 - z_2). \qquad (5.6)$$

[It can be proven by integrating z_2 over the plane minus a small disk around z_1. Multiplying both sides with $\partial_{\bar{z}_1}$, one obtains on the left-hand side the derivative of the θ function $\theta\left[(z_1 - z_2)(\bar{z}_1 - \bar{z}_2) - \epsilon^2\right]$, which is proportional to a delta function.] Using this identity we find

$$\left(\delta\Gamma^{W_3}_{\mathrm{ind}} \right) \exp \Gamma^{W_3}_{\mathrm{ind}} = \langle \left[\left(\frac{-c}{12\pi} \int \partial^3_- \epsilon\, h\, d^2 z \right) + \left(-\frac{1}{\pi} \int 2\partial_- \epsilon\, h\, T\, d^2 z \right) \right.$$
$$\left. + \left(-\frac{1}{\pi} \int \epsilon\, h\, \partial_- T\, d^2 z \right) \right] e^{S_{\mathrm{int}}} \rangle. \qquad (5.7)$$

All terms with T can be canceled by adding two suitable extra terms to δh

$$\delta h = \partial_+ \epsilon + \epsilon \partial_- h - h \partial_- \epsilon. \qquad (5.8)$$

We are then left with 'the minimal anomaly' containing the h field. If the sources T and W are realized by a local lagrangian scalar field theory as in (4.2), the Feynman diagram which yields this minimal anomaly is given by

δh ∼ ∂₊ε

$$= -\frac{c}{12\pi} \int h \partial^3_- \epsilon\, d^2 z \qquad (5.9)$$

We still have to deal with the TW contractions in (5.4). They yield

$$\langle \left(\frac{1}{\pi^2} \int \partial_+ \epsilon(z_1) b(z_2) \left[\frac{3W(z_2)}{(z_1 - z_2)^2} + \frac{W'(z_2)}{(z_1 - z_2)} \right] d^2 z_1 d^2 z_2 \right) e^{S_{int}} \rangle$$

$$= \langle -\frac{1}{\pi} \int (3\partial_- \epsilon\, b\, W + \epsilon\, b\, W') d^2 z\, e^{S_{int}} \rangle. \tag{5.10}$$

All these variations are canceled by adding suitable terms to δb

$$\delta b = \epsilon \partial_- b - 2b \partial_- \epsilon. \tag{5.11}$$

Defining variables u and v proportional to the effective currents

$$T_{eff} = -\pi \frac{\delta \Gamma^{W_3}_{ind}}{\delta h} \equiv \frac{c}{12} u, \quad W_{eff} = -\pi \frac{\delta \Gamma^{W_3}_{ind}}{\delta b} \equiv \frac{c}{360} v, \tag{5.12}$$

so that

$$u = \frac{\partial^3_-}{\partial_+} h + \cdots, \quad v = \frac{\partial^5_-}{\partial_+} b + \cdots, \tag{5.13}$$

the final Ward identity corresponding to ϵ symmetry for the induced action reads

$$\partial_+ u = D_1 h + \frac{1}{30} (3v\partial_- + 2\partial_- v)\, b, \tag{5.14}$$

where

$$D_1 \equiv \partial^3_- + 2u\partial_- + u'. \tag{5.15}$$

This is a (complicated) current conservation law ($\partial_+ u \sim \partial^\mu T_{\mu-}$) with an anomaly ($\partial^3_- h$). It is suggestive to consider all terms except the anomaly to constitute a covariant derivative on the doublet of currents u and v, but at present no geometry is known which can explain this. We observe that this Ward identity is local: the non-local operator $\frac{1}{\partial_+}$ does not appear explicitly but is hidden inside the currents u and v.

We now turn to the λ symmetry. Here, the nonlinearities in the OPE of W with itself are expected to lead to interesting complications. We begin as in the case of ϵ symmetry, and vary b as $\delta b = \partial_+ \lambda$. This yields

$$(\delta \Gamma^{W_3}_{ind}) \exp \Gamma^{W_3}_{ind} = \langle \left(-\frac{1}{\pi} \int \partial_+ \lambda\ W d^2 z_1 \right) \left(-\frac{1}{\pi} \int (hT + bW) d^2 z_2 \right) e^{S_{int}} \rangle. \tag{5.16}$$

The WT OPE yields terms with a W, namely $3W(z_2)/(z_1 - z_2)^2 + 2W'(z_2)/(z_1 - z_2)$, which are treated with (5.6) and are canceled by a suitable b transformation law

$$\delta_\lambda b = \partial_+ \lambda + 2\lambda \partial_- h - h \partial_- \lambda. \tag{5.17}$$

From the WW contraction we get a central charge term, terms linear in T and terms with Λ. The central charge term $(c/3)(z_1 - z_2)^{-6}$ yields the minimal λ anomaly proportional to $\partial_-^5 b$, which is the counterpart of the minimal ϵ anomaly proportional to $\partial_-^3 h$. The terms linear in T are all canceled by a suitable law for h

$$\delta_\lambda h = \frac{1}{30} \left(2\lambda \partial_-^3 b - 3\partial_- \lambda \partial_-^2 b + 3\partial_-^2 \lambda \partial_- b - 2\partial_-^3 \lambda b\right). \tag{5.18}$$

The remaining terms with Λ appear in $\delta_\lambda \Gamma_{\text{ind}}^{W_3}$ as $\langle \Lambda(z) \exp S_{\text{int}} \rangle \equiv \Lambda_{\text{eff}}(z)$, and will be studied in more detail in the next section. Thus the response of $\Gamma_{\text{ind}}^{W_3}$ under $\delta_\lambda h$ and $\delta_\lambda b$ given above contains the minimal anomaly and the terms with Λ_{eff}. The corresponding λ Ward identity reads

$$\partial_+ v = D_2^{lin} b + (3v\partial_- + \partial_- v) h + \frac{360}{c} \beta (b\partial_- + 2\partial_- b) \Lambda_{\text{eff}} \tag{5.19}$$

with

$$D_2^{lin} = \partial_-^5 + \left(10u\partial_-^3 + 15u'\partial_-^2 + 9u''\partial_- + 2u'''\right) \tag{5.20}$$

and β as in (2.3).

In the next section we shall show that all h-dependent but b-independent terms in Λ_{eff} are local when written in terms of u, and can be transferred to D_2^{lin}, which then becomes the Gelfand-Dickey operator D_2

$$D_2 = D_2^{lin} + 16uu' + 16u^2\partial_- . \tag{5.21}$$

The b^2 terms in Λ_{eff}, however, can not all be written as local expressions in terms of the variables h, b, u, v, and they lead to nonlocal b^3 terms in the λ Ward identity. The exact form of these b^3 terms is not an easy matter to obtain, but we shall need it in order to later show that they cancel against other nonlocal terms due to integrating the induced action over h and b. A simple dimensional argument already shows that nonlocalities must appear in the λ Ward identity. (By dimension we mean the usual notion according to which a coordinate has dimension -1. In conformal field theory this corresponds to the sum of left and right conformal dimensions.) To see this, note that Λ_{eff} has dimension 4, h has dimension 0, b has dimension -1, u has dimension 2, while v has dimension 3. The h^2 terms in Λ_{eff} may appear as u^2 as, in fact, they do (see (6.7)), but the b^2 terms cannot be produced by a local expression in h, b, u, v [v is odd in b, hence $b^2 v^2$ and $\partial_- v$ are out, while buv produces at best an hb^2 term but no b^2 term. Further, $\partial_-^2 u$ is out because it would also produce a term with a simple h field, whereas Λ_{eff} contains no terms linear in h or b. Finally, a term with $\partial_-^2 bv$ is out, as the dual requirements of dimensions and Lorentz covariance (the correct left- and right-conformal dimensions) imply nonlocality]. Thus, nonlocalities in Λ_{eff} cannot be avoided.

6 Computing $\Lambda_{\text{eff}}(h, b)$.

We now study the function $\Lambda_{\text{eff}}(h, b)$ in more detail. From its definition

$$\Lambda_{\text{eff}} = \langle \Lambda(z) \exp -\frac{1}{\pi} \int (hT + bW) d^2 z_1 \rangle,$$

$$\Lambda(z) = (TT)(z) - \frac{3}{10} T''(z) \tag{6.1}$$

we see that we shall need the singular terms in the OPE of $\Lambda(z)$ with $T(z_1)$ and $W(z_1)$. As we shall discuss in a moment, the first result is (compare with (2.6))

$$T(z_1)\Lambda(z) = \frac{5c + 22}{5} \frac{T(z)}{(z_1 - z)^4} + \left(\frac{4}{(z_1 - z)^2} + \frac{\partial_z}{z_1 - z} \right) \Lambda(z). \tag{6.2}$$

Were it not for the term with T on the right hand side, Λ would be a primary field with dimension 4. For the contraction of $\Lambda(z)$ with $W(z_1)$ one finds

$$W(z_1)\Lambda(z) = \left\{ \frac{6(TW)(z)}{(z_1 - z)^2} + \frac{4(TW')(z)}{(z_1 - z)} + \frac{15\, W(z)}{(z_1 - z)^4} + \frac{8\, W'(z)}{(z_1 - z)^3} + \frac{\frac{3}{2}\, W''(z)}{(z_1 - z)^2} \right\}$$

$$- \frac{3}{10} \left\{ \frac{18W(z)}{(z_1 - z)^4} + \frac{16\, W'(z)}{(z_1 - z)^3} + \frac{7\, W''(z)}{(z_1 - z)^2} + \frac{2\, W'''(z)}{(z_1 - z)} \right\}. \tag{6.3}$$

We have given the contractions of $W(z_1)$ with $(TT)(z)$ and $T''(z)$ separately (compare with (6.1)).

We now restrict our attention to the terms in Λ_{eff} with only one or two h and/or b fields. There are no terms with zero or one field, as in the expansion

$$\Lambda_{\text{eff}} = \langle \Lambda(z) + \Lambda(z) \left(-\frac{1}{\pi} \int (hT + bW) d^2 z \right) + \ldots \rangle \tag{6.4}$$

the terms explicitly written vanish due to $< \Lambda >= 0$, $< T >= 0$ and $< W >= 0$. The contributions to Λ_{eff} with two fields are given by

$$\langle \Lambda(z) \frac{1}{2!} \left(-\frac{1}{\pi} \int (hT + bW) d^2 z_1 \right) \left(-\frac{1}{\pi} \int (hT + bW) d^2 z_2 \right) \rangle. \tag{6.5}$$

We shall evaluate these correlators by putting $T(z_1)$ and $W(z_1)$ in front. Cross terms with hb vanish since $\langle \Lambda TW \rangle$ vanishes.

The h^2 terms come from the contractions

$$\frac{1}{2} \langle T(z_1) \Lambda(z) T(z_2) \rangle + \frac{1}{2} \langle \Lambda(z) T(z_1) T(z_2) \rangle \tag{6.6}$$

where only the $T\Lambda \sim T$ in the first term contributes, while the second term vanishes. It yields

$$\begin{aligned}
\Lambda_{\text{eff}}(h^2) &= \frac{5c + 22}{10} \int \frac{1}{(z_1 - z)^4} \frac{c/2}{(z - z_2)^4} h(z_1) h(z_2) d^2 z_1 d^2 z_2 \\
&= \frac{(5c + 22)c}{720} \left[\frac{\partial^3}{\partial_+} h(z) \right]^2 = \frac{c}{45\beta} u^2 + \mathcal{O}(h^3)
\end{aligned} \tag{6.7}$$

since the effective field equation u is given by $u = \frac{\partial^3}{\partial_+} h + \ldots$. Thus we see that due to the fact that we can hide all nonlocalities of $\left[\frac{\partial^3}{\partial_+} h \right]^2$ in u^2 to the order in h we work, the h^2 terms of Λ_{eff} do not produce nonlocalities in the Ward identity. It can be shown [4] that the relation $\Lambda_{\text{eff}}(h, b = 0) = \frac{c}{45\beta} u^2(h, b = 0)$ is exact to all orders in h.

The situation with the b^2 terms is more complicated. Now we must evaluate

$$\frac{1}{2} \langle W(z_1) \Lambda(z) W(z_2) \rangle + \frac{1}{2} \langle \Lambda(z) W(z_1) W(z_2) \rangle \tag{6.8}$$

and all the TW and W terms in $W\Lambda$ contribute. From WW, however, we only need the terms with Λ. To evaluate $\langle \Lambda(z_1)\Lambda(z_2)\rangle$ we replace $\Lambda(z_2)$ by $(TT)(z_2)$ and use the result for $\Lambda(z)T(z_2)$. The final result for the b^2 terms in Λ_{eff} reads

$$\Lambda_{\text{eff}}(b^2) = \frac{8\,c}{5\cdot 7!}\frac{1}{\partial_+}\left[b\frac{\partial_-^8}{\partial_+}b + 6\,\partial_-b\frac{\partial_-^7}{\partial_+}b + 14\,\partial_-^2 b\frac{\partial_-^6}{\partial_+}b + 14\,\partial_-^3 b\frac{\partial_-^5}{\partial_+}b\right]. \qquad (6.9)$$

As we already mentioned, for the h^2 terms in Λ_{eff} we find the square of (the leading part of) the h effective field equation (namely $u = \frac{\partial_-^3}{\partial_+}h + \ldots$). This came about because in the OPE only one pole structure contributed, namely $(z - z_1)^{-4}(z - z_2)^{-4}$, which factorized into $u^2 = \left[\frac{\partial_-^3}{\partial_+}h\right]^2 + \ldots$. For the b^2 terms several pole structures contribute. Those coming from $W(z_1)\Lambda(z)W(z_2)$ give terms with $(z_1 - z)^{-p}(z - z_2)^{-q}$ with $p + q = 10$ and $p = 1, 2, 3, 4$, while from $\Lambda(z)W(z_1)W(z_2)$ one obtains terms with $(z_1 - z_2)^{-2}(z - z_2)^{-8}$ and $(z_1 - z_2)^{-1}(z - z_2)^{-9}$. The former produce products of the form $\left[\frac{\partial_-^{p-1}}{\partial_+}b\right]\left[\frac{\partial_-^{q-1}}{\partial_+}b\right]$, while the latter yield structures of the form $\frac{\partial_-^8}{\partial_+}\left[b\frac{1}{\partial_+}b\right]$ and $\frac{\partial_-^7}{\partial_+}\left[b\frac{\partial_-}{\partial_+}b\right]$. One should be careful with such nonlocal expressions, as they can be rewritten in various other forms which look very different. For example, if we would have started to evaluate the OPE with $\Lambda(z)$ in front, and computed the three point function $\langle \Lambda WW\rangle$ as $\langle \Lambda(z)W(z_1)W(z_2)\rangle + \langle W(z_1)\Lambda(z)W(z_2)\rangle$, we would have found a sum of terms of the form $\frac{\partial_-^s}{\partial_+}\left[b\frac{\partial_-^t}{\partial_+}b\right]$ with $s + t = 8$, $s = 0, 1, 2, 3$. This is the result given in (6.9). To show that both results are equivalent, one could multiply the first result by $\frac{1}{\partial_+}\partial_+$, and act with ∂_+ to the right.

Let us finally show how one goes about to compute the (singular) OPE of elementary fields with composites such as $\Lambda(z)$. As an example, we will show how to derive the OPE $T\Lambda$ given in (6.2).

We begin by recalling that the normal ordering of two operators A and B can in general be written as [36]

$$(AB)(z) = \frac{1}{2\pi i}\oint\frac{dx}{x - z}A(x)B(z)\,, \qquad (6.10)$$

where the x-contour runs around z. For contractions of operators with such composites, there is a Wick theorem [36], which in our case can be written as

$$
\underbrace{T(z)(TT)(w)}_{} = \frac{1}{\pi} \int \frac{dx}{x-w} \left[\underbrace{T(z)T(x)T(w)}_{} + T(x)\underbrace{T(z)T(w)}_{} \right] .
\tag{6.11}
$$

Substituting the TT OPE one obtains

$$
\frac{1}{2\pi i} \oint \frac{dx}{x-w} \left[\frac{c/2}{(z-x)^4} + \frac{2T(x)}{(z-x)^2} + \frac{T'(x)}{z-x} \right] T(w)
$$
$$
+ \frac{1}{2\pi i} \oint \frac{dx}{x-w} T(x) \left[\frac{c/2}{(z-w)^4} + \frac{2T(w)}{(z-w)^2} + \frac{T'(w)}{z-w} \right]
\tag{6.12}
$$

The second line gives straightaway

$$
\frac{(c/2)\,T(w)}{(z-w)^4} + \frac{2\,(TT)(w)}{z-w)^2} + \frac{(TT')}{z-w}
\tag{6.13}
$$

but in the first line we must expand $(z-x)^{-2}$ and $(z-x)^{-1}$ in x about w since any positive powers of $(x-w)$ can be overcome by the singularities in the OPE of $T(x)T(w)$. The leading terms of the first line give

$$
\frac{(c/2)\,T(w)}{(z-w)^4} + \frac{2\,(TT)(w)}{(z-w)^2} + \frac{(T'T)(w)}{z-w}
\tag{6.14}
$$

while the terms obtained by expanding the poles yield

$$
\frac{3c}{(z-w)^6} + \frac{8T(w)}{(z-w)^4} + \frac{3T'(w)}{(z-w)^3} .
\tag{6.15}
$$

One can cancel the central term by adding to $(TT)(w)$ a term proportional to $T''(w)$. Indeed,

$$
T(z)T''(w) = \frac{10c}{(z-w)^6} + \frac{12\,T(w)}{(z-w)^4} + \frac{10\,T'(w)}{(z-w)^3} + \frac{4\,T''}{(z-w)^2} + \frac{T'''(w)}{z-w} + \ldots
\tag{6.16}
$$

so that $\Lambda = (TT) - \frac{3}{10}T''$ has an OPE with T without central term. It was given in (6.2).

7   The b^4 terms in the induced action.

The Ward identities for the induced action can be expanded in terms of $1/c$, and their solution can also be expanded in a $1/c$ series. The Ward identities read (see (5.14) and (5.19))

$$\partial_+ u \;=\; D_1 h + \frac{1}{30}\left(3v\partial_- + 2v'\right)b$$

$$\partial_+ v \;=\; D_2 b + (3v\partial_- + v')h + \frac{360}{c}\beta\,(b\partial_- + 2b')\left(\Lambda_{\text{eff}} - \frac{c}{45\beta}u^2\right) \qquad (7.1)$$

where

$$\Lambda_{\text{eff}} - \frac{c}{45\beta}u^2 = \frac{8c}{5\cdot 7!}\frac{1}{\partial_+}Q^8_{bb} + \mathcal{O}(b^2 h^{\ge 1}) + \mathcal{O}(b^4 h^{\ge 0})\dots \qquad (7.2)$$

with

$$Q^8_{bb} = b\frac{\partial^8_-}{\partial_+}b + 6\,\partial_- b\frac{\partial^7_-}{\partial_+}b + 14\,\partial^2_-\,b\frac{\partial^6_-}{\partial_+}b + 14\,\partial^3_-\,b\frac{\partial^5_-}{\partial_+}b. \qquad (7.3)$$

The contributions due to $\Lambda_{\text{eff}} - \frac{c}{45\beta}u^2$ are all suppressed by at least one factor $1/c$ as compared to the pure u terms in Λ_{eff}. For example, in ΛTT each contraction can give a factor c, but in ΛWW the ΛW contraction is of order c^o. The leading terms in $1/c$ in the Ward identity, namely, the c-independent terms, read therefore

$$\partial_+ u \;=\; D_1 h + \frac{1}{30}\left(3v\partial_- + 2v'\right)b$$

$$\partial_+ v \;=\; D_2 b + (3v\partial_- + v')\,h. \qquad (7.4)$$

These equations generalize the KdV equations, and are called the Boussinesq equations. They characterize the induced action in the limit $c \to \pm\infty$. We have solved these coupled nonlinear partial differential equations for u and v in terms of h and b. The solution is

$$u_L(h,b) = -\pi\frac{\delta}{\delta h}\Gamma_L, \qquad v_L(h,b) = -30\pi\frac{\delta}{\delta b}\Gamma_L, \qquad (7.5)$$

where $\Gamma_L[h,b]$ is a reference functional which, up to a change of variables, we obtained in closed, explicit form in [5]. It starts out with

$$\Gamma_L[h,b] = \frac{-1}{2\pi}\int d^2 z\left(h\frac{\partial^3_-}{\partial_+}h + \frac{1}{30}b\frac{\partial^5_-}{\partial_+}b + \dots\right). \qquad (7.6)$$

A related functional $W_L[u,v]$ is obtained from $\Gamma_L[h,b]$ by a Legendre transformation

$$W_L[u,v] = \Gamma_L[h(u,v),b(u,v)] + \frac{1}{\pi}\int\left(hu + \frac{1}{30}bv\right), \tag{7.7}$$

where $h(u,v)$ and $b(u,v)$, which we denote by $h_L(u,v)$ and $b_L(u,v)$ for later reference, are determined through the relations in (7.5) and we have that

$$h_L(u,v) = \pi\frac{\delta}{\delta u}W_L, \qquad b_L(u,v) = 30\,\pi\frac{\delta}{\delta v}W_L. \tag{7.8}$$

Let us now consider the original problem and consider the induced action for finite c. We decompose

$$\Gamma_{\text{ind}}^{W_3}[h,b] = \frac{c}{12}\Gamma_L[h,b] \;+\; \left(b^4\text{ terms, plus } b^4h,\, b^4h^2,\ldots, b^6,\, b^6h,\ldots\text{ terms}\right)$$

$$+\;\frac{1}{c}\text{ terms }+\frac{1}{c^2}\text{ terms }+\ldots \tag{7.9}$$

and expand $u(h,b)$ and $v(h,b)$ (as defined in (5.12)) as

$$u(h,b) = u_L(h,b) + \frac{1}{c}\left(b^4\text{ terms } + b^4h,\ldots, b^6,\ldots\right) + \frac{1}{c^2}\left(b^6,\ldots\right)$$

$$v(h,b) = v_L(h,b) + \frac{1}{c}\left(b^3\text{ terms } + b^3h,\ldots, b^5,\ldots\right) + \frac{1}{c^2}\left(b^5,\ldots\right). \tag{7.10}$$

(The b^2 terms in u are accounted for by u_L.)

The full result for the b^4 terms in the induced action was computed in [4], where we directly computed the four-point correlator $\langle WWWW\rangle$. The result is

$$\Gamma_{\text{ind}}[h=0,b^4] = -\frac{c}{60\cdot 6!}\frac{1}{\pi}[I] - \frac{2\beta c}{5\cdot 7!}\frac{1}{\pi}[II], \tag{7.11}$$

where the two structures $[I]$ and $[II]$ are given by

$$[I] = \int\left(2b\frac{\partial_-^3}{\partial_+}b - 3\,\partial_-b\frac{\partial_-^2}{\partial_+}b + 3\,\partial_-^2 b\frac{\partial_-}{\partial_+}b - 2\,\partial_-^3 b\frac{1}{\partial_+}b\right)\frac{1}{\partial_+}\left(2b\frac{\partial_-^6}{\partial_+}b + 3\,\partial_-b\frac{\partial_-^5}{\partial_+}b\right),$$

$$[II] = \int\left(b\frac{\partial_-}{\partial_+}b - \partial_-b\frac{1}{\partial_+}b\right)\frac{1}{\partial_+}\left(b\frac{\partial_-^8}{\partial_+}b + 6\,\partial_-b\frac{\partial_-^7}{\partial_+}b + 14\,\partial_-^2\,b\frac{\partial_-^6}{\partial_+}b + 14\,\partial_-^3\,b\frac{\partial_-^5}{\partial_+}b\right).$$

$$\tag{7.12}$$

We remark that the last term in (7.11) corresponds to the lowest nonlocal contribution (which is of order b^3) to the λ Ward identity (see (7.1)-(7.3)). To prove this, one may use that taking $\frac{\delta}{\delta b}$ of the expression $[II]$ is equivalent to 4 times varying w.r.t. the b at the second position.

To lowest order in $1/c$, the last term of (7.11) is $-\frac{32}{25.7!}\frac{1}{\pi}[II]$. In the following sections we will explicitly show that this term will get cancelled when we pass from the *induced* to the *effective* action.

8   The effective action.

We now turn to the effective action, and will argue that its Ward identities are local in terms of h, b, u and v. The effective action is obtained as follows. First consider the path-integral

$$\exp W_{\text{conn}}^{W_3}[t, w] = \int DhDb \exp\left(\Gamma_{\text{ind}}^{W_3}[h, b] + \frac{1}{\pi}\int(ht + bw)d^2z\right). \qquad (8.1)$$

It represents the connected Feynman diagrams with propagating h and b fields. Then take the Legendre transform

$$\Gamma_{\text{eff}}^{W_3}[h, b] = W_{\text{conn}}^{W_3}[t, w] - \frac{1}{\pi}\int(ht + bw)d^2z, \qquad (8.2)$$

where $t(h, b)$ and $w(h, b)$ are the solutions of

$$\pi\frac{\delta W_{\text{conn}}^{W_3}}{\delta t} = h, \qquad \pi\frac{\delta W_{\text{conn}}^{W_3}}{\delta w} = b. \qquad (8.3)$$

Then $\Gamma_{\text{eff}}^{W_3}$ corresponds to all graphs which cannot be split into two parts by cutting one internal h or b line. We shall from now on deal with $W_{\text{conn}}^{W_3}[t, w]$. If one wishes $\Gamma_{\text{eff}}^{W_3}$ instead, one must make the above Legendre transformation.

We evaluate $W_{\text{conn}}^{W_3}[t, w]$ in an h, b-loop expansion as follows. First we take the saddle point approximation. The saddle point are those functions $h_0(t, w)$ and $b_0(t, w)$ where the integrand is stationary

$$\left.\begin{array}{l} \frac{\delta}{\delta h}\Gamma_{\text{ind}}^{W_3}[h, b] + \frac{1}{\pi}t = 0 \\[2ex] \frac{\delta}{\delta b}\Gamma_{\text{ind}}^{W_3}[h, b] + \frac{1}{\pi}w = 0 \end{array}\right\} \text{solution}: \begin{array}{l} h = h_0(t, w) \\ b = b_0(t, w) \end{array} \qquad (8.4)$$

Comparing with (5.12), (7.5) and (7.9), we find

$$
\begin{aligned}
h_o(t,w) &= h_L(u = \frac{12}{c}t, v = \frac{360}{c}w) + \mathcal{O}(\frac{1}{c}), \\
b_o(t,w) &= b_L(u = \frac{12}{c}t, v = \frac{360}{c}w) + \mathcal{O}(\frac{1}{c}).
\end{aligned}
$$

$$(8.5)$$

Then we decompose h and b as follows

$$
h = h_0(t,w) + h', \quad b = b_0(t,w) + b'
$$

$$(8.6)$$

and expand the exponent in terms bilinear in the quantum fields h', b', terms trilinear in h', b', etc. Hence,

$$
e^{W^{W_3}_{\text{conn}}[t,w]} = e^{W^{W_3}_{\text{saddle}}[t,w]} \int Dh' Db' \, e^{I^{(2)} + I^{(3)} + \cdots},
$$

$$(8.7)$$

where

$$
\begin{aligned}
W^{W_3}_{\text{saddle}}[t,w] &= \Gamma^{W_3}_{\text{ind}}[h_0, b_0] + \frac{1}{\pi} \int (h_0 t + b_0 w) d^2 z, \\
I^{(2)} &= \frac{1}{2!} \int (h'\, b') \begin{pmatrix} \frac{\delta u}{\delta h} & \frac{\delta u}{\delta b} \\ \frac{\delta v}{\delta h} & \frac{\delta v}{\delta b} \end{pmatrix} \begin{pmatrix} h' \\ b' \end{pmatrix} d^2 z
\end{aligned}
$$

$$(8.8)$$

We have dropped the factors $\frac{c}{12}$ and $\frac{c}{360}$ to the power infinity (see (5.12)), as they are additive constants in $W^{W_3}_{\text{conn}}$. Clearly, the saddle point approximation of $W^{W_3}_{\text{conn}}$ is the Legendre transform of $\Gamma^{W_3}_{\text{ind}}$.

The leading order terms in $W^{W_3}_{\text{conn}}[t,w]$ are known; they are given by the Legendre transform of $\frac{c}{12}\Gamma_L[h,b]$. We are interested in the order $1/c$ corrections to $W^{W_3}_{\text{conn}}[t,w]$ since we have seen that similar terms lead to non-localities in the Ward identity (7.1) for the induced action. $W^{W_3}_{\text{conn}}[t,w]$ depends on t and w through the explicit t and w appearing in $(h_0 t + b_0 w)$ and through $h_0(t,w)$ and $b_0(t,w)$. [A little thinking shows that, for the purpose of considering $1/c$ corrections to the saddle-point result, we may replace $h_o(t,w)$ by $h_L(u = \frac{12}{c}t, v = \frac{360}{c}w)$ and $b_o(t,w)$ by $b_L(u = \frac{12}{c}t, v = \frac{360}{c}w)$ (compare with (8.5). Conversely, one may replace the saddle-point value for t and w, to be denoted by $t_0(h,b)$ and $w_0(h,b)$ by $\frac{c}{12}u_L(h,b)$ and $\frac{c}{360}v_L(h,b)$. We will use this in sections 11, 12.]

Since the kinetic terms for h and b are proportional to c, whereas the interactions are proportional to c or down by powers of $1/c$, we can interpret $1/c$ as the Planck's constant. The action $\Gamma_{\text{ind}}^{W_3}$ is thus: a 'classical action' ($\frac{c}{12}\Gamma_L$) plus '$\hbar$ corrections' (the term with c^0) plus $\hbar^2$ corrections (term with $1/c$), etc. It follows that the leading 1-loop results give order $1/c$ corrections to the saddle-point result.

In what follows we will focus on the $c^0 b_0^4$ terms in $W_{\text{conn}}^{W_3}$. They come from two sources

(i) from the b^4 term in $\Gamma_{\text{ind}}^{W_3}$ in (7.11), with b replaced by b_0,

(ii) from the 1-loop corrections to the path-integral.

9 1-Loop corrections to $W_{\text{conn}}^{W_3}$.

The complete 1-loop correction to the saddle-point approximation is given by

$$K = -\frac{1}{2}\ln\det\begin{pmatrix}\frac{\delta u}{\delta h} & \frac{\delta v}{\delta h} \\ \frac{\delta u}{\delta b} & \frac{\delta v}{\delta b}\end{pmatrix} \quad \text{at } h = h_0, b = b_0. \tag{9.1}$$

In here we may replace u, v by u_L and v_L since we only want the c^0 terms (*i.e.*, we keep only the vertices in $\Gamma_{\text{ind}}^{W_3}$ which are of order c). Since u_L and v_L satisfy the Ward identities (7.4) we can find an equation for the entries $\frac{\delta u_L}{\delta h}$ etc., by differentiating each of these Ward identities w.r.t. h or b. One finds then an equation of the form

$$M\begin{pmatrix}\frac{\delta u_L}{\delta h} & \frac{\delta u_L}{\delta b} \\ \frac{\delta v_L}{\delta h} & \frac{\delta v_L}{\delta b}\end{pmatrix} = N \tag{9.2}$$

and hence

$$K = \frac{1}{2}\ln\det(N^{-1}M) = \frac{1}{2}\ln\det M - \frac{1}{2}\ln\det N. \tag{9.3}$$

The explicit form of the operator-valued matrices M and N is

$$M = \begin{bmatrix}\nabla_+^{(2)} & -\left(\frac{1}{10}b' + \frac{1}{15}b\partial_-\right) \\ -L & \nabla_+^{(3)}\end{bmatrix}\begin{bmatrix}\delta^2(z-w) & 0 \\ 0 & \delta^2(z-w)\end{bmatrix}$$

$$N = \begin{bmatrix} D_1 & (\frac{1}{10}v\partial_- + \frac{1}{15}v') \\ 3v\partial_- + v' & D_2 \end{bmatrix} \begin{bmatrix} \delta^2(z-w) & 0 \\ 0 & \delta^2(z-w) \end{bmatrix}, \tag{9.4}$$

where

$$\nabla_+^{(j)} = \partial_+ - h\partial_- - jh' \tag{9.5}$$

and

$$L = \left(10b''' + 15b''\partial_- + 9b'\partial_-^2 + 2b\partial_-^3\right) + \left(32ub' + 16bu' + 16bu\partial_-\right). \tag{9.6}$$

Our task is to compute each of these two determinants. This is similar to the evaluation of the Jacobians for anomalies, and we must similarly regulate these expressions. We will use the well-known representation of determinants as Gaussian integrals of anticommuting variables and write the determinants as the partition functions of certain 'b-c' systems.[9] We introduce anticommuting fields b_1, b_2, c_1, c_2 for M, and B_1, B_2, C_1, C_2 for N, and write

$$\det M = \int dc_1 db_1 dc_2 db_2 \, e^{\frac{1}{\pi} \int (b_1 \, b_2) M \begin{pmatrix} c_1 \\ c_2 \end{pmatrix} d^2 z} \tag{9.7}$$

and a similar expression for N.

We now write the exponent (the 'action') in this expression as

$$I^{(M)} = \frac{1}{\pi} \int (b_1 \, b_2) M \begin{pmatrix} c_1 \\ c_2 \end{pmatrix} d^2 z = I_{\text{free}}^{(M)} + I_{\text{int}}^{(M)}, \tag{9.8}$$

where

$$I_{\text{free}}^{(M)} = \frac{1}{\pi} \int (b_1 \partial_+ c_1 + b_2 \partial_+ c_2) \, d^2 z$$

$$I_{\text{int}}^{(M)} = -\frac{1}{\pi} \int (hT_h + bT_b + buT_{bu} + b'uT_{b'u}) d^2 z, \tag{9.9}$$

and

$$T_h = -b_1 c_1' - 2 b_1' c_1 - 2 b_2 c_2' - 3 b_2' c_2$$

---

[9]Equivalently, we could have taken a commuting b-c system to obtain the inverses of the determinants.

$$T_b = -\frac{1}{30}(b_1 c_2' + 3b_1' c_2) + (10\, b_2''' c_1 + 15\, b_2'' c_1' + 9\, b_2' c_1'' + 2\, b_2 c_1''')$$

$$T_{bu} = -16\, b_2' c_1$$

$$T_{b'u} = 16\, b_2 c_1. \tag{9.10}$$

Hence

$$\det M = \langle e^{-\frac{1}{\pi}\int d^2 z(h\, T_h + b\, T_b + bu\, T_{bu} + b'u\, T_{b'u})} \rangle. \tag{9.11}$$

The action for N analogous to (9.8) is given by

$$I^{(N)} = \frac{1}{\pi}\int (B_1\, B_2)\, N\begin{pmatrix} C_1 \\ C_2 \end{pmatrix} d^2 z. \tag{9.12}$$

It can be written as $I = I_{free} + I_{int}$, where

$$I_{free}^{(N)} = \frac{1}{\pi}\int (B_1 \partial_-^3 C_1 + B_2 \partial_-^5 C_2) d^2 z$$

$$I_{int}^{(N)} = \frac{1}{\pi}\int (u H_u + v H_v + u^2 H_{uu}) d^2 z \tag{9.13}$$

and

$$H_u = (B_1 C_1' - B_1' C_1) + 10\, B_2 C_2''' - 15\, \partial_-(B_2 C_2'')$$

$$+9\, \partial_-^2(B_2 C_2') - 2\, \partial_-^3(B_2 C_2)$$

$$= (B_1 \overset{\leftrightarrow}{\partial_-} C_1) + (2 B_2 C_2''' - 3 B_2' C_2'' + 3 B_2'' C_2' - 2 B_2''' C_2)$$

$$H_v = \frac{1}{30}(B_1 C_2' - 2 B_1' C_2) + (2 B_2 C_1' - B_2' C_1)$$

$$H_{uu} = 8\,(B_2 C_2' - B_2' C_2). \tag{9.14}$$

Hence the expression for $W_{conn}^{W_3}[t, w]$ through 1-loop reads

$$e^{W_{conn}^{W_3}[t, w]} = e^{W_{saddle}^{W_3}[t, w]}\left(\langle e^{-\frac{1}{\pi}\int (h\, T_h + b\, T_b + bu\, T_{bu} + b'u\, T_{b'u})\, d^2 z}\rangle\right)^{1/2}$$

$$\times \left(\langle e^{\frac{1}{\pi}\int (u H_u + v H_v + u^2 H_{uu})\, d^2 z}\rangle\right)^{-1/2}. \tag{9.15}$$

It would be nice if we could write down a pair of Ward identities for the partition functions $\ln \det M$, and $\ln \det N$ and then solve them exactly. For the $c \to \pm\infty$ limit of the Ward identities for $\Gamma_{\text{ind}}^{W_3}$ we were able to achieve this, because in this limit the Ward identities became local (albeit nonlinear) differential equations (anyhow, it was a hard job). For the present case, it seems hopeless to follow this path, as the Ward identities are expected to be nonlocal (and nonlinear): their nonlocalities are, after all, supposed to cancel the original nonlocalities in $\Gamma_{\text{ind}}^{W_3}$.

To get an idea how to proceed to obtain $\ln \det M$ and $\ln \det N$, we now first consider the truncation to pure gravity. In this case,

$$
\begin{aligned}
M^{\text{grav}} &= \nabla_+^{(2)} \delta^2(z - w), & \nabla_+^{(2)} &= \partial_+ - h\partial_- - 2h', \\
N^{\text{grav}} &= D_1 \delta^2(z - w), & D_1 &= \partial_-^3 + 2u\partial_- + u'.
\end{aligned}
\tag{9.16}
$$

The results for $\det M^{\text{grav}}$ and $\det N^{\text{grav}}$ will then suggest what the leading terms in the result for $\det M$ and $\det N$ will be.

10 The determinants in pure gravity.

Let us first consider the most complicated case, which is the truncation of N to pure gravity: $N^{\text{grav}} = D_1 \delta^2(z - w)$. We have

$$
I_{\text{free}}^{(N)} = -\frac{1}{\pi} \int B \partial_-^3 C, \quad I_{\text{int}}^{(N)} = -\frac{1}{\pi} \int uJ, \quad J = BC' - B'C.
\tag{10.1}
$$

We claim that the propagators are given by[10]

$$
\langle C(z, \bar{z}) B(w, \bar{w}) \rangle = \langle B(z, \bar{z}) C(w, \bar{w}) \rangle = -\frac{1}{2} \frac{(z - w)^2}{\bar{z} - \bar{w}}.
\tag{10.2}
$$

Assuming for a moment this to be the case, we obtain the OPE

$$
J(z, \bar{z})J(w, \bar{w}) = -\frac{(z - w)^2}{(\bar{z} - \bar{w})^2} - \frac{(z - w)}{\bar{z} - \bar{w}} J(w, \bar{w}) - \frac{1}{2}\frac{(z - w)^2}{\bar{z} - \bar{w}} J'(w, \bar{w}).
\tag{10.3}
$$

---

[10]Since B and C have total conformal dimension -1 and total conformal spin $+3$, the propagator must be of the form $(z - w)^2/(\bar{z} - \bar{w})$.

The truncation of this OPE at the position shown here should be compared to the 'contraction' or 'singular OPE' and is such that Wick-decomposition of correlation functions holds. The truncation of course corresponds to the on-shell condition $\partial_-^3 J = 0$.

Let us now come back to the propagators (10.2). In general, propagators are obtained by adding the source terms $\frac{1}{\pi}\int(j_C C + j_B B)d^2z$ to the free action and completing squares. This yields $K = -\frac{1}{\pi}\int j_C \frac{1}{\partial_-^3}j_B d^2z$. Then $\langle C(z,\bar{z})B(w,\bar{w})\rangle = -\pi^2 \frac{\delta}{\delta j_C(z)}\frac{\delta}{\delta j_B(w)}$ $\exp K = -\pi\frac{1}{\partial_-^3}\delta^2(z-w)$. Comparing with (10.2), we should thus prove the distribution identity[11]

$$\pi\frac{1}{\partial_-^3}\delta^2(z-w) = \lim_{\epsilon\downarrow 0}\frac{1}{2}\frac{(z-w)^2}{(\bar{z}-\bar{w})}\theta\left[(z-w)(\bar{z}-\bar{w})-\epsilon^2\right]. \tag{10.4}$$

To prove this, we multiply by a test function $\partial_-^3 f(w,\bar{w})$ and integrate over d^2w. On the l.h.s. this gives $\pi f(z,\bar{z})$, while on the r.h.s. we find

$$\text{r.h.s.} = \lim_{\epsilon\downarrow 0}\int d^2w\frac{1}{2}f(w,\bar{w})\partial_w^3\left\{-\frac{(z-w)^2}{\bar{z}-\bar{w}}\theta\left[(z-w)(\bar{z}-\bar{w})-\epsilon^2\right]\right\}. \tag{10.5}$$

Since $(z-w)(\bar{z}-\bar{w}) = \rho^2$ is larger than ϵ^2, the ∂_w derivatives annihilate $(\bar{z}-\bar{w})^{-1}$, and we get terms with 1,2 or 3 ∂_w derivatives on the θ function. They yield, respectively,

$$6\left(\frac{1}{\bar{z}-\bar{w}}\right)(\bar{z}-\bar{w})\delta(\rho^2-\epsilon^2) - 6\frac{z-w}{\bar{z}-\bar{w}}\partial_w\left\{(\bar{z}-\bar{w})\delta(\rho^2-\epsilon^2)\right\}$$

$$+\frac{(z-w)^2}{(\bar{z}-\bar{w})}\partial_w^2\left\{(\bar{z}-\bar{w})\delta(\rho^2-\epsilon^2)\right\}$$

$$= 6\delta(\rho^2-\epsilon^2) - 6(z-w)\partial_w\delta(\rho^2-\epsilon^2) + (z-w)^2\partial_w^2\delta(\rho^2-\epsilon^2). \tag{10.6}$$

Putting this into (10.5), and integrating by parts, it yields indeed $\int d^2w\frac{1}{2}f(w,\bar{w})2\delta(\rho^2-\epsilon^2) = \pi f(z,\bar{z})$.

We derive, as before, a Ward identity for $\det N^{\text{grav}} = \langle\exp\frac{1}{\pi}\int uJd^2z\rangle \equiv \exp\Gamma_{\text{ind}}^{(N)}[u]$. We recall that $u = \frac{\partial_-^3}{\partial_+}h + \ldots$ and begin by varying $\delta u = \partial_-^3\epsilon$. This yields

$$\left(\delta\Gamma_{\text{ind}}^{(N)}\right)\exp\Gamma_{\text{ind}}^{(N)} = \langle\left(\frac{1}{\pi}\int\partial_-^3\epsilon\,J\,d^2z_1\right)\left(\frac{1}{\pi}\int u\,J\,d^2z_2\right)e^{\frac{1}{\pi}\int uJd^2z}\rangle. \tag{10.7}$$

---

[11]Note that one recovers (5.6) for $n = 1$ by acting with ∂_-^2.

Using the contraction for JJ given above and distribution identities similar to (10.4) we obtain

$$\left(\delta\Gamma_{\text{ind}}^{(N)}\right)\exp\Gamma_{\text{ind}}^{(N)} = \langle\left[\frac{1}{\pi}\int(-u\epsilon' + \epsilon u\partial_-)J + \frac{2}{\pi}(\partial_+\epsilon)u\right]d^2z_2\, e^{\frac{1}{\pi}\int uJd^2z}\rangle.$$

(10.8)

By adding a suitable term to δu, we remove the terms with J, after which only the minimal anomaly (i.e., with one u) remains. Discarding the local parameter ϵ, we arrive at the following Ward identity

$$\left(\partial_-^3 + 2u\partial_- + u'\right)\frac{\delta}{\delta u}\Gamma_{\text{ind}}^{(N)} = \frac{2}{\pi}\partial_+ u.$$

(10.9)

Do we recognize this identity? The induced action of pure gravity, defined by

$$e^{\Gamma_{\text{ind}}^{\text{grav}}[h]} = \langle e^{-\frac{1}{\pi}\int hT d^2z}\rangle$$

(10.10)

satisfies

$$(\partial_+ - h\partial_- - 2h')\frac{\delta}{\delta h}\Gamma_{\text{ind}}^{\text{grav}} = -(\frac{c}{12\pi})\partial_-^3 h.$$

(10.11)

Let us turn this relation 'inside out' by using the Legendre transform,

$$t = -\pi\frac{\delta}{\delta h}\Gamma_{\text{ind}}^{\text{grav}}, \qquad h = \pi\frac{\delta}{\delta t}W_{\text{ind}}^{\text{grav}}$$

$$W_{\text{ind}}^{\text{grav}}[t] = \Gamma_{\text{ind}}^{\text{grav}}[h(t)] + \frac{1}{\pi}\int ht\, d^2z,$$

(10.12)

To make contact with (10.9), we consider suitably normalized functionals as in (7.7)

$$\Gamma_L^{\text{grav}}[h] = \frac{12}{c}\Gamma_{\text{ind}}^{\text{grav}}[h], \qquad W_L^{\text{grav}}[u] = \frac{12}{c}W_{\text{ind}}^{\text{grav}}[t = \frac{c}{12}u]$$

(10.13)

with

$$W_L^{\text{grav}}[u] = \Gamma_L^{\text{grav}}[h] + \frac{1}{\pi}\int hu\, d^2z.$$

(10.14)

Then (10.11) becomes

$$\frac{1}{\pi}\partial_+ u - [u' + 2u\partial_-]\frac{\delta}{\delta u}W_L^{\text{grav}} = \partial_-^3\left[\frac{\delta}{\delta u}W_L^{\text{grav}}\right]. \tag{10.15}$$

Comparison of (10.15) and (10.11) reveals that

$$\ln\det N = \Gamma_{\text{ind}}^{(N)}[u] = 2\,W_L^{\text{grav}}[u]. \tag{10.16}$$

Thus we see that, up to a constant, $\ln\det N$ is given by the Legendre transform of the induced action of pure gravity.

Next we consider the truncation of M to pure gravity. We write

$$\det(\partial_+ - 2h' - h\partial_-) = \int dc\,db\,e^{\frac{1}{\pi}\int(b\partial_+ c - hT_h)d^2z}, \tag{10.17}$$

where

$$T_h = -bc' - 2b'c. \tag{10.18}$$

The propagators are $\langle c(z)b(w)\rangle = -\frac{1}{z-w}$ and T_h is just the stress tensor for coordinate ghosts

$$T_h(z)T_h(w) = \frac{-13}{(z-w)^4} + \frac{2\,T_h(w)}{(z-w)^2} + \frac{T_h'(w)}{z-w}. \tag{10.19}$$

The Ward identity for

$$\det M^{\text{grav}} = e^{\Gamma_{\text{ind}}^{(M)}} = \langle e^{-\frac{1}{\pi}\int hT_h d^2z}\rangle \tag{10.20}$$

follows from varying $\delta h = \partial_+\epsilon$ and then using T_hT_h to obtain extra δh terms which cancel everything except the minimal anomaly. One finds with (5.6)

$$(\partial_+ - h\partial_- - 2h')\frac{\delta}{\delta h}\Gamma_{\text{ind}}^{(M)} = \frac{26}{12\pi}\partial_-^3 h. \tag{10.21}$$

Comparison with (10.11) immediately yields

$$\ln \det M = \Gamma^{(M)}_{\text{ind}}[h] = -\frac{13}{6}\Gamma^{\text{grav}}_L[h]\,. \tag{10.22}$$

Hence, $\det M$ is proportional to the induced action itself.

For later reference we mention the results for the determinants of the operators $\nabla^{(j)}_+$ and D_j, which are the spin-j analogs of the operators $M = \nabla^{(2)}_+$ and $N = D_1$ discussed here. We have [31, 37]

$$
\begin{aligned}
\ln \det \nabla^{(j)}_+ &= -\frac{(6j^2 - 6j + 1)}{6}\,\Gamma^{\text{grav}}_L[h]\,, \\
\ln \det D_j &= \frac{j(2j+1)(2j+2)}{6}\,W^{\text{grav}}_L[u]\,.
\end{aligned}
\tag{10.23}
$$

11   The determinants in W_3 gravity.

Based on the results in the previous section, we can easily find some first results for the W_3 determinants $\det M$ and $\det N$. If we put $b = 0$ in $\det M$ and $v = 0$ in $\det N$, the matrices M and N reduce to diagonal matrices, whose determinant we easily find [37]:

$$
\begin{aligned}
\ln \det M[h, b = 0] &= \ln \det \nabla^{(2)}_+ + \ln \det \nabla^{(3)}_+ = \frac{(-13 - 37)}{6}\Gamma^{\text{grav}}_L[h] \\
\ln \det N[u, v = 0] &= \ln \det D_1 + \ln \det D_2 = (2 + 10) W^{\text{grav}}_L[u]\,.
\end{aligned}
\tag{11.1}
$$

One may expect that putting in the b and v dependences will simply extend these results to $\ln \det M = -\frac{50}{6}\Gamma_L[h, b]$ and $\ln \det N = 12\,W_L[u, v]$. However, we will find below that things are not as simple as that. Let us now work out these determinants in more detail.

We first focus on $\ln \det M$. Up to a change in notation, the currents T_h and T_b in (9.10) are very similar to the the currents T^{gh} and W^{gh} that one finds in a BRST treatment of critical ($c = 100$) W_3 gravity (see section 4). However, in this case the current T_b is *not* a primary current with respect to T_h, and the T_b-T_b OPE is modified accordingly. One finds

$$T_h(z)T_h(w) = \frac{-50}{(z-w)^4} + \frac{2\,T_h(w)}{(z-w)^2} + \frac{\partial_- T_h(w)}{(z-w)} + \cdots$$

$$T_h(z)T_b(w) = \frac{388\,b_2 c_1(w)}{(z-w)^5} + \frac{96\,(b_2 \partial_- c_1 + 2\,\partial_- b_2 c_1)(w)}{(z-w)^4} + \frac{3\,T_b(w)}{(z-w)^2} + \frac{\partial_- T_b(w)}{(z-w)} + \ldots$$

$$T_b(z)T_b(w) = -\frac{348}{5}\frac{1}{(z-w)^6} + \frac{2\,T_h(w)}{(z-w)^4} + \frac{\partial_- T_h(w)}{(z-w)^3}$$

$$+ \frac{1}{(z-w)^2}\left[\frac{3}{10}\partial_-^2 T_h(w) - \frac{16}{15}\partial_-^3 b_1\,c_1(w)\right]$$

$$+ \frac{1}{(z-w)}\left[\frac{1}{15}\partial_-^3 T_h(w) - \frac{8}{15}\partial_-(\partial_-^3 b_1\,c_1)(w)\right] + \ldots. \tag{11.2}$$

The T_h-T_h OPE shows that the functional $-\frac{50}{6}\Gamma_L[h,b]$ indeed reproduces the b independent terms in the logarithm of the determinant. However, since the current algebra (11.2) is significantly different from the exact W_3 algebra at $c = -100$, we can not expect that $-\frac{50}{6}\Gamma_L[h,b]$ is the exact result to all orders in b. For example, in the exact $c = -100$ W_3 algebra, the coefficient of the pole $(z-w)^{-6}$ in the OPE T_b-T_b is $-100/3$, rather than the value $-348/5$ that we find here. This implies that already at the level of the terms quadratic in b, the logarithm of $\det M$ deviates from $-\frac{50}{6}\Gamma_L[h,b]$. The extra terms quadratic in b can be written in the following suggestive way

$$\frac{1}{240}\left(\frac{348}{5} - \frac{100}{3}\right)\frac{1}{\pi}\int d^2z\,b\frac{\partial_-^5}{\partial_+}b = \frac{272}{5\pi c}\int d^2z\,w_0(h,b)\,b + \ldots, \tag{11.3}$$

where we used (see section 8) that $w_0(h,b) = \frac{c}{360}v_L(h,b) + \ldots = \frac{c}{360}\frac{\partial_-^5}{\partial_+}b + \ldots$ at the saddlepoint. We will later see that also certain higher orders in b can be reproduced by the right hand side of (11.3).

If we now look at the terms of order $b^2 h$, we find contributions from (i) the 3-point function $< T_b T_b T_h >$ and (ii) the 2-point functions $< T_b T_{b'u} >$ and $< T_b T_{bu} >$, where in the latter two cases we use that u is given by $u_L(h,b) = \frac{\partial_-^3}{\partial_+}h + \ldots$. It so turns out that the terms (ii) precisely cancel the terms in (i) that come from the anomalous part of the OPE T_h-T_b. The remaining terms are precisely described by $-\frac{50}{6}\Gamma_L[h,b]$ plus the $b^2 h$ terms in the correction term (11.3).

In a similar way one can compute, order by order, higher terms in the logarithm of $\det M$. As announced in section 8, we now discuss the terms of order b^4. These arise from (i) the 4-point function $< T_b T_b T_b T_b >$ and (ii) the 2-point functions $< T_b T_{b'u} >$ and $< T_b T_{bu} >$, where in the latter we now use the part of the saddle-point value $u_L(h,b)$

that is quadratic in b [4]:

$$u(h, b) = \ldots + \frac{1}{30} \frac{1}{\partial_+} \left(3 \partial_- b \frac{\partial_-^5}{\partial_+} b + 2 b \frac{\partial_-^6}{\partial_+} b \right) + \ldots . \tag{11.4}$$

After a straightforward though rather lengthy computation one finds that all the b^4 terms can be collected in two terms proportional to the strucutures $[I]$ and $[II]$ given in (7.12)! One finds

$$\frac{397}{75 \cdot 6!} \frac{1}{\pi} [I] + \frac{64}{25 \cdot 7!} \frac{1}{\pi} [II] . \tag{11.5}$$

The terms proportional to $[I]$ are precisely reproduced by the two terms in $\ln \det M$ which we identified above, leaving us with the terms proportional to structure $[II]$.

In summary, we find the following result, which is exact through the orders h^n, b^2, $b^2 h$ and b^4

$$\ln \det M = \ln \det \begin{pmatrix} \nabla_+^{(2)} & -\frac{1}{10}(\partial_- b) - \frac{1}{15} b \partial_- \\ -L & \nabla_+^{(3)} \end{pmatrix} =$$

$$-\frac{50}{6} \Gamma_L[h, b] + \frac{272}{5\pi c} \int d^2 z \, w_0(h, b) b + \frac{64}{25 \cdot 7!} \frac{1}{\pi} [II] + \ldots , \tag{11.6}$$

where the dots will not affect the abovementioned orders in h, b.

For the computation of $\ln \det N$, we consider the current algebra of the currents H_u and H_v. We find the following OPE's[12]

$$H_u(z, \bar{z}) H_u(w, \bar{w}) = -6 \frac{(z-w)^2}{(\bar{z}-\bar{w})^2} - \frac{(z-w)}{(\bar{z}-\bar{w})} H_u(w, \bar{w}) - \frac{1}{2} \frac{(z-w)^2}{(\bar{z}-\bar{w})} \partial_- H_u(w, \bar{w})$$

$$- 2 \delta^2(z-w) H_{uu}(w, \bar{w}) + \ldots$$

$$H_u(z, \bar{z}) H_v(w, \bar{w}) = -2 \frac{(z-w)}{(\bar{z}-\bar{w})} H_v(w, \bar{w}) - \frac{1}{2} \frac{(z-w)^2}{(\bar{z}-\bar{w})} \partial_- H_v(w, \bar{w}) + \ldots$$

$$H_v(z, \bar{z}) H_v(w, \bar{w}) = \frac{-1}{60} \frac{(z-w)^4}{(\bar{z}-\bar{w})^2} - \frac{1}{180} \frac{(z-w)^3}{(\bar{z}-\bar{w})} H_u(w, \bar{w})$$

$$- \frac{1}{360} \frac{(z-w)^4}{(\bar{z}-\bar{w})} \partial_- H_u(w, \bar{w}) + \ldots \tag{11.7}$$

---

[12] To obtain the terms with $\delta^2(z-w) H_{uu}(w)$, use that $\partial_-^5 \left((z-w)^4/(\bar{z}-\bar{w})\right)$ equals $24\pi \delta^2(z-w)$. This comes about because further terms with derivatives of delta functions cancel each other after partial integration.

Note that the singularities in these OPE's are not just functions of the form $\frac{(z-w)^n}{(\bar{z}-\bar{w})^m}$ (as is the case for pure gravity), but also take the form of bare delta-functions. These lead to terms in the determinant containing a factor u^2, which combine with similar terms coming from the coupling $u^2 H_{uu}$ in (9.13). One already encounters this complication when evaluating the determinant of the operator D_2 defined in (5.21), which is a part of our more complicated operator M. It was suggested in [31] that in the computation of $\ln \det D_2$ the different u^2 terms precisely cancel, so that the final result is simply proporional to $W_L^{grav}[u]$ as in (10.23). We checked this claim for the contributions of the form $\int u^2 \frac{\partial}{\partial \bar{z}} u$ and $\int u^2 \frac{\partial}{\partial \bar{z}} u^2$ and found it to be correct. In our case we do not expect a complete cancellation of the u^2 terms, since the Ward identities (7.4) that determine the form of the reference functional $W_L[u,v]$ explicitly contain u^2 terms.

We used the OPE's (11.7) to explicitly compute the leading terms in the logarithm of $\ln \det N$, which are all consistent, up to a factor of 12, with the WI's (7.4) of the reference functional $W_L[u,v]$, so that

$$\ln \det N = \ln \det \begin{pmatrix} D_1 & \frac{1}{10}v\partial_- + \frac{1}{15}(\partial_- v) \\ 3\,v\partial_- + (\partial_- v) & D_2 \end{pmatrix} = 12\,W_L[u,v] + \dots \quad (11.8)$$

Using the saddle-point expressions $u_L(h,b)$ and $v_L(h,b)$, we can express the result in terms of h and b; we checked that (11.8) is exact through the orders h^n, b^2, b^2h and b^4.

12 All order result for the effective action.

We are now ready to combine the results (7.9), (7.11), (11.6) and (11.8) into an expression for the effective action, which is exact through the orders h^n, b^2, b^2h and b^4 in the leading $1/c$ correction to the saddle-point result. To our great satisfaction, we find that the explicit non-local structure $[II]$ precisely cancels between the induced action and the determinant corrections. The remaining terms are

$$W_{conn}^{W_3}[t,w] =$$
$$\frac{c}{12}W_L[\frac{12}{c}t, \frac{360}{c}w] - 6\,W_L[\frac{12}{c}t, \frac{360}{c}w] - \frac{50}{12}\Gamma_L[h_0, b_0] + \frac{1}{\pi c}\frac{272}{10}\int d^2z\, wb_0 + \dots$$
$$(12.1)$$

Once more using the saddle-point equations, we can rewrite this as

$$W_{\text{conn}}^{W_3}[t, w] =$$

$$\frac{c}{12}\left(1 - \frac{122}{c} + \ldots\right) W_L \left[\frac{12}{c}\left(1 + \frac{50}{c} + \ldots\right) t, \frac{360}{c}\left(1 + \frac{386}{5c} + \ldots\right) w\right].$$

$$(12.2)$$

We thus find that the computed result for $W_{\text{conn}}^{W_3}[t, w]$ can be summarized by the simple formula (12.2). We now propose that the exact, all-order result for this functional can be gotten by simply completing the $1/c$ expansions indicated by the dots in (12.2). This leads to the formula

$$W_{\text{conn}}^{W_3}[t, w] = 2\, k\, W_L \left[Z^{(t)}t, Z^{(w)}w\right], \qquad (12.3)$$

where k, $Z^{(t)}$ and $Z^{(w)}$ are functions of c that allow the $1/c$ expansions

$$k = \frac{c}{24}\left(1 - \frac{122}{c} + \ldots\right)$$

$$Z^{(t)} = \frac{12}{c}\left(1 + \frac{50}{c} + \ldots\right)$$

$$Z^{(w)} = \frac{360}{c}\left(1 + \frac{386}{5c} + \ldots\right). \qquad (12.4)$$

We remark that the result for k is consistent (in the classical limit $c \to -\infty$) with the formula

$$k = -\frac{1}{48}\left(50 - c + \sqrt{(c-2)(c-98)}\right) - 3, \qquad (12.5)$$

which is the *conjectured* outcome of a KPZ type analysis of constraints in a more covariant formulation of W_3 gravity [38, 20]. For the two Z factors, the following all-order results have been proposed ([39])

$$Z^{(t)} = \frac{1}{2(k+3)}, \qquad Z^{(w)} = \frac{\sqrt{30}}{\sqrt{\beta}(k+3)^{3/2}}, \qquad (12.6)$$

with β given in (2.3). They correctly reproduce the singularity structure that one expects, and are in agreement with the expansions (12.4).

For pure gravity formulas similar to (12.3), (12.5) and (12.6) were proposed in [31] on the basis of the results in [30]. It would be most interesting to try and prove the proposal (12.6) on the basis of a similar analysis of W_3.

In general quantum field theories, an extrapolation from (partial) 1-loop results to an all-order result such as (12.3) for the effective action is not at all possible. Clearly, the quantum theory of chiral W_3 gravity enjoys miraculous features that are a consequence of its quantum integrability. It is especially appealing that the rather complicated underlying symmetry structure, being nonlinear and infinite dimensional and falling outside the traditional class of finitely generated Lie algebras, in the end leads to transparent results such as (12.3) for the effective action of the fully quantized theory.

13   Divergence equations and Tini Veltman.

We have extensively used (anomalous) conservation equations for the currents u and v to obtain the quantum theory of W_3 gravity. It is perhaps appropriate to recall how Tini founded his research of quantum Yang-Mills theory on such current divergences.

Tini, in 1966 at Brookhaven, wrote an important paper, of which we reproduce the first page, in which he replaced current commutation rules by divergence equations. This eliminated Schwinger term difficulties. He was able to reproduce the known collection of results from current algebra, in particular the Adler-Weisberger relation. A few months later John Bell showed that these divergence equations can be derived from a theory with a gauge invariance. That was later a reason for Tini to start working on Yang-Mills theories.

In 1966, in London, Tini demonstrated that as a consequence of the divergence equations the decay of the neutral pion in photons was forbidden. John Bell, in the audience, started to work on that. First he stimulated Sutherland to prove this result using current algebra. Since then this forbiddenness is known as the Veltman-Sutherland theorem. Continued work from Bell with Jackiw led to the discovery of the anomaly, independently discovered by Adler as well.

Tini's first paper on massive Yang-Mills theories was published in 1968. It was thought at the time that the theory was not renormalizable, for example Salam and Komar had demonstrated in 1960 that the one-loop three point function contained non-renormalizable infinities. These divergences were caused by the kk/M^2 term in the vector boson propagator; no one knew how to get rid of these kk/M^2 terms. Tini invented a technique (a free field technique, or as called frivolously by Tini, the 'Bell-Treiman transformation'), to get rid of these kk/M^2 terms, and thus to implement the cancellation of many divergences. This technique consisted in taking a free scalar field, and performing a canonical transformation (which does not change the S-matrix, an obvious but important fact) such that it became interacting in such a way that many divergences in loop diagrams canceled. Since the Yang-Mills bosons were still massive, this was not a gauge transformation, but it looked very much like going from the unitary to the renormalizable gauge. The crucial observation by Tini was that one could go to different (renormalizable) formulations ("gauges", although they were really canonical transformations). All this was still at the one-loop level. The breaking of the gauge invariance by the W mass terms led to non-renormalizable divergences at two loops.

Thus renormalizability (by power counting) was proven up to one loop diagrams. This paper revived interest in field theory of vector bosons and stimulated a number of authors (GIM for example), in particular Boulware, Fradkin and Tyutin in 1969. They reformulated the techniques in the context of path integrals, rederiving the same results. Like Tini, they saw one-loop renormalizability at the one-loop level and non-renormalizable divergences at the two-loop level. (Actually, Fradkin and Tyutin thought the theory was completely renormalizable).

A subsequent very important contribution of Fradkin and Tyutin was the application of the free field technique (in path integral setting) to the massless theory. This enabled them to formulate Feynman rules in various gauges. The earlier famous paper of Faddeev and Popov (1967) derives the rules in the Landau gauge. (Tini received this paper as an editor to Physics Letters B. He did not understand the paper, and hesitated for a while, but then decided to accept it anyhow. Fortunately. Just imagine...) Through the work of De Witt (1964) and Mandelstam (1968) the rules were also known in the Feynman gauge. It is this paper of Fradkin and Tyutin that led 't Hooft later to his general gauge formulation for the massless case; he derived the recipe with gauge-fixing and ghost lagrangian as we know today.

The issue of massless versus massive Yang-Mills theory was initially (1968) not well understood. Intuitively one thought that the massless theory would obtain from the massive in the limit of zero mass. This suggested that the kk/M^2 terms were probably harmless. The situation was clarified in 1969-1970 by Faddeev and Slavnov, and Tini with Henk Van Dam; the limit of zero mass of the massive theory turns out to be discretely different from the massless theory. This made it clear why the massless theory was renormalizable while the massive was not (at least at two loops and beyond).

So by the beginning of the seventies many of the ingredients for a successful future quantum field theory of the weak and strong interactions had been found: Yang-Mills theory, currents, diagrammatic techniques, one-loop renormalizability. In addition there was the paper of Gell-Mann and Levy on the sigma model, PCAC and spontaneous symmetry breaking. Many students in Utrecht had to study this article. Yet it was not clear how to combine all these concepts, and most physicists did not appreciate the fundamental importance of the results obtained.

We now reach the beginning of the seventies, when Tini and Gerard did their important discoveries on regularization, renormalizability and unitarity of the Yang-Mills theory. Their work, together with the classical theory of Yang and Mills, has become the basis of particle physics. The radiative corrections which follow from their work, are observed in large accelerators around the world.

References

[1] K. Schoutens, A. Sevrin and P. van Nieuwenhuizen, Phys.Lett. **243B** (1990) 245

[2] K. Schoutens, A. Sevrin and P. van Nieuwenhuizen, Nucl.Phys. **B349** (1991) 791

[3] K. Schoutens, A. Sevrin and P. van Nieuwenhuizen, in Proceedings of the Trieste conference on 'Topological methods in field theory', Int.Jour.Mod.Phys. **A6** (1991) 2891

[4] K. Schoutens, A. Sevrin and P. van Nieuwenhuizen, Nucl.Phys. **B364** (1991) 584

[5] H. Ooguri, K. Schoutens, A. Sevrin and P. van Nieuwenhuizen, *The Induced Action of W_3 Gravity*, preprint ITP-SB-91-16, RIMS-764, Comm.Math.Phys. to be published

[6] K. Schoutens, A. Sevrin and P. van Nieuwenhuizen, *On the Effective Action of Chiral W_3 Gravity*, preprint ITP-SB-91-21, Nucl.Phys. **B** to be published

[7] A. van de Ven, *Two-loop Quantum Gravity*, preprint ITP-SB-91-52

[8] M.H. Goroff and A. Sagnotti, Phys.Lett. **160B** (1985) 81

[9] K. Schoutens, A. Sevrin and P. van Nieuwenhuizen, Phys.Lett. **255B** (1991) 549

[10] A.B. Zamolodchikov, Theor.Math.Phys. **63** (1985) 1205

[11] C.M. Hull, Phys.Lett. **240B** (1990) 110; Nucl.Phys. **B353** (1991) 707

[12] A. Belavin, Adv.Stud. in Pure Math. **19** (1989) 117

[13] A. Bilal and J.-L. Gervais, Phys.Lett. **206B** (1988) 412; Nucl.Phys. **B314** (1989) 646; Nucl.Phys. **B318** (1989) 579; in 'Infinite dimensional Lie algebras and Lie groups', V.G. Kac editor (World Scientific, 1988), p. 483

[14] K. Li, Nucl.Phys. **B346** (1990) 329; Phys.Lett. **251B** (1990) 54

[15] E. Bergshoeff, C.N. Pope, L.J. Romans, E. Sezgin, X. Shen and K.S. Stelle, Phys.Lett. **243B** (1990) 350;

[16] K. Schoutens, A. Sevrin and P. van Nieuwenhuizen, Phys.Lett. **251B** (1990) 355

[17] E. Bergshoeff, P.S. Howe, C.N. Pope, E. Sezgin, X. Shen and K.S. Stelle, Nucl.Phys. **B363** (1991) 163.

[18] C.N. Pope, L.J. Romans and X. Shen, Phys.Lett. **236B** (1990) 173; Nucl.Phys. **B339** (1990) 191.

[19] F. Bastianelli, *An Action for Super-W_3 Gravity*, preprint ITP-SB-90-91

[20] Y. Matsuo, Phys.Lett. **227 B** (1989) 209; in proceedings of the meeting 'Geometry and Physics', Lake Tahoe, July 1989

[21] C.M. Hull, *W Gravity Anomalies 1: Induced Quantum W Gravity*, preprint QMW/PH/91/2; Phys.Lett. **265B** (1991) 347.

[22] K. Schoutens, A. Sevrin and P. van Nieuwenhuizen, in Proceedings of the Jan. 1991 Miami Workshop on 'Quantum Field Theory, Statistical Mechanics, Quantum Groups and Topology' (Plenum, 1991)

[23] A.T. Ceresole, M. Frau, J. McCarthy and A. Lerda, Phys.Lett. **265B** (1991) 72

[24] K. Li and C. Pope, in Proceedings of the Trieste Summer High-Energy Workshop, July 1990

[25] C.M. Hull, *W Gravity Anomalies with Ghost Loops and Background Charges*, preprint QMW/PH/91/14; C.M. Hull and L. Palacios, *W algebra Realisations and W Gravity Anomalies*, preprint QMW/PH/91/15.

[26] J. Thierry-Mieg, Phys.Lett. **197B** (1987) 368; K. Schoutens, A. Sevrin and P. van Nieuwenhuizen, Comm.Math.Phys. **124** (1988) 87;

[27] L.J. Romans, Nucl.Phys. **B352** (1991) 829

[28] C.N. Pope, L.J. Romans and K.S. Stelle, Phys.Lett. **268B** (1991) 167

[29] A.M. Polyakov, Mod.Phys.Lett. **A2** (1987) 893; in Proceedings of the Les Houches 1988 meeting on 'Fields, Strings and Critical Phenomena' (Elsevier, 1989)

[30] V.G. Knizhnik, A.M. Polyakov and A.B. Zamolodchikov, Mod.Phys.Lett. **A3** (1988) 819

[31] Al.B. Zamolodchikov, *Liouville Action in the Cone Gauge*, preprint ITEP 84-89 (1989)

[32] A. Bilal and J.-L. Gervais, Nucl.Phys. **B326** (1989) 222

[33] S.R. Das, A. Dhar and K.S. Rama, *Physical properties of W gravities and W strings*, preprint TIFR/TH/91-11; *Physical states and scaling properties of W gravities and W strings*, preprint TIFR/TH/91-20

[34] C.N. Pope, L.J. Romans and K.S. Stelle, *On W_3 strings*, preprint CERN-TH.6188/91; C.N. Pope, L.J. Romans, E. Sezgin and K.S. Stelle, *The W_3 string spectrum*, preprint CPT TAMU-68/91

[35] A.A. Belavin, A.M. Polyakov and A.B. Zamolodchikov, Nucl.Phys. **B241** (1984) 333

[36] F.A. Bais, P. Bouwknegt, K. Schoutens and M. Surridge, Nucl.Phys. **B304** (1988) 348

[37] J. Pawelczyk, Phys.Lett. **255B** (1991) 330

[38] M. Bershadsky and H. Ooguri, Comm.Math.Phys. **126** (1989) 49

[39] J. de Boer, private communication

thesis of Ref. 5). Note, however, that (5) is not quite the same as the SU(6) result[9] which in our notation reads

$$(f_\rho{}^2/4\pi) = (G_{\pi NN}{}^2/4\pi)(g_V/g_A)^2(m_M{}^2/m_B{}^2)$$

$$= (9/25)(G_{\pi NN}{}^2/4\pi)(m_M{}^2/m_B{}^2), \qquad (6)$$

where m_M and m_B, respectively, stand for the mean masses of the meson 35-plet and the baryon 56-plet of SU(6).

It is a pleasure to thank Professor Peter G. O. Freund for stimulating conversations which led to this investigation.

---

*This work supported in part by the U. S. Atomic Energy Commission.

†Alfred P. Sloan Foundation Fellow.

[1]M. Gell-Mann, Phys. Rev. 125, 1067 (1962).

[2]M. Gell-Mann and M. Lévy, Nuovo Cimento 16, 705 (1960); Y. Nambu, Phys. Rev. Letters 4, 380 (1960); J. Bernstein, S. Fubini, M. Gell-Mann, and W. Thirring, Nuovo Cimento 17, 757 (1960).

[3]Y. Tomozawa, to be published.

[4]S. Weinberg, to be published.

[5]J. J. Sakurai, Ann. Phys. (N.Y.) 11, 1 (1960).

[6]See also M. Gell-Mann and F. Zachariasen, Phys. Rev. 124, 953 (1961). These authors define a constant γ_ρ, which is equivalent to our $(\frac{1}{2}f_\rho)$.

[7]Equation (5) gives a ρ width of 160 MeV. Possible off-the-mass-shell corrections are discussed in footnote 10 of Ref. 8. It may also be mentioned that if we use $G_{\pi NN}{}^2/4\pi$, which gives the correct pion lifetime when inserted in the Goldberger-Treiman relation, then the predicted ρ width is changed to 120 MeV in good agreement with observation.

[8]K. Kawarabayashi and M. Suzuki, Phys. Rev. Letters 16, 255 (1966). See also Riazuddin and Fayyazuddin, Phys. Rev. 147, 1071 (1966).

[9]F. Gürsey, A. Pais, and L. A. Radicati, Phys. Rev. Letters 13, 299 (1964).

DIVERGENCE CONDITIONS AND SUM RULES*

M. Veltman†

Brookhaven National Laboratory, Upton, New York

(Received 29 July 1966)

Recently several sum rules have been derived employing current commutation[1] rules and divergence conditions for those currents. As is well known, the application of commutation rules involves the manipulation of the so-called Schwinger terms,[2] and where some of these calculations avoid such complications, others may be criticized in this respect. An alternative derivation of these sum rules, based on assumptions other than current commutation relations may, therefore, be of help in understanding the mechanism involved.

Consider the vector current of hadrons that is coupled to leptons and photons. Neglecting higher-order electromagnetic (em) and weak interactions one customarily assumes, following Feynman and Gell-Mann,[3]

$$\partial_\mu \vec{J}_\mu{}^V = 0. \qquad (1)$$

As is well known, em interactions break this law for the charged components of $\vec{J}_\mu$. Similarly the weak interactions break (1) for the neutral component because they carry off a nonzero charge. (Remember that $\vec{J}_\mu$ is the hadron current only.) We will try to find the

first-order em and weak effects on (1).

According to the principle of minimality, we find the em effect on (1) by substituting $\partial_\mu - ieA_\mu$ for ∂_μ applying to a (negative) charged field. Thus, neglecting here the case that $\vec{J}_\mu{}^V$ itself contains derivatives, we find

$$\partial_\mu \vec{J}_\mu{}^V = +ie\vec{A}_\mu \times \vec{J}_\mu{}^V,$$

$$(\vec{A}_\mu \times \vec{J}_\mu)^i = i\epsilon_{ijk}A_\mu{}^j J_\mu{}^k, \qquad (2)$$

where $\vec{A}_\mu$ is an isotopic vector whose first two components are zero. Equation (2) is already sufficient to derive the Cabibbo-Radicati[4] sum rule.

In accordance with the observations made above, we generalize (2) to include also first-order weak interaction effects:

$$\partial_\mu \vec{J}_\mu{}^V = ie\vec{A}_\mu \times \vec{J}_\mu{}^V + ig\vec{l}_\mu \times \vec{J}_\mu{}^V. \qquad (3)$$

Here l_μ represents the lepton current.[5] Equation (3) is valid if no axial currents are present. The generalization to include axial currents also requires some care. Let us intro-

Tini and the Cosmological Constant*

Leonard Susskind
Department of Physics
Stanford University
Stanford, CA 94305-4060

*Lecture given at the 60th birthday celebration of Tini Veltman, University of Michigan, 1991.

Tini and the Cosmological Constant

In addition to his many technical accomplishments in physics, Tini Veltman was the first to clearly call attention to what is undoubtably one of the greatest paradoxes in the history of physics, namely why does the vacuum energy not gravitate. According to the standard theory of elementary particles, the vacuum is not the state in which all fields vanish but even at the classical level, at least one field, the Higgs, is spontaneously shifted from zero, thus giving rise to a vacuum energy $V(\phi_0) \sim (250 \text{ GeV})^4$. Why this energy does not manifest itself as a source of gravitational field is a mystery. Einsteins' equations in the presence of such a source are entirely consistent

$$R_{\mu\nu} - \frac{1}{2} g_{\mu\nu} R = k T_{\mu\nu}$$
$$= \Lambda_1 g_{\mu\nu}$$

with the cosmological constant Λ being the vacuum energy density itself. The situation is scandalous. The overwhelming bulk of the energy of the world is not seen as a source of gravity while only the minute left over bit in fluctuations is gravitationally potent.

The only known answer is that the zero of energy can be arbitrarily cancelled by adding an additional term to the left side of the gravitational equations. A new coupling constant Λ_0 with dimensions of $(\text{length})^{-4}$ is introduced.

$$R_{\mu\nu} - \frac{1}{2} g_{\mu\nu} R + \Lambda_0 g_{\mu\nu} = k T_{\mu\nu}$$

The constant Λ_0 is of course Einsteins cosmological term but now it is introduced not to create an effect but rather to cancel one. Furthermore, the cancellation must be of almost perfect precision. The difference $\Lambda = \Lambda_0 - \Lambda_1$ is known to be less than $10^{-12} (\text{ev})^4$ and may well be exactly zero. The knowledge that Λ is so small comes not from high energy physics but from cosmology at the very largest scales.

The problem is that we are inclined to believe that the origin of fundamental constants such as Λ_0 and the various constants which enter into Λ_1 is at very small microscopic scales. From this point of view it seems a completely mysterious conspiracy that the various microscopic constants should combine in just such a way as to lead to a simple and special cosmic behavior.

The situation only becomes more uncomfortable when quantum flucuations are accounted for. The expression for the vacuum energy of a quantum field receives contributions from fluctuations at every length scale. The weak scale is only one of many scales which appear in particle physics and in general, the smaller the distance scale the larger the contributions. Typically fluctuations with wave length ℓ give rise to vacuum energy of order ℓ^{-4}. The preceise contribution is proportional to a complicated function of coupling constants which can only be determined perturbatively as an infinite series. If we assume that scales of distance down to the planck scale contribute, then the vacuum energy is of order 10^{76}GeV^4 with the exact numerical value being determined from the infinitely complicated solution of quantum field theory including quantum gravity. What principles guarantee that the resulting vacuum energy cancels to one part in 10^{120} is at present totally unknown.

Needless to say there have been no lack of attempts to explain the vanishing of Λ. Most have been foolish, a few interesting and probably all are wrong. I will discuss three.

1. Supersymmetry

Supersymmetry is a remarkable symmetry which interchanges fermions and bosons. Its simplest consequence is that it predicts what particles come in fermion-boson pairs which are degenerate in mass, change color but not spin. It is easy to see that if the particles are free then the vacuum energy is zero. Recall that the zero point energy of a free boson field is $\frac{1}{2}\hbar\omega$ for each mode of the field with frequency ω. Similarly for a fermi field the answer is $-\frac{1}{2}\hbar\omega$. If the particles come in degenerate pairs the energies cancel.

What about radiative corrections? For example, the masses of the particles can be shifted by interactions. The miracle of supersymmetry is that order by order in perturbation theory the symmetry persists and the vacuum energy is exactly zero. In fact, the contribution of each scale of wavelengths and frequencies cancel.

But supersymmetry can not be an exact symmetry. It must be either explicitly or spontaneously broken. In either case the degeneracy of superpartners is lifted. We know this because the superpartners of the usual particles have mass of at least a hundred GeV or more. The breaking of supersymmetry destroys the delicate balance which cancelled Λ. Nevertheless, the natural magnitude of Λ will be much smaller in a broken supersymmetric theory than in other kinds of theories. The reason is that the splitting of masses does not affect scales of momentum much larger than the splitting. The contribution to vacuum energy from momentum scales larger than the typical supersymmetry breaking scale still vanishes. Unfortunately since the symmetry breaking scale must be at least ~ 100 GeV, we are very far from an explaination of the smallness of Λ.

II. Wormholes

Sidney Coleman's wormhole ideas are to my mind the most interesting of the failures. Of course not everyone agrees that wormholes failed but I am convinced.

Wormholes are planck scale configurations which connect otherwise distant space-time regions. They can even connect otherwise disconnected regions. According to Coleman, they should be thought of as configurations in the euclidean path integral formulation of quantum gravity. They can be thought of as microscopic tubes connecting large smooth regions of space time. Such configurations are thought to be part of the sum over histories of the gravitational field which defines quantum gravity.

This lecture is not the place to review the technical aspects of

Coleman's proposal. However, I will try to give a picture in words. First of all the wormholes have a small distance or microscopic importance. They are continually popping open and closing off so that our smooth space time is in reality full of wormhole ends. The effect of these insertions was argued by Coleman to renormalize coupling constants including Λ_0.

Secondly there is a long distance aspect to wormholes. They connect regions of arbitrarly large separation. Thus the source of a wormhole end that appears on earth in 1991 could be near the Andromeda galaxy 6 billion years ago. In other words, the distribution of coupling-constant-renormalizing-worm-hole-ends could depend on the global structure of space-time.

Coleman then argues that summing over such configurations gives rise to a "probabiltiy" for the cosmological constant which he computes to be

$$\exp\left[\exp\tfrac{1}{\Lambda}\right]$$

In other words, the probability for a given Λ is violently peaked at $\Lambda = 0$.

I do not believe that Coleman's logic survives careful scrutiny. In addition to a number of lesser problems of a phenomonological type and the so called giant worm hole catastrophe which might be overcome, I believe the basic logic is incorrect. The Euclidean path integral is a notoriously ill defined concept for gravity and can be made to yield any result. It is therefore important to use methods which do not rely on it. Klebanov, Fischler, Polchinsky and I attempted to analyse the quantum mechanics of a system of large universes coupled by worm-holes. We found a number of things. First of all the quantity which Coleman identifies as the probability for the cosmological constant is not a probability. It is the amplitude for no universes to occur. Furthermore it is not $\exp\left[\exp\tfrac{1}{\Lambda}\right]$ but more likely $\exp[-\exp\tfrac{1}{\Lambda}]$.

Secondly we found a result that had previously been independently found by Rubakov that the mean number of large universes formed is $\exp\tfrac{1}{\Lambda}$ so that far more universes occur when Λ is vanishing by small. These universes are however not viable candidates for our own universe because they are completely devoid of matter. The number of viable universes is not peaked as $\Lambda \to 0$.

More recently Adrian Cooper, Larus Thorlacius and I have have studied the case of two-dimensional space-time. The mathematics in this case is identical to string theory and a good deal is known. Polchinsky had previously shown that all the conditions needed for Coleman's arguement were present in the 2-d case and so if Coleman is right the two-dimensional Λ should vanish in this theory.

The structure of string theory and its relation to two dimensional gravity is through the renormalization group. The theory we are renormalizing is two-dimensional quantum gravity coupled to some matter fields called X_i. There are an infinite set of coupling constants which describe all possible covariant interactions and these couplings are functions of scale size. Among them is the two-dimensional cosmological constant.

Long ago Ken Wilson suggested that the equations governing the running of couplings were like equations for a mechanical system's time evolution. For example we could consider the field potential V(x) of some scalar field x. In quantum field theory the potential runs with scale so we consider it to be a function of x and the log of the scale called t by Wilson. The remarkable fact discovered by string theorists is that the renormalization group equations in 2-d gravity have the form of coupled relativistic wave equations for a collection of "target space fields" like V(x,t).

The two-dimensional cosmological constant is the value of V(x,t) at its minimum in x-space. Its dependence on t is just the usual dependence of a running coupling on scale. What then is the problem of the cosmological constant from this viewpoint? The answer is simple: The target space field is a tachyon. It is unstable and tends to roll away from V = 0. The initial condition at t = 0 (short distance behavior) must be fine tuned to insure that at large scale (t → ∞) the field V arrives at exactly V = 0, the point of unstable equalibrium.

Do wormholes help? To answer this we go back to string theory where wormholes are higher topologies such as handles on the worldsheet. The effect of such configurations is to quantize the equations for the target space fields. Note that I am not talking about quantizing 2-d gravity. That was already assumed, we are talking about 3rd quantization in which the target space fields become quantum fields.

The effect of quantizing the theory of target space fields turns out to be concentrated at small scales (early target space time t). This is roughly equivalent to Coleman's assumption that only small wormholes are important. However, their effect does not in any way seem to stabilize the 2-d cosmological constant at zero. Instead it introduces quantum fluctuations in the initial conditions which would even frustrate the fine tuning option.

Perhaps there is some kind of truth to Coleman's idea, but I think it has yet to be found.

III. Anthropic Arguements

My impression is that there is no explanation of the vanishing of Λ within any more or less conventional context. We probably need a totally new framework for thinking about the parameters of nature. One such very radical idea is the anthropic principle proposed by Dicke. The anthropic principle has a bad reputation among most physicists who view it as a metaphysical antiscientific cop-out. They may be quite right. But maybe not. Let me try to convince you that it is worth thinking about.

A Fish Story

It seems that long ago in a very ancient sea a big brained race of fish evolved intelligence. Because of their physiology, they were confined to a fairly narrow range of depths and therefore knew neither of the sea's surface nor its bottom. They believed the universe to be a homogeneous sea of water-ether.

Their scientists became puzzled about one point. How is it that the temperature of the sea is so finely adjusted that it is neither frozen nor steam or ionized plasma or something even worse. Of all the possible values of temperature from absolute zero to the planck temperature, how is it that the universe is liquid.

Then came The Fish Dicke who proclaimed the ichthropic principle. The universe is liquid because only in a liquid can fish exist. The fish Dicke was essentially correct. As they might have found out, had they ever reached the surface, the universe is a very big place with a very wide range of temperatures from 3° in intergalactic space to extrodinary temperatures in the hearts of quasars. In fact the conditions for liquid water are very very rare. But because fish can only evolve in water that is in fact where they are found.

Can we apply similar reasoning to the cosmological constant fine tuning? To do so we would first of all need a theory in which an enormous range of values of Λ could occur in different regions of space time. In other words an incredible number of stable or almost stable vacua should exist and that in the formation of the universe domains formed with almost any value of Λ. The second condition is that life can only form if Λ is at most an order of magnitude or two bigger than the observational bound.

As to the first of these conditions, I will only note that string theory seems to have an incredible number of vacuua. Unfortunately, we know very little about them. As to the second question, Weinberg has discussed it in some detail. He finds that if Λ were a factor of ~ 100 bigger than the empirical bound, galaxies could not form.

Does the anthropic principle lead to an observable prediction? I think it does. Since there is no reason for Λ to be smaller than the threshold for like to form it should not be zero. In fact, it should be about at the limits of observation.

Is Λ Exactly Zero?

This brings me to the final point in my talk. Is Λ zero? It is too early to say with any confidence, but it is possible that a non-zero Λ is needed to repair a serious cosmological inconsistency. The most recent measurements of the Hubble constant are converging at about 80 km/sec/Mpc. For a coventionally matter dominated universe this would correspond to an age of about 8 billion years. However, the oldest observed stars are believed to be about 15 billion years old. Most likely a pedestrian way out of this inconsistency will be found. But if not, a non zero Λ can fix things. With a $\Lambda \sim (10^{-3} eV)^4$ the age of the universe can be pushed to 15 billion years.

In conclusion I would like to thank Tini for providing us with so interesting a puzzle and maybe even the "Dark Cloud On The Horizon" which will lead us to the next revolution in Physics.

THE VELTMAN VERTEX
(in celebration of Martinus Veltman's 60th birthday)

R. Thun

University of Michigan

Let me begin by saying that I enjoy immensely Tini Veltman's presence in our Physics Department. He's generally the first guy I go to when a deeply puzzling question bounces around in my brain. If it's a good question, I can be sure he's already thought about it. Tini is always generous with his time, and even when there is no answer, the ensuing discussions are always lively and illuminating. What better compliment can one give one's colleague? Tini is a man with deep insights into the human condition who rarely hides his convictions. His remarks, jokes, opinions, and judgements are at times outrageous, but they rarely miss the truth of a situation. I'm not surprised that Tini would love to see strongly interacting W bosons: the Veltman Vertex is anything but feeble.

A few years ago, when life in Randall Lab was a bit sleepy, we had afternoon coffee and cookies in the tiny alcove next to the preprint shelves of our seminar room. It's still not clear to me how the faculty and students of our great department were all expected to participate in this social event. Tini alone used up about 25% of the available floor space. However, the resulting short-range interactions led to some deeply philosophical discussions. I recall one conversation in which Tini remarked in his usual blunt way that we walk around this Earth for a few years, and then we are dead forever. I don't remember the context of this observation, but it surely provides some insight into Tini's passion for science and his provocative view of life and of the various people that participate in its adventure.

Ideally, a talk like this ought to reflect deeply on the work of the man we are celebrating. For me this represents a slight difficulty which is illustrated by the following story. A few years ago I sat next to Tini during a dinner associated with a ritual familiar to many of you: the annual DoE review. For some reason, which now escapes me, the conversation turned to the topic of renormalization. Tini leaned over to me in his inimitable way, looked across the top of his glasses, and said: "You know Rudi, there are only about five people in the world who really understand renormalization." Now, it was absolutely clear then, and it shall remain so forever, that I'm not one of those five. I have no profound insight into dimensional regularizations [1]. I have trouble enough dealing with 4 dimensions, let alone 3.999 or 4.001. Tini's virtuosity with computers and calculations is well known, but personally I have little understanding of the intricacies that he has mastered in these areas.

However, over the years I've been intrigued by a number of fundamental physics questions toward whose understanding Tini has made profound contributions. For example, the repetition of fermions completely eludes our understanding. The existence of the muon is as mysterious now as it was at the time of its discovery. The simplest question one can ask in this regard concerns the number of lepton and quark families. A number of years ago, I became excited by the prospect of measuring the number of neutrino species in electron-positron annihilations above the Z resonance by measuring the photon spectrum in the process $e^+e^- \rightarrow \gamma\nu\bar{\nu}$ [2]. I remember talking to Tini about this, and hearing him proclaim with great conviction that there could only be three families of fermions. His reasoning was based on a few simple arguments. Tini had calculated the shift of the W and Z masses arising from fermion loops in the propagators of these bosons [3]. By the mid-1980's the ratio of $M_W/(M_Z cos\theta_w)$ was known well enough to constrain the mass splitting of any fermion weak isodoublet to less than about 300 GeV. But as shown in fig. 1, the pattern of the known (and quasi-known) quark masses, extrapolated in a reasonable way, required the mass splitting of a fourth generation of quarks to be substantially larger than this. Tini's insight proved to be correct. The first measurements of the number of light neutrinos were not made by observing radiative electron-positron annihilation as I had originally hoped, but rather by measuring the cross section at the peak of the Z resonance [4] as shown in fig. 2. It is amusing to note that this determination was made by possible by the fact that the cross section is maximal for a minimal Z, one that couples only to the three neutrinos and to the known charged fermions. Tini's daughter Helene made major contributions to the construction of the Mark II luminosity monitor, a device that was essential for making this measurement. The Old Man's desire to turn his daughter into an experimentalist was therefore not entirely in vain, although my best effort at facilitating his wish were insufficient to overcome the Veltman genetic heritage.

Of course, one of the key themes of Tini's work has been the enigma of the so-called

Higgs mechanism [5]. The thing is a little bit like the white whale in Moby Dick. These days, with LEP, LHC, and SSC either here or on the horizon, even kindergarden kids know that the Higgs boson is THE physics goal of future high-energy experiments. Tini already worked on the Higgs problem during ancient times when even the detection of W's and Z's seemed to be just a dream for the distant future. When accelerators are inadequate to materialize the particles of their dreams, physicists attempt to glimpse the outline of very-high energy phenomena through the murky window of radiative effects. Tini's investigation of Higgs-induced radiative effects showed, unfortunately, that the Higgs mechanism eludes this form of spying by producing only highly suppressed logarithmic deviations from lowest-order scattering at small energies. As shown in fig. 3, vertex corrections involving fermions include a factor M_f^2/M_W^2 which has a very small value for all fermions except the top quark. Tini called the suppression of Higgs radiative effects the "screening theorem" [6]. One way to reduce this suppression is to produce massive particles in the final state, and Tini pointed to the annihilation process $e^+e^- \rightarrow W^+W^-$ as a promising reaction for sensing the existence of Higgs fields [7]. Depending on Higgs mass, the effect on the cross section can be as large as 5%, and may be observable as LEP energies exceed the W-pair threshold in a few years.

At the time, this "screening theorem" must have been a depressing finding. However, the remarkable ambition of those who build accelerators and measure particle collisions will allow us to perform a much more frontal assault on the problem of electroweak symmetry breaking in the near future. Already at LEP, the lower limit on the Higgs mass has been pushed up to about 50 GeV [8]. By the mid-nineties we can expect the mass reach of Higgs searches to be extended to about 80 GeV when LEP operates at c.m. energies up to 200 GeV [9]. At the turn of the century, if all goes as planned, the two large hadron colliders LHC at CERN and SSC at Texas will stretch the accessible mass range out to about 1 TeV [10]. Tini was one of the first to recognize that this is a critical demarcation energy for the Higgs mechanism [11]. A Higgs particle with a mass well below 1 TeV leads to a reasonably self-consistent, perturbative theory of electroweak interactions. If the mass is above 1 TeV, the picture of electroweak symmetry breaking is much less clear, but is likely to involve strong interactions of some kind. Despite more than a decade of intense speculation on these matters, and despite the great success of the Standard Model at lower energies, it is really remarkable how little we know about particle physics above the several-hundred GeV level.

Electroweak symmetry breaking is characterized by the existence of massive particles. As simple-minded as it sounds, the mechanism for this symmetry breaking is investigated most effectively by studying the most massive particles available to us in the laboratory. It is therefore not surprising that Higgs searches conducted at LEP and as proposed for LHC and SSC involve processes with either real or virtual W bosons, Z bosons, or t quarks. At a fundamental level we are searching for a fifth force which, like gravity, has a strength dependent on mass.

During the past few years, I have participated in the design and planning of an SSC detector which has as one of its primary goals an elucidation of this mass-dependent fifth force. The detector goes by the acronym SDC (for Solenoidal Detector Collaboration [12]) and is shown schematically in fig. 4. Very few people outside of high-energy physics appreciate the remarkable complexity of the instrumentation and human interactions required to perform this kind of science. The SSC will generate interactions at a rate of one hundred million or more per second, of which only a few can be recorded and analyzed. A detector such as SDC will have of order one million or more elements for sensing collision products. Effectively, 10^{14} pieces of information must be evaluated each second to decide which are the precious few events that might contain the Higgs particle or other equivalently interesting physics.

The design, construction, and eventual operation of this detector will involve more than a thousand physicists who are scattered all over the world, and who belong to hundreds of independent organizations. It has been fascinating to observe the interactions between so many participants during the birth process of this one-of-a-kind enterprise. It reminds me of quantum mechanics with its many amplitudes, interference effects, Heisenberg uncertainties, and occasional collapses of wave functions. The sequence of realities emerging through this process can be understood in retrospect but is not easily anticipated. This sort of non-algorithmic, collaborative activity also raises interesting questions which have application far beyond high-energy physics. Clearly, individuals working in a complex but directed group activity have some similarity to neurons in a brain engaged in problem solving. Do groups develop a supra-individual intelligence, consciousness, and awareness?

Such philosophical speculations are amusing, but let's return to the question at hand.

What are the prospects for observing a Higgs boson, assuming that it indeed exists? The expected cross section for producing Higgs particles at the SSC is shown in fig. 5. To convert this to an expected signal, one must take into account that the only obviously "clean" decay mode is $H \rightarrow ZZ \rightarrow 4$ leptons (electron or muon) with a branching fraction of about 0.0011. By "clean" we mean a mode with reasonably small backgrounds. It would be nice if the Higgs searches could be extended to decay channels with final-state quark jets so that one could avoid the penalty of such small branching fractions. However, backgrounds from $pp \rightarrow Z+$ jets are expected to be both severe and not known with great precision. Figure 6 illustrates the kind of Higgs signals that might be observed with the SDC detector during a year of running at the SSC, corresponding to an integrated luminosity of $10^{40} cm^2$. What is interesting about these simulated results is that the ability to distinguish a signal fades as the Higgs mass approaches a TeV, a magnitude at which the theoretical concept of a Higgs particle itself begins to lose meaning.

Assuming that a Higgs particle exists, do we have any clue about its mass? I know of three arguments, none of them completely compelling, which point to a mass of several hundred GeV. On a most general level, the weak interactions are characterized by a mass scale given by $1/\sqrt{G_F} = 293$ GeV. It is, of course, not obvious what numerical factor multiplies this scale to obtain the Higgs mass. In a paper examining the stability of electroweak interactions with respect to radiative effects [13], Tini derived a tantalizing mass relationship given by:

$$(1) \qquad \sum M_f^2/M_W^2 = 3/2 + 3/(4cos^2\theta_w) + 3M_H^2/4M_W^2$$

where the sum over fermions includes distinct color states, and is completely dominated by the top quark. Equation (1) expresses the assumption that the quadratically divergent contributions to the Higgs self energy cancel at the one-loop level. The relevant diagrams are shown in fig. 7. Neglecting all quarks except the t yields:

$$(2) \qquad 3M_t^2/M_W^2 = 3/2 + 3/(4cos^2\theta_w) + 3M_H^2/4M_W^2$$

or

$$(3) \qquad M_H^2 = 4M_t^2 - 2M_W^2 - M_Z^2$$

which relates the masses of the heavy particles: W, Z, t-quark, and Higgs! For M_t=150 GeV one obtains M_H=260 GeV and for M_t=200 GeV, M_H=370 GeV. A third approach, by Bardeen et al. [14], postulates a $t\bar{t}$ condensate as the origin of electroweak symmetery breaking, and obtains mass predictions for the t quark and Higgs boson using renormalization group techniques. The result is M_t=229 GeV and M_H=256 GeV if the scale of the interaction triggering the condensate is 10^{15} GeV. These predicted mass values depend only weakly on this scale. If indeed there is a Higgs particle with a mass around 300 GeV, it will be readily observed at LHC and SSC.

I believe that Tini would be somewhat disappointed by the discovery of a 300 GeV Higgs boson, although he and the rest of us would have to accept whatever nature has invented. One reason for Tini's uncomfortable feeling about the Higgs mechanism is its impact on cosmology [15]. The basic problem is that the Higgs condensate introduces a stupendous contribution to the cosmological constant which then must be cancelled by some other, unknown term to give agreement with the observation that our universe has very little intrinsic curvature. So in the end we somehow come back to understanding the nature of the physical vacuum. What kinds of experiments can one do with the vacuum? Pursuing this question a few years ago, I looked into the work of one of Tini's compatriots, H. B. G. Casimir, who derived the result that two perfectly conducting but uncharged parallel plates attract each other with a force per unit area given by:

$$(4) \qquad F = \hbar c \pi^2/(240a^4) = 0.013/a^4 \; dyne/cm^2 \; (a \; in \; microns)$$

where a is the separation of the plates [16]. This prediction was subsequently confirmed within the limitations of dealing with imperfect conductors [17]. The boundary conditions introduced into the vacuum by plates of finite area shift the zero-point energy of electromagnetic field oscillations by a finite amount U that depends on the separation a. The resulting force is given by -dU/da. Equation (4) is really remarkable because it implies that anyone possessing conducting plates can extract energy out of the vacuum. When discussing the Casimir effect with Tini, he pointed out that the dressing of the vacuum by the Higgs field is very different from the phenomenon of zero-point fluctuations. Spontaneous symmetry breaking produces a finite, non-zero field value everywhere, whereas the Casimir effect arises from fields fluctuating around zero mean values. Nevertheless, it seems to me that if the Higgs mechanism is to be understood in terms of a modification of the vacuum, a complete understanding of the mechanism must include a consideration of all dynamical aspects of the vacuum. I believe that Tini would agree that the problem of the cosmological constant will be with us for a long time.

In this talk I have touched on a few of the central questions of modern particle physics towards whose formulation and answers Tini Veltman has made major contributions. During the preparation of the talk I began to notice a remarkable thing. I was never tempted to proceed with a systematic literature search of Veltman's papers. Those quoted in the references below simply accumulated in my office as a result of many years of impromptu interactions with Tini. Retrieving these haphazardly collected papers reminded me that the joy of doing science is a living thing which somehow seems to survive the jealousies, false ambitions, endless reviews, and constant institutional imperatives that are part of the business of doing science. What I like most about Tini is that he never loses sight of the most important reason for doing physics: to examine and understand to the best of our ability the universe which we are privileged to experience during our brief sojourn on Earth.

REFERENCES

[1] G.'t Hooft and M. Veltman, Nucl. Phys. B44 (1972) 189

[2] G. Barbiellini, B. Richter, J. L. Siegriest, Phys. Lett. 106B (1981) 414

[3] M. Veltman, Nucl. Phys. B123 (1977) 89

[4] G. S. Abrams et al., Phys. Rev. Lett. 63 (1989) 2173
   D. Decamp et al., Phys. Lett. B235 (1990) 399
   B. Adeva et al.,Phys. Lett. B237 (1990) 136
   M. Z. Akrawy et al., Phys. Lett. B240 (1990) 497
   P. Abreu et al., Phys. Lett. B241 (1990) 435

[5] P. W. Higgs, Phys. Lett. 12 (1964) 132

[6] M. Veltman, Acta Phys. Pol. B8 (1977) 475
   M. Veltman, Phys. Lett. 70B (1977) 253

[7] M. Lemoine and M. Veltman, Nucl. Phys. B164 (1980) 445

[8] D. Decamp et al., Phys. Lett. B246 (1990) 306 and CERN-PPE/91-19 (ALEPH Collaboration) which gives a 95% C.L. lower limit on the Higgs mass of 48 GeV.

[9] Physics at LEP, CERN-86-02 (1986)

[10] Conceptual Design of the Superconducting Super Collider, SSC-SR-2020 (1986)
The Large Hadron Collider in the LEP Tunnel, CERN 87-05 (1987)

[11] see references in [6]
and B. W. Lee, C. Quigg, and H. B. Thacker, Phys. Rev. Lett. 38 (1977) 883

[12] Letter of Intent by the Solenoidal Detector Collaboration (1990)

Expression of Interest, Solenoidal Detector Collaboration (1990)

[13] M. Veltman, Acta Phys. Pol. B12 (1981) 437

[14] W. A. Bardeen, C. T. Hill, and M. Lindner, Phys. Rev. D41 (1990) 1647

[15] M. Veltman, Phys. Rev. Lett. 34 (1975) 777

[16] H. B. G. Casimir, Proc. kon. Ned. Akad. Wetenschap 51 (1948) 793
for a recent review of the Casimir effect see C. Plunien, B. Müller, and W. Greiner, Phys. Rep. 134 (1986) 87.

[17] M. J. Sparnaay, Physica 24 (1958) 751

FIGURES

1) Pattern of quark masses.

2) a) First detailed measurement of the Z line shape in electron-positron annihilations by the Mark II collaboration (G. S. Abrams et al., ref [4]). The number of light neutrinos deduced from the peak cross section is 2.8 ± 0.6.

   b) High-statistics measurement of the Z-line shape performed by the L3 collaboration (B. Adeva et al., ref [4]).

3) Some diagrams contributing to Higgs radiative effects.

4) Proposed SSC detector of the Solenoidal Detector Collaboration.

5) Expected cross section for producing Higgs bosons at the SSC as a function of the Higgs and t quark masses.

6) Expected Higgs signals for a one year run at the SSC for Higgs masses of 200, 400, and 800 GeV.

7) Diagrams contributing to the Higgs self energy.

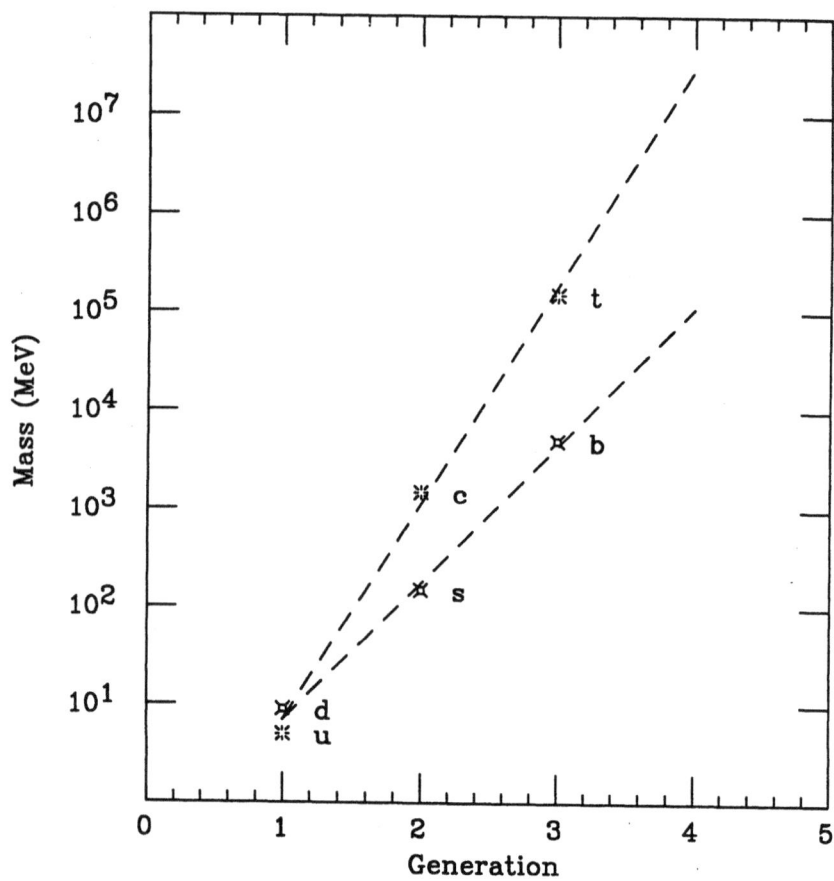

Fig. 1

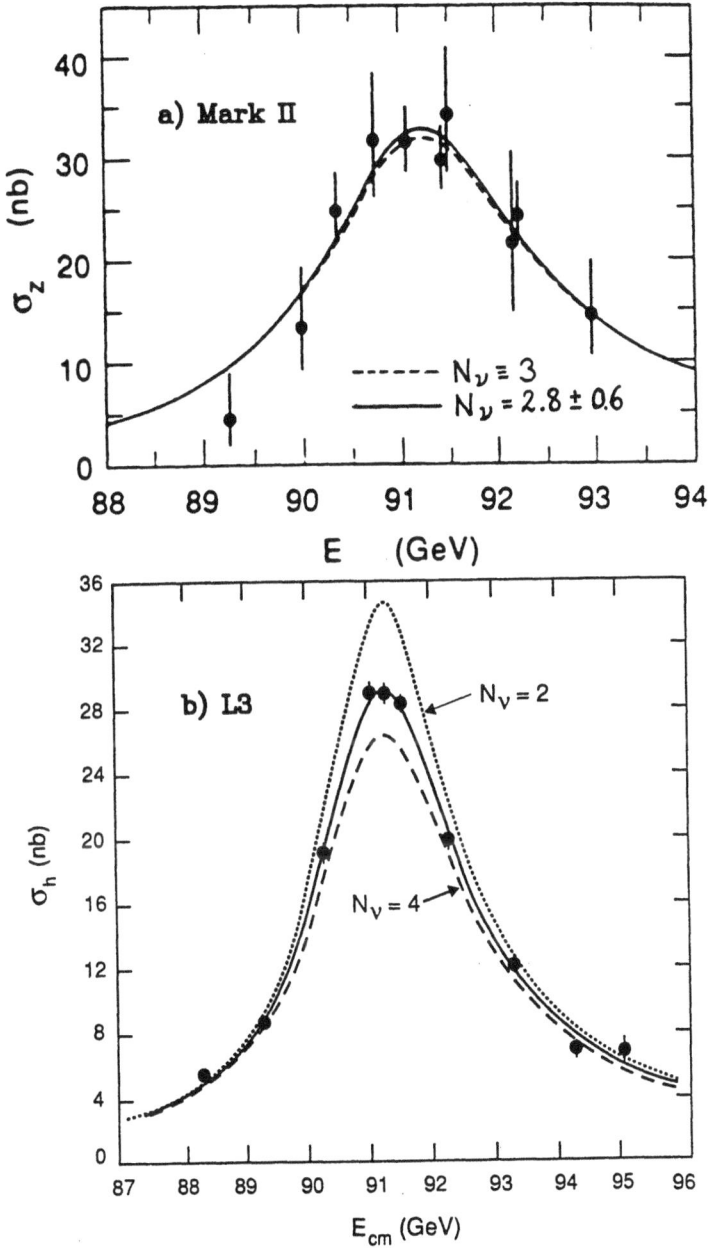

Fig. 2

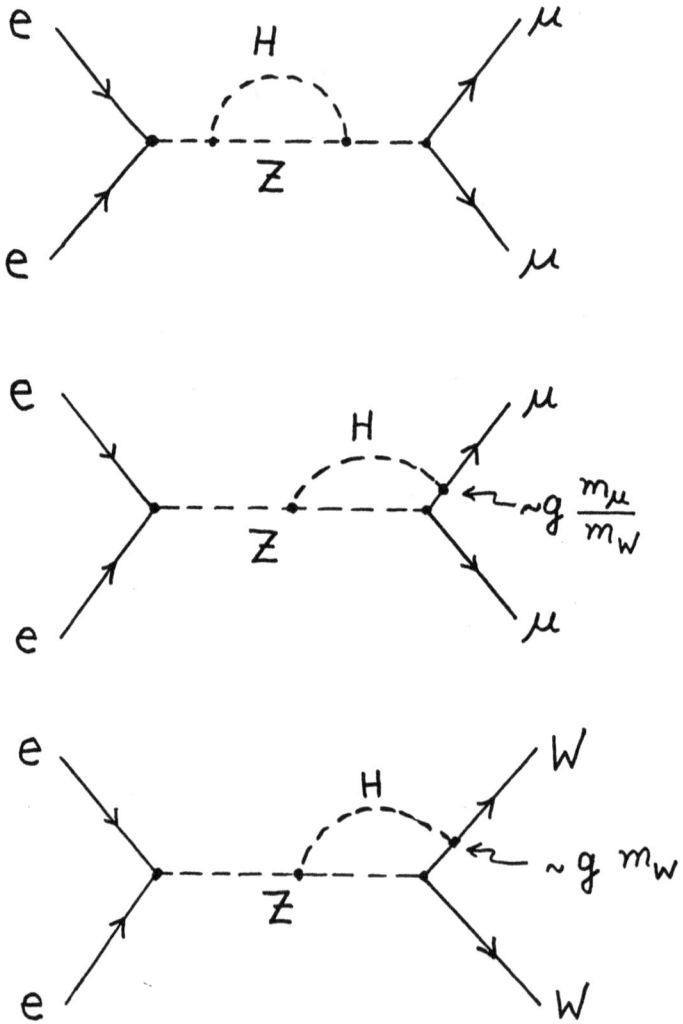

Fig. 3

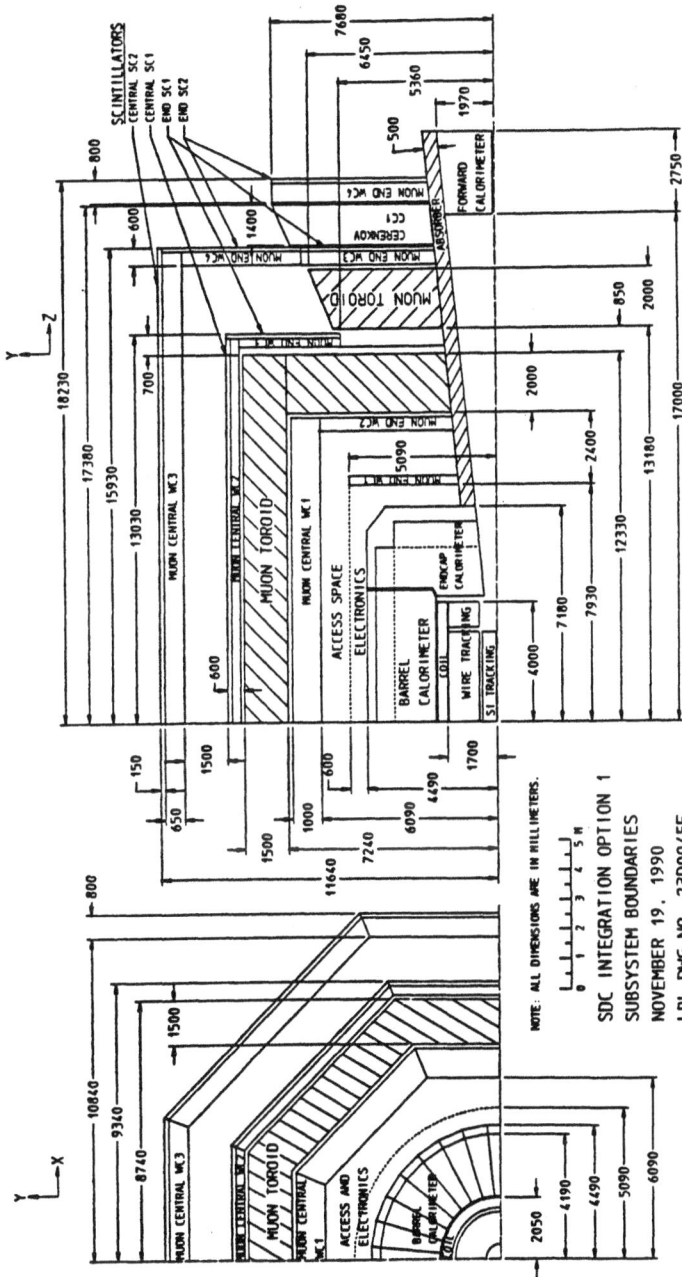

Fig. 4 Quadrant plan view of SDC detector with lead/scintillator calorimeter option.

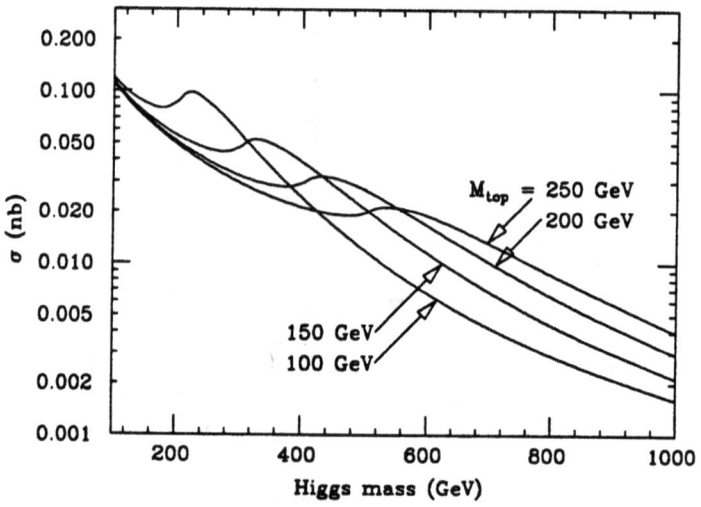

Fig. 5

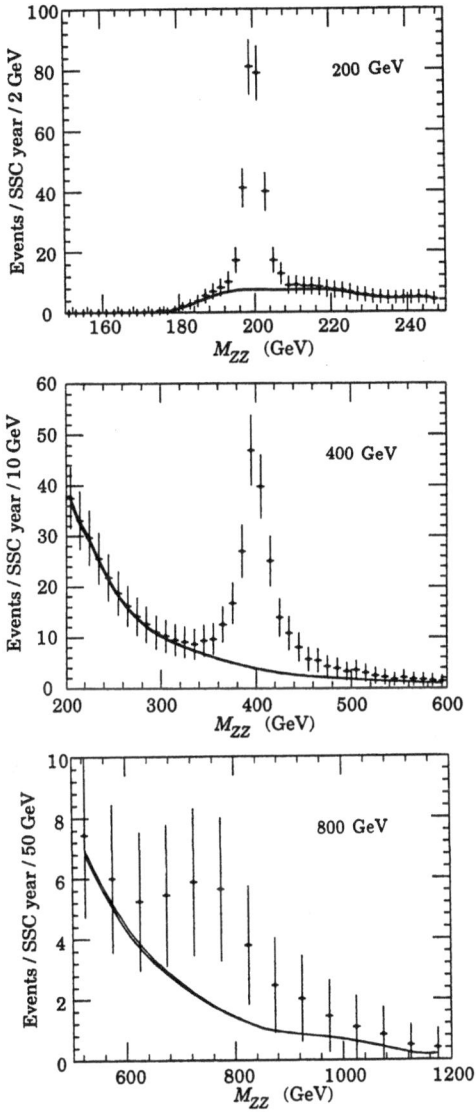

Fig. 6

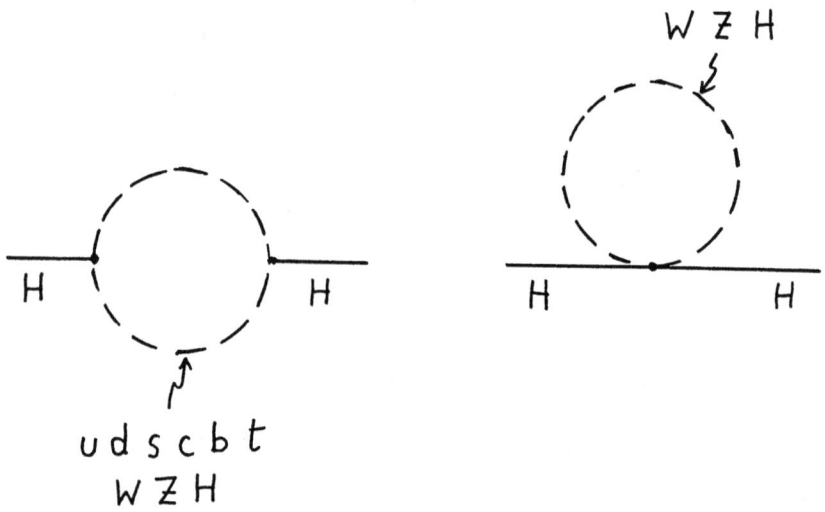

Fig. 7

Third threshold in the weak interactions ?

J.J. van der Bij
Universität Freiburg
Hermann Herder strasse 3
7800 Freiburg i. B.
Germany

&

H. Veltman
DESY
Notkestrasse 85
2 Hamburg 52
Germany

Abstract

We study the possibility of the existence of non-standard four vector boson couplings induced by strongly interacting Higgs particles. We show that such interactions can exist if there is a hierarchy of strong interactions and masses in the Higgs sector, thereby circumventing Veltman's screening theorem. We show that resonances can be formed in this case.

1 Introduction

The study of the Higgs sector of the Standard Model in the large Higgs mass limit has been pioneered by M. Veltman. He developed elegant techniques in formulating certain physical consequences in this limit, like the screening theorem. His work on the possibility of a second threshold in the weak interactions is the basis of many subsequent works, such as the one that is described here.

As the experiments at LEP indicate, the Standard Model of the weak interactions is in agreement with all data. What has not been tested so far is the existence of the vector boson self couplings. While the three vector boson coupling will be studied at LEP200, the four vector boson coupling can only be studied at future accelerators like the LHC, SSC or the NLC. Within the Standard Model the vectorboson self couplings are completely determined by the gauge structure of the theory and deviations of its values are only possible through radiative corrections, which tend to be small. An exception to this situation can in principle arise if the Higgs particle is very heavy, because then strong couplings are present. Since the Higgs particle has not been found so far and since problems like the generation of a large cosmological constant are associated with the Higgs sector it motivates us to consider the Standard Model without the Higgs particle. This model is then non-renormalizable and divergences arise. An alternative way is to study the situation where the Higgs mass becomes infinite, which at least at the tree level is equivalent to having no Higgs at all. This limit was first studied by Veltman [1]. He found that radiative corrections indeed blow up but only logarithmically. As a consequence the vector boson couplings stay small unless one makes the Higgs mass unreasonably large. This observation is now known as Veltman's screening theorem. In order to generate large self couplings of the vector bosons one therefore has to find a way to avoid this theorem. In section 2 we discuss in some detail how Veltman's theorem arises and we try to put limits on the generality of the theorem. A careful study shows that the theorem can be circumvented if extra strong interactions are present in the Higgs sector. This conclusion was also reached in a previous paper [2], where an explicit model was constructed. More precisely in that paper it was shown that strong interactions among the vector bosons can be generated if there is a hierarchy of strong interactions in the Higgs sector. This feature is likely to be generally valid and we therefore call it Hill's theorem. The model is discussed in section 3. In section 4 we then study the dynamics of Hill's model in more detail and we find that resonances exist in the four vector boson scattering following an argument presented in [3].

2 Veltman's screening theorem

It is known that the weak interactions are mediated by massive vector bosons. In the Standard Model the mass of the vector bosons arises through the mechanism of spontaneous symmetry breaking. A doublet of bosonic fields is introduced, which receives a vacuum expectation value, because the minimum of the potential is not located at zero. As a consequence the vector bosons receive a mass, but there is also a particle left, the so called Higgs boson. As there is no experimental evidence of the Higgs boson sofar it is natural to study the Standard Model without the Higgs boson. At the tree level this is equivalent to studying the Standard Model with the Higgs boson mass becoming very large.

At the one-loop level one can take two approaches. In the first approach one takes the Standard Model and calculates the Higgs mass dependence of the radiative corrections to vector boson interactions. In the limit of large Higgs mass the Higgs sector becomes strongly interacting and therefore the decoupling theorem [4] is not valid. As a consequence corrections growing with the Higgs mass are present. It was found in [1] that these corrections grow only logarithmically with the Higgs mass, which make them very hard to see experimentally. The Higgs effects are screened. A precise study of the way these infinities arise was done in a non-gauged model in [5].

In the second approach one looks at the Standard Model without the Higgs boson altogether. This model is nonrenormalizable and therefore infinities arise. The possible infinities were classified in [6] for the SU(2) theory and in [7] for the full SU(2)xU(1) model. Since the theory is nonrenormalizable a precise definition of the infinities is needed in this case. Dimensional regularization [8] was used and it was found that infinities arise in the form of poles in $n-4$, where n is the dimension of space-time.

A priori it is not clear that these approaches should give the same result. In order to see how this comes about we describe here both the Standard Model and the model without the Higgs boson in a gauge-invariant way. The Standard Model is a gauged linear σ-model

$$\mathcal{L} = -\frac{1}{2}(D_\mu \Phi)^\dagger (D^\mu \Phi) - \frac{\lambda}{8}(\Phi^\dagger \Phi - f^2)^2 \qquad (2.1)$$

$$\Phi = (\sigma + i\vec{\tau} \cdot \vec{\phi})\begin{pmatrix} 1 \\ 0 \end{pmatrix} \qquad (2.2)$$

At the tree level, the Higgs particle can be removed from (2.1) by taking the limit $\lambda \to \infty$ or, equivalently, $m_H^2 \to \infty$. The Standard Model then turns into a gauged nonlinear σ-model

$$\mathcal{L} = \frac{f^2}{4}Tr(D_\mu U)^\dagger (D^\mu U) \qquad (2.3)$$

$$U = \sqrt{1 - \vec{\pi}^2} + i\vec{\tau} \cdot \vec{\pi} \qquad (2.4)$$

$$\vec{\pi} = \frac{\vec{\phi}}{f}$$

which is equivalent to massive Yang-Mills theory. It can be seen on the formal level that the Standard Model reduces to (2.3) in the limit $\lambda \to \infty$ by noticing that the potential acts like a constraint in this case. Explicit calculations show that at the one-loop level the effects of a heavy Higgs particle are given by terms involving higher order derivative terms of U. In the unitary gauge these terms correspond to extra interactions beyond the ordinary gauge couplings. If we limit ourselves to the SU(2) model, which we do here and in the following for simplicity, the extra effects are summarized by the following effective Lagrangian

$$\mathcal{L}_{eff} = \alpha_1 Tr(V_\mu V^\mu) Tr(V_\nu V^\nu) + \alpha_2 Tr(V_\mu V^\nu) Tr(V_\mu V^\nu) + g\alpha_3 Tr(F_{\mu\nu}[V^\mu, V^\nu]) \quad (2.5)$$

where

$$V_\mu = (D_\mu U) U^\dagger \quad (2.6)$$

and

$$F_{\mu\nu} = (\partial_\mu - \frac{ig}{2}\vec{W}_\mu \cdot \vec{\tau}) \frac{\vec{W}_\nu \cdot \vec{\tau}}{2i} - (\mu \leftrightarrow \nu) \quad (2.7)$$

Explicit calculation in the linear model gives

$$\alpha_1 = \frac{1}{384\pi^2} \ln(m_H^2/M_W^2) + \mathcal{O}(1),$$

$$\alpha_2 = \frac{1}{192\pi^2} \ln(m_H^2/M_W^2) + \mathcal{O}(1), \quad (2.8)$$

$$\alpha_3 = -\frac{1}{384\pi^2} \ln(m_H^2/M_W^2) + \mathcal{O}(1).$$

In the nonlinear model one gets the same coefficients with the following replacement, for the moment not worrying about the $\mathcal{O}(1)$ terms,

$$\text{linear model}: \ln(m_H^2/M_W^2) \longleftrightarrow \text{nonlinear model}: -\frac{2}{n-4} \quad (2.9)$$

The possibility of calculating the divergences directly in the nonlinear model seems to indicate that the coefficients are universal. We wish to study whether this is indeed the case. While in the calculation of α_2 and α_3 only irreducible graphs contribute it is not too surprising that the relation (2.9) is valid. For the coefficient α_1 also reducible graphs involving Higgs exchange (fig.1) appear and a delicate cancellation between the graphs and the renormalization counterterm for the Higgs mass is present in this case. Therefore it is natural to consider extensions involving extra strong interactions in the Higgs sector, which may change the coefficient α_1 strongly. This will be done in the next chapter. Since we will later be considering the possibility of resonances in the longitudinal vector boson scattering, we remark here that only the coefficients α_1 and α_2 contribute in this case. In the end the possibility of resonance formation in the isospin $I = 1$ channel will be determined by the following combination of parameters

$$\beta = 128\pi^2(\alpha_2 - 2\alpha_1).$$

Note that β is independent of $\ln(m_H^2)$.

3 Hill's theorem; avoiding the screening theorem

In this chapter we study the possibility of generating strong interactions in the gauge sector via radiative corrections in the Higgs sector. As argued in the previous sector to get large effects one has to find a way to change the Higgs propagator in order to avoid the cancellations of the Standard Model. A simple way to do this is to add a strongly interacting singlet field to the Higgs sector [2]. We call this field the X field. The corresponding Lagrangian is

$$\mathcal{L} = -\frac{1}{2}(D_\mu \Phi)^\dagger (D^\mu \Phi) - \frac{1}{2}(\partial_\mu X)^2 - \frac{\lambda_1}{8}(\Phi^\dagger \Phi - f_1^2)^2 - \frac{\lambda_2}{8}(2f_2 X - \Phi^\dagger \Phi)^2 + \mathcal{L}_{gauge} \quad (3.1)$$

Even though no X self couplings are present this model is renormalizable, because the X field is a singlet under the gauge group. Therefore no divergent X^4 interactions can be induced via gauge boson loops. Explicit power counting of divergences bears out this argument [2].

Within this model we are interested in the limit

$$\lambda_2 \gg \lambda_1 \gg 0, \quad (3.2)$$

because here there is a hierarchy of coupling strengths and one can expect effects to be present even at lower energies. In this limit the mass eigen values in the boson sector become

$$m_+^2 \approx \lambda_2(f_1^2 + f_2^2), \quad (3.3)$$

$$m_-^2 \approx \frac{\lambda_1 f_1^2 f_2^2}{f_1^2 + f_2^2}. \quad (3.4)$$

We note that the mass of the Higgs and the mass of the X are each some combination of m_+ and m_-. In the absence of the X particle, which corresponds to $\lambda_2 = 0$, we have $m_+ = 0$ and $m_- = m_H$.

It is clear from these formulae that there is in this limit also a hierarchy of mass scales in the higgs boson sector $m_+ \gg m_- \gg m_W$, so that one can speak of a third threshold in the weak interactions. For future reference we define here

$$\alpha = \frac{f_1^2}{f_1^2 + f_2^2}. \quad (3.5)$$

In this limit the one loop corrections to the α_i's, due to the contribution of the X particle, become

$$\alpha_1 = \frac{1}{384\pi^2}\{\ln(m_-^2/m_W^2) - \frac{1}{2}\alpha(6 + \alpha)\ln(m_+^2/m_-^2) + 3\delta_H\} + \mathcal{O}(1), \quad (3.6)$$

$$\alpha_2 = \frac{1}{192\pi^2}\{\ln(m_-^2/m_W^2) + \alpha^2\ln(m_+^2/m_-^2)\} + \mathcal{O}(1), \quad (3.7)$$

$$\alpha_3 = -\frac{1}{384\pi^2}\{\ln(m_-^2/m_W^2) + \alpha\ln(m_+^2/m_-^2)\} + \mathcal{O}(1). \tag{3.8}$$

where we defined

$$\begin{aligned}
\delta_H = \frac{\lambda_2}{\lambda_1} \Bigg\{ &-\frac{9}{2}\alpha^2(2 - \frac{\pi}{\sqrt{3}}) - \frac{1}{2}(4 - 2\alpha + \alpha^2)\ln(1-\alpha) \\
&+ \alpha^2(1 - \sqrt{\frac{3+\alpha}{1-\alpha}}\mathrm{arctg}\sqrt{\frac{1-\alpha}{3+\alpha}}) + \alpha^2\ln\alpha - \frac{1}{6}(1-\alpha)(3+17\alpha) \\
&+ 4\pi\sqrt{\frac{\lambda_2}{\lambda_1}}\alpha^{3/2}(1-\alpha)^{3/2} \Bigg\} + O(\lambda_1).
\end{aligned} \tag{3.9}$$

We see therefore that deviations from the Standard Model are present. For the coefficients α_2 and α_3 these deviations are only logarithmic, but for α_1 they grow with the ratio of the coupling constants λ_2/λ_1 and therefore arbitrarily large effects can be expected. A further simplification appears when one takes $\alpha \to 0$, i.e. $f_2 >> f_1$. In that case the whole effect of the extra interactions beyond the Standard Model can be summarized by the parameter

$$\beta = 128\pi^2(\alpha_2 - 2\alpha_1) = \lambda_2/\lambda_1. \tag{3.10}$$

Therefore a large β can appear if there is a hierarchy of coupling strenghts, thereby violating the screening theorem. A similar effect appears in a related model [9] where a slightly different scalar is introduced, the so called U particle. Therefore we claim that the following statement holds: "The screening theorem can be avoided if a hierarchy of strong interactions exists in the Higgs sector". We call this Hill's theorem.

In the next chapter we will study the consequence of Hill's theorem on the formation of resonances for longitudinally polarized vector boson scattering.

4   Resonances in vector boson scattering

Let us consider the amplitude for longitudinally polarized vector boson scattering in the large Higgs mass limit, where the interacting vector bosons have an energy much larger than their mass. We thus consider the energy region

$$m_W \ll \sqrt{s} \ll m_{\mathrm{Higgs}}.$$

The tree amplitude grows like the energy squared and the unitarity limit is reached around 1 TeV. Thus if no Higgs has been found below 1 TeV, new physics will have to show up in the TeV region. This new physics could manifest itself for example through the occurrence of resonances in the different isospin channels for $W_L W_L$ scattering. In ref.[3] it was shown that the occurrence of a resonance in the isospin I=1 channel is sensitive to the details of the Higgs sector. As an example of such model dependence we consider in the following the effect that the X-particle may have.

In the energy region below 1 TeV we assume that physics is described by some effective Lagrangian, for example the Standard Model in the large Higgs mass limit. Furthermore we assume that in this energy region the one loop correction, being of the order of 10 %, is a reasonable approximation. Using partial wave analysis, it is then possible to describe vector boson dynamics in the TeV region.

In general the amplitude for $W_L W_L$ scattering, including the one loop correction, may be written as

$$A(W_L^a W_L^b \to W_L^c W_L^d) = \frac{s}{f_1^2}$$
$$- \frac{1}{96\pi^2 f_1^4} \cdot \{3s^2(\ln s - \beta_1) + t(t - u)(\ln t - \beta_2) + u(u - t)(\ln u - \beta_2)\}. \quad (4.1)$$

Here all model dependence is contained in the parameters β_1 and β_2. Expressed in terms of α_1 and α_2 we have

$$\beta_1 = 64\pi^2(4\alpha_1 + \alpha_2),$$
$$\beta_2 = 192\pi^2\alpha_2. \quad (4.2)$$

Different models give different values for β_1 and β_2. The question is now how the occurrence of a resonance in the different isospin channels depends on these two parameters. Using partial wave analysis, it was derived in refs. [3,10] that in the isospin I=1 channel a resonance is located at

$$s = \frac{32\pi^2 v^2}{\beta - \frac{1}{9}}, \quad (4.3)$$

where $\sqrt{s}$ is the center of mass energy and

$$\beta = \beta_2 - \beta_1$$
$$= 128\pi^2(\alpha_2 - 2\alpha_1). \quad (4.4)$$

For example, when $\beta = 5$, a resonance occurs around 2 TeV. For $\beta < 0$, no resonance appears at any energy. Similarly, the occurrence of a resonance in the I=0 channel for a particular value of $\sqrt{s}$ depends on another linear combination of β_1 and β_2.

Let us summarize what values we have for β according to various models.

(1) Standard Model in the large Higgs mass limit, linear σ model;

$$\beta_1 = \ln(m_H^2) + \frac{4}{3} + 9\left(\frac{\pi}{\sqrt{3}} - 2\right),$$
$$\beta_2 = \ln(m_H^2) - \frac{2}{3},$$
$$\beta = -2 - 9\left(\frac{\pi}{\sqrt{3}} - 2\right) = -0.32. \quad (4.5)$$

We see that here no resonance can be formed in the I=1 channel at any energy.

(2) Standard Model in the large Higgs mass limit, non-linear σ model;

$$\beta_1 = \Delta + \frac{11}{6},$$
$$\beta_2 = \Delta + \frac{13}{6},$$
$$\beta = \frac{1}{3}. \tag{4.6}$$

In this case we find that, according to eq.(4.3) a resonance in the I=1 channel will occur around 9 TeV.

(3) Contribution of the U-particle [9]; to the Standard Model add a piece :

$$\mathcal{L}(U) = -\frac{1}{2}(\partial_\mu U)^2 - \frac{1}{2}m_U^2 U^2 - g g_U r m_W U^2 H - \frac{1}{4}g^2 g_U r U^2 (H^2 + \phi^2) \tag{4.7}$$

with $r = m_H^2/4m_W^2$ and g_U is the parameter associated with the U-particle. Thus the U is coupled to the Higgs with a strength proportional to m_H^2. In the limit $m_U = m_H \to \infty$, we obtain

$$\beta_1 = g_U^2 \left(\frac{\pi}{\sqrt{3}} - 2\right),$$
$$\beta_2 = 0,$$
$$\beta = -g_U^2 \left(\frac{\pi}{\sqrt{3}} - 2\right). \tag{4.8}$$

The contribution to β is always positive, and depending on the value of g_U, a resonance can occur at any value of s.

(4) Contribution of the X particle. Eq.(4.2) gives β_1 and β_2 in terms of α_1 and α_2, which in the limit $\lambda_2 \gg \lambda_1 \gg 0$ are given by eqs. (3.6) and (3.7). According to eq.(4.4) we find

$$\beta = \alpha(2 + 3\alpha)\ln(m_+^2/m_-^2) - 2\delta_H. \tag{4.9}$$

In the limit $f_2 \gg f_1$ we have

$$\beta = \frac{\lambda_2}{\lambda_1}. \tag{4.10}$$

Just like in the case of the U-particle, β is completely arbitrary.

We can now make the following remarks. First of all the reason that the isospin I=1 channel is interesting is now obvious; β is independent of $\ln(m^2)$ and thus sensitive to the model considered. This is not the case for example for the I=0 channel, where one finds that the location of a resonance depends on $\ln(m_H^2)$. Next, from eqs.(4.5) and (4.6), if we consider the Standard Model in the large Higgs mass limit, we find that the result depends on which way the large Higgs mass limit is taken and therefore β, although a calculable number, is an arbitrary parameter. Indeed, through the U or the X particle, the value for β can be altered and can in principle be any positive or negative number. In order for a resonance to occur in the I=1 channel, β must be > 0. Hill's Theorem states that a large and positive value for β is only possible if there are extra strong interactions present in the Higgs sector. For example for the case of the U particle, eq.(4.8) was derived given that $m_U = m_H \to \infty$, resulting in a positive contribution to β. On the other hand, we find that $\beta < 0$, when the mass m_U is well below a given choice of m_H. For the case of the X particle, we have plotted in fig. 2 β as a function of λ_2/λ_1 for various values of α. We notice that for example for $\alpha \approx 1$, which corresponds to $f_2/f_1 << 1$ β is always negative when $\lambda_2 > \lambda_1$. In this case no resonance occurs in the I=1 channel.

In fig. 3 we have plotted the I=1 amplitude as a function of $\sqrt{s}$ for various values of β. We observe that the resonance is located at lower energies for higher values of β. We can therefore put an upper limit on β. Up to $\sim$ 100-200 GeV no bound state of two vector bosons has been observed. Therefore $\beta_{max} \sim 500$. This can be roughly translated into a ratio $m_+/m_- = 50$. Therefore if the second threshold is at 1 TeV the new physics corresponding to the strong interactions between singlet and doublet fields would appear below 50 TeV.

References

[1] M. Veltman, Acta Phys. Polon. B8, 475 (1977).

[2] A. Hill and J. J. van der Bij, Phys. Rev. D36, 3463 (1987).

[3] H. Veltman and M. Veltman, DESY 91-050 (1991).

[4] T. Appelquist and J. Carazzone, Phys. Rev. D11, 2856 (1975).

[5] R. Akhoury and Y. P. Yao, Phys. Rev. D25, 3361 (1982).

[6] T. Appelquist and C. Bernard, Phys. Rev. D22, 200 (1980).

[7] A. Longhitano, Phys. Rev. D22, 1166 (1980); Nucl. Phys. B188, 118 (1981).

[8] M. Veltman and G. 't Hooft, Nucl. Phys. B44, 189 (1972).

[9] M. Veltman and F. Yndurain, Nucl. Phys. B325, 1 (1989).

[10] H. Lehmann, Acta Phys. Austriaca, Suppl. XI, 139 (1973).

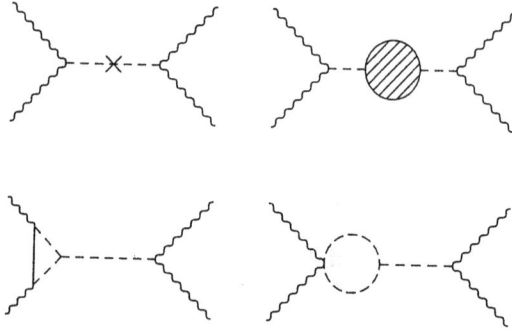

Figure 1:   Various Higgs reducible graphs contributing to the four vector boson vertex.

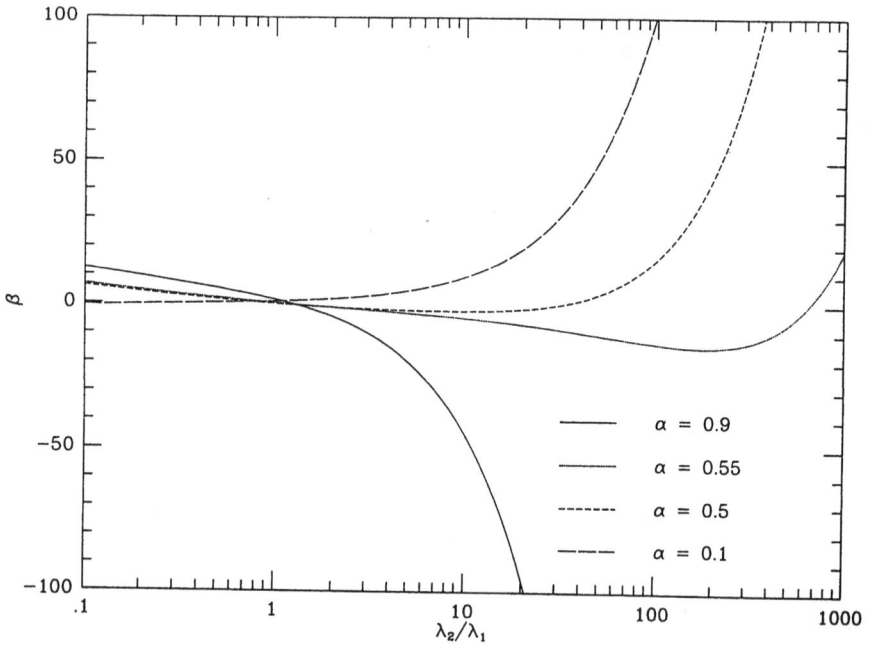

Figure 2: β as a function of λ_2/λ_1 for various values of α.

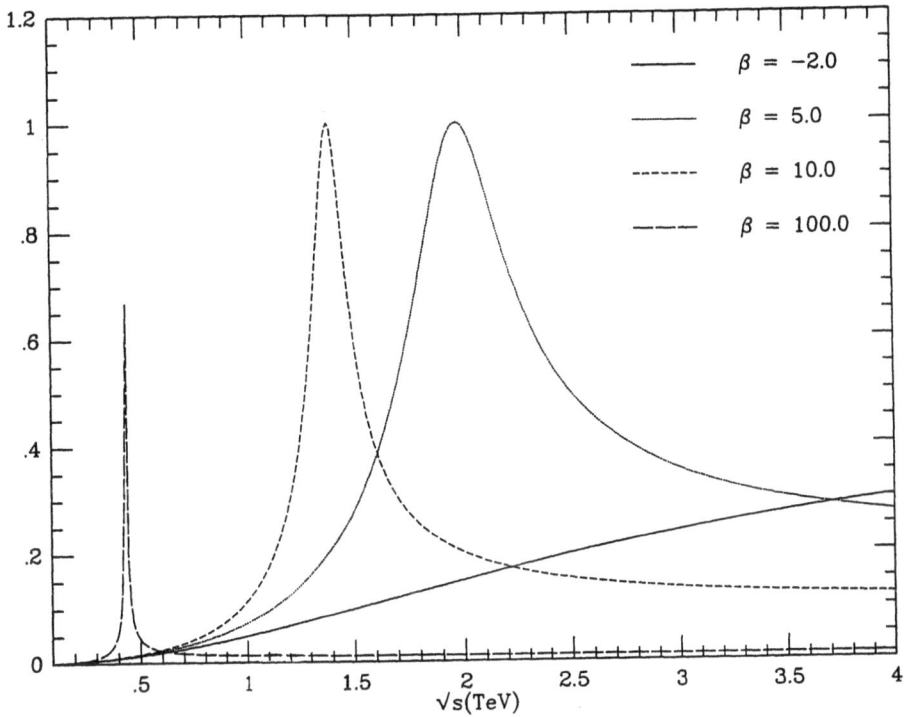

Figure 3:   Absolute value of the I=1 partial wave amplitude as a function of $\sqrt{s}$ for various values of α.

RELATION BETWEEN THE PRACTICAL RESULTS
OF CURRENT ALGEBRA TECHNIQUES AND
THE ORIGINATING QUARK MODEL[*]

M. Veltman

Institute for Theoretical Physics,
University of Utrecht,
Utrecht, the Netherlands

1. Introduction

In recent years much work has been devoted to the properties of weak and e.m. currents. First of all there is the problem of finding the correct equations obeyed by the currents (CVC, PCAC, divergence equations, current commutators) and to derive methods to verify these equations experimentally; secondly one must try to understand the origin of those equations.

In Sec. 2 we will write down the various equations for the currents, without any attempt to prove them from any model. Some low energy theorems on Compton scattering on nucleons and on η-decay will be derived in Sec. 3. In Sec. 4 some high energy results will be investigated while in Sec. 5, finally, we will consider the question of the origin of the current equations.

2. Current Equations

Historically the first equation concerns the electric current, coupled to the e.m. field. The Maxwell equations are

$$\partial_\mu F_{\mu\nu} = -ej_\nu, \qquad \left(\partial_\mu = \frac{\partial}{\partial x_\mu}\right) \tag{1}$$

with

$$F_{\mu\nu} = (\partial_\mu A_\nu - \partial_\nu A_\mu), \qquad F_{\mu\nu} = -F_{\nu\mu} . \tag{2}$$

The $F_{\mu\nu}$ are the e.m. field strengths $\mathbf{E}$ and $\mathbf{H}$:

$$F_{4\mu} = iE_\mu, \qquad \mu = 1, 2, 3 ,$$
$$F_{12} = H_3, \qquad F_{31} = H_2, \qquad F_{23} = H_1 .$$

Further e = electric charge, $e^2/4\pi = \alpha = 1/137$. The A_μ are the vector potentials and the j_μ are the charge and current densities. Specifically $j_4(x) = i\rho(x)$, where $\rho(x)$ is the charge density at the place $\mathbf{x}$ at time $x_0 = -ix_4$.

---

[*] Copenhagen Lectures July 1968

From (1) and the antisymmetry of $F_{\mu\nu}$ (2) we find

$$\partial_\nu j_\nu(x) = 0 . \tag{3}$$

This equation expresses local conservation of charge. To see this consider a small cube with sides of length a. We take our coordinate system as in the figure. The total charge Q contained in this cube C is

$$Q = -i \int_C d_3x\, j_4(x) .$$

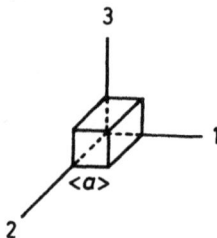

This charge will vary with time as current flows in and out through the surfaces of the cube. We have

$$\frac{\partial Q}{\partial t} = i\frac{\partial Q}{\partial x_4} = \int_C d_3 j_4(x)$$

Using (3)

$$\frac{\partial Q}{\partial t} = -\int_C d_3x\, \partial_i j_i(x) = -\left[\int dx_2 dx_3 j_1(a, x_2, x_3)\right.$$
$$\left. -\int dx_2 dx_3 j_1(0, x_2, x_3) + \int dx_1 dx_3 j_2(x_1, a, x_3) \ldots \right]$$

The first term is the total charge per unit time flowing through the rightmost surface ($x_1 = a$, x_2 and x_3 between 0 and a) of the cube. Analogously for the other sides. We see that (3) implies that any charge decrease or increase in C comes about by charge flowing in or out through the surface of C. This is valid for an arbitrary small cube, which is why we speak of *local* conservation of charge. In principle, from the experimental point of view it could be that charge is not locally conserved for very small distances, i.e. that charge vanishes at some point and comes up at some very nearby point. For this reason one must consider (3) as an equation to be tested for higher and higher energies.

From (2) we see that the replacement (gauge transformation)

$$A_\mu \to A_\mu + \partial_\mu \Lambda , \tag{4}$$

with Λ any function, does not affect the $F_{\mu\nu}$, because of the antisymmetry of $F_{\mu\nu}$. It was just this antisymmetry that led to Eq. (3), and in this simple way we see that gauge invariance and local conservation of charge are intimately connected. Later on we will consider this connection in much more detail.

Let us now consider processes between elementary particles. Isospin is a well defined concept in strong interactions, and we are interested in the I-spin properties of the e.m. current operator j_μ. If the e.m. current were an isoscalar photons would couple equally strong to protons as the neutrons. This is evidently not so, the neutron having no charge. In fact the combination that couples to the photon is

$$\frac{1}{2}[\bar{N}(1+\tau_3)N] \,,$$

where

$$N = \begin{pmatrix} 1 \\ 0 \end{pmatrix} \quad \text{for proton,} \quad \tau_3 = \begin{pmatrix} 1 & 0 \\ 0 & -1 \end{pmatrix} ,$$

$$N = \begin{pmatrix} 0 \\ 1 \end{pmatrix} \quad \text{for neutron,} \quad 1 = \begin{pmatrix} 1 & 0 \\ 0 & 1 \end{pmatrix} ,$$

$$\bar{N} = \begin{pmatrix} 1 \\ 0 \end{pmatrix} \quad \text{for outgoing} \quad \begin{pmatrix} 0 \\ -1 \end{pmatrix} , \quad \text{for outgoing}$$
$$\text{proton,} \qquad\qquad\qquad\qquad\qquad \text{neutron.}$$

τ_3 represents the third component of the isovector current

$$(\bar{N}\tau_i N)$$

$$\tau_1 = \begin{pmatrix} 0 & 1 \\ 1 & 0 \end{pmatrix}, \quad \tau_2 = \begin{pmatrix} 0 & -i \\ i & 0 \end{pmatrix}, \quad \tau_3 = \begin{pmatrix} 1 & 0 \\ 0 & -1 \end{pmatrix} .$$

We will *assume* that the e.m. current contains an isoscalar and an isovector part, but no more. This assumption is experimentally well verified. Thus we write

$$j_\mu = j_\mu^s + j_\mu^3 \,, \tag{5}$$

where j_μ^s transforms like an isoscalar and j_μ^3 as the third component of an isovector. Eq. (3) goes over into two separate equations:

$$\partial_\mu j_\mu^s(x) = 0 \,, \tag{6}$$
$$\partial_\mu j_\mu^3(x) = 0 \,. \tag{7}$$

As already indicated, j_μ is now considered an operator, and Eqs. (6) and (7) are operator equations.

As a consequence of isospin invariance there exist two other currents, $j_\mu^1(x)$ and $j_\mu^2(x)$ that can be obtained from $j_\mu^3(x)$ by rotations in isospin space. To the extent that isospin is conserved these currents will obey a law similar to (7), and we write:

$$\partial_\mu j_\mu^i(x) \simeq 0, \qquad i, 1, 2, 3, \ . \tag{8}$$

We now make the further assumption that I-spin invariance is broken only by e.m. forces (in this section we disregard the breaking of I-spin invariance in weak interactions). Then (8) will be true up to terms containing e, the electric charge. It turns out to be possible to find, from gauge invariance, the precise form of those terms; one finds

$$\partial_\mu j_\mu^i(x) = -e\varepsilon_{i3k}A_\mu^3(x)j_\mu^k(x) , \tag{9}$$

where $A_\mu^3(x)$ is the e.m. vector potential coupled to the third component of the isovector current. If we define j_μ^+ and j_μ^- by:

$$j_\mu^1 = j_\mu^+ + j_\mu^- ,$$
$$j_\mu^2 = i(j_\mu^+ - j_\mu^-) ,$$

we find the more familiar expressions:

$$(\partial_\mu - ieA_\mu)j_\mu^+ = 0 ,$$
$$(\partial_\mu + ieA_\mu)j_\mu^- = 0 ,$$
$$\partial_\mu j_\mu^3 = 0 .$$

Equation (9) leads in itself already to consequences that can be verified experimentally. Eqs. (8) and (9) gain enormously in interest if we adopt the C.V.C. hypothesis. This hypothesis tells us that the currents j_μ^+ and j_μ^- as defined above are experimentally observable in leptonic weak interactions. In particular, consider neutron decay

$$n \to p + e + \bar{\nu} .$$

It is well established that this interaction contains a vector and an axial vector coupling between the nucleon part and the lepton part. The C.V.C. hypothesis suggests that the vector part of the nucleon current is nothing but $j_\mu^-(x)$ multiplied by the appropriate weak interaction coupling constant. Thus, by this hypothesis the nucleon vector part of

$$n \to p + e + \bar{\nu}$$

becomes intimately related to the couplings

$$p \to p + \gamma ,$$
$$n \to n + \gamma .$$

Stated somewhat differently: consider any process involving an isovector photon. Performing a rotation in I-spin space and replacing the photon by a lepton pair gives the vector part of a genuine weak process. This observation suggests a further generalization of (9); ignoring axial currents we must have

$$\partial_\mu j_\mu^i(x) = -\varepsilon_{ijk}\{\varepsilon A_\mu^j(x) + G l_\mu^j(x)\}j_\mu^k(x) , \tag{10}$$

where $A_\mu^j = 0$, unless $j = 3$, and where l_μ^j stands for the lepton fields, with $l_\mu^3 = 0$ and $l_\mu^1 = l_\mu^+ + l_\mu^-$ etc. The lepton current l_μ^+ is the current as found in the β-decay of the neutron. $G = G_v = 1.02 \cdot 10^{-5}/(\sqrt{2}m_p^2)$, $m_p =$ proton mass.

Inspection of (10) shows that $\partial_\mu j_\mu^3$ is no longer zero, but proportional to G. That means that charge is no longer exactly conserved. The reason is that in writing (10) we neglected the coupling of photons to leptons, and seemingly nonconservation of charge results. Thus the right-hand side of (10), for $i = 3$, gives us essentially the amount of charge carried away by leptons in a weak process. Remember that $j_\mu(x)$ is the vector current of hadrons alone.

Including axial currents becomes easy, one must supplement (10) with a term giving the amount of charge carried away by leptons in a weak process involving axial currents. One finds:

$$\partial_\mu j_\mu^i(x) = -\varepsilon_{ijk}\{\varepsilon A_\mu^j(x)j_\mu^k(x) + Gl_\mu^j(x)[j_\mu^k(x) + j_\mu^{Ak}(x)]\} , \tag{11}$$

where j_μ^{Ak} is the axial vector hadron current.

We wish to emphasize that the essential ingredients necessary to derive (11) are local conservation of charge, or rather gauge invariance, and the C.V.C. hypothesis.

Consider now the axial vector current of hadrons. Its four-divergence can not be zero, as testified by the pion decay. The pion decay matrix element is for a pion of four-momentum k_μ given by

$$\langle \mu^- \bar{\nu}|l_\mu^+|0\rangle\langle 0|j_\mu^{A-}(0)|\pi^-\rangle .$$

The matrix-element $\langle 0|j_\mu^{A-}(0)|\pi^-\rangle$ transform like a four-vector, and since the only four-vector available is the pion four-momentum k_μ, it must be proportional to k_μ:

$$\langle 0|j_\mu^{A-}(0)|\pi^-\rangle = Ck_\mu \cdot \frac{1}{\sqrt{2k_0 V}} .$$

V is the volume in which the system is enclosed. By actually working out the decay rate of the π^- in terms of C, the constant C may be deduced from the pion lifetime. On the other hand, one has

$$\langle 0|\partial_\mu j_\mu^{A-}(0)|\pi^-\rangle = ik_\mu\langle 0|j_\mu^{A-}(0)|\pi^-\rangle$$
$$= \frac{iCk^2}{\sqrt{2V k_0}} = -iCm_\pi^2 \cdot \frac{1}{\sqrt{2V k_0}} .$$

As $C \neq 0$ we see that $\partial_\mu j_\mu^{A-} \neq 0$.

From the above it is clear that $\partial_\mu j_\mu^{A-}$ must contain a part that can absorb a negative pion, just as one finds in the ordinary pion field $\pi^-(x)$. The simplest assumption that we can make is therefore that $\partial_\mu j_\mu^{A-}$ equals the pion field (apart from some constant):

$$\partial_\mu j_\mu^{A-} = -iCm_\pi^2\pi^-(x) .$$

Apart from a sign the constant C can be determined from π-decay. In fact, with

$$l_\mu = i\{\bar{\nu}\gamma^\mu(1+\gamma^5)\mu\}\ ,$$

$$G = \frac{1.02}{\sqrt{2}}\cdot 10^{-5}m_p^{-2}\ ,$$

$$\frac{C^2}{\pi}\frac{m_\mu^2}{m_\pi}\left(\frac{m_\pi^2-m_\mu^2}{2m_\pi}\right) = \Gamma_\pi\quad(=2.524\cdot 10^{-14}\mathrm{MeV})\ ,$$

where m_μ = muon mass, m_π = pion mass and Γ_π = pion decay width, we find

$$iC = \pm 1.057\cdot 10^{-9}\ \mathrm{MeV}^{-1}$$

and

$$\frac{iC}{G} = \pm 129\ \mathrm{MeV} = \pm 0.924\,m_{\pi^-}\ .$$

Thus we have

$$\partial_\mu j_\mu^{A-} = \lambda m_\pi^3 \pi^-(x) \tag{12}$$

with $\lambda = \pm 0.924$ and where the charged pion mass is intended. Note that we divided by a factor G.

Analogously to the vector current, Eq. (12) may be extended to include higher order e.m. and weak interaction effects:

$$\partial_\mu j_\mu^{Ai} = \lambda m_\pi^3 \pi^i(x) - \varepsilon_{ijk}\{eA_\mu^j(x)j_\mu^{Ak}(x)$$
$$+ Gl_\mu^j(x)[j_\mu^{Ak}(x)+j_\mu^k(x)]\}\ . \tag{13}$$

The second and third term follow essentially from gauge-invariance and C.V.C. The last term cannot be derived unless some further assumptions are made concerning the axial current. A gauge principle, as first introduced by Gell-Mann and Levy in connection with the σ-model, may be used to this effect.[a]

We rewrite the equations found so far in somewhat different notation:

$$\partial_\mu v_\mu^i = \varepsilon_{ijk}\{eA_\mu^j v_\mu^k + Gl_\mu^j(v_\mu^k+a_\mu^k)\}\ , \tag{14}$$

$$\partial_\mu a_\mu^i = \lambda m_\pi^3 \pi^i - \varepsilon_{ijk}\{eA_\mu^j a_\mu^k + Gl_\mu^j(a_\mu^k+v_\mu^k)\}\ , \tag{15}$$

$$\lambda = \pm 0.924,\quad e^2/4\pi = 137,\quad G = 1.02\cdot 10^{-5}/(\sqrt{2}m_p^2)\ .$$

v_μ^i, a_μ^i, A_μ^i, l_μ^i and π^i represent vector current, axial vector current, e.m. vector potentials, lepton fields and pion fields respectively. To lowest order in e and G one has:

$$\partial_\mu v_\mu^i = 0\ , \tag{16}$$

$$\partial_\mu a_\mu^i = \lambda m_\pi^3 \pi^i\ . \tag{17}$$

[a] *Il. Nuovo. Cimento* **26**, 705 (1960).

The latter relation is often called P(artially) C(onserved) A(xial) C(urrent) hypothesis. Everywhere we suppressed the dependence on x : $v_\mu^i = v_\mu^i(x)$ etc.

We must now fill in a gap. The right-hand side of (17) has a well defined meaning when taken between the vacuum and the one-pion state, but is meaningless in other situations unless we define what we understand by $\pi^i(x)$. Customarily one introduces so-called smoothness assumptions, which are assumptions concerning the behaviour of any matrix-element of $\pi^i(x)$ as function of the four-momentum q_μ associated with x. In fact one usually assumes dependence of the form

$$a + bq ,$$

where a and b are independent of q. To illustrate this we consider the derivation of the Goldberger–Treiman relation.

Consider the axial-vector current in neutron β-decay ($g_a = G_A/G_v \simeq 1.18$)

$$\langle p|a_{\mu(0)}^-|n\rangle = ig_a\bar{u}(p)\gamma^\mu\gamma^5 u(k) . \tag{18}$$

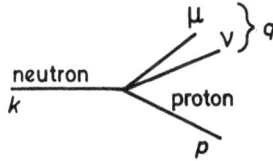

We neglected the induced pseudo-scalar term that is important if $q^2 \simeq -m_\pi^2$ ($q = k - p =$ lepton pair four-momentum) but very small otherwise, in particular for this process. Consider now

$$\langle p|\partial_\mu a_\mu^-(0)|n\rangle = ig_a i(k - p)_\mu\bar{u}(p)\gamma^\mu\gamma^5 u(k)$$
$$= 2im_p g_a\bar{u}(p)\gamma^5 u(k) , \tag{19}$$

where m_p = mass nucleon. In deriving the last line we used the Dirac equation, for instance

$$(i\gamma k + m)u(k) = 0 .$$

On the other hand we have

$$\langle p|\partial_\mu a_\mu^-(0)|n\rangle = \lambda m_\pi^3\langle p|\pi^-(0)|n\rangle . \tag{20}$$

Now, with $q = k - p$:

$$\langle p|\pi^-(0)|n\rangle = \frac{1}{q^2 + m_\pi^2}\langle p|j^-(0)|n\rangle , \tag{21}$$

which is nothing but the Fourier transform of the definition of the pion source current

$$(\Box - m^2)\pi^-(x) = -j^-(x) \ .$$

The matrix element at the right-hand side describes the interaction between pions and nucleons. One has

$$\langle p|j^-(0)|n\rangle = iG_{\pi N}(q^2)\bar{u}(p)\gamma^5 u(k) \ .$$

We make now the following smoothness assumption

$$(q^2 + m_\pi^2)\langle\pi\rangle$$

depends at most linearly on q. This implies $G_{\pi N}(q^2)$ is constant. We will need this smoothness assumption only for values of q^2 between 0 and $-m_\pi^2$. This assumption is essential in order to connect $G_{\pi N}$ with the pion-nucleon coupling constant determined from $\pi - N$ scattering; we get

$$G_{\pi N} \simeq 19.24 \ .$$

For $q^2 \simeq 0$ [for $n \to p + e + \bar{\nu}$ we have $|q^2| \lesssim (0.5 \text{ MeV})^2$] we find

$$\langle p|\pi^-(0)|n\rangle = \frac{iG_{\pi N}}{m_\pi^2}\bar{u}(p)\gamma^5 u(k) \ .$$

With the help of (19) and (20) we find

$$2m_p g_a = \lambda m_\pi G_{\pi N}$$

or

$$g_a = \frac{\lambda G_{\pi N} m_\pi}{2m_p} \sim 1.34 \ . \tag{21}$$

This is the Goldberger–Treiman relation, connecting $g_a = G_A/G_\nu$ and the pion lifetime, represented by λ.

To see the essential ingredients here suppose for a moment that instead of (17) we had

$$\partial_\mu a_\mu^i = \lambda m_\pi^3 \pi^i + X \ ,$$

where X is some pseudo-scalar operator with $\langle 0|X|\pi\rangle = 0$, that is X has no one pion pole. This ensures again $\lambda = 0.924$, from the pion lifetime. Instead of (20) and (21) we would have obtained

$$\langle p|\partial_\mu a_\mu^-(0)|n\rangle = \frac{1}{q^2 + m_\pi^2}\langle p|j^-(0)|n\rangle + \langle p|X(0)|n\rangle \ .$$

Smoothness for the pion field implies again that for $-m_\pi^2 \le q^2 \le 0$ the matrix element $\langle p|j^-(0)|n\rangle$ may be taken as constant. But due to the occurrence of the unknown operator X we cannot derive the Goldberger–Treiman relation.

However, introducing nonsmoothness we can compensate the X-term. Suppose $\langle p|X|n\rangle$ is a constant as function of q^2

$$\langle p|X|n\rangle = \beta \ .$$

Suppose the following behaviour

$$\langle p|j^-(0)|n\rangle = \text{const.} - \beta(q^2 + m_\pi^2) \ .$$

Again, from πN scattering one may derive the value of the constant. The second factor is not experimentally accessible since for any physical pion one has $q^2 + m_\pi^2 = 0$, and thus the second factor is a pure assumption on the behaviour of this matrix element. This factor exactly cancels the term $\langle p|X|n\rangle$ and we get once more the Goldberger–Treiman relation.

The above demonstrates that the name P(ion pole) D(ominated) D(ivergence of the) A(xial) C(urrent), i.e. PDDAC is a more appropriate name than PCAC.

We now turn to Current Commutation rules. They are:

$$\left.\begin{array}{l}
\delta(x_0 - y_0)[v_4^i(x), v_\mu^j(y)] = -\varepsilon_{ijk}v_\mu^k(x)\delta^4(x - y) \ , \\[4pt]
\delta(x_0 - y_0)[a_4^i(x), v_\mu^j(y)] = -\varepsilon_{ijk}a_\mu^k(x)\delta^4(x - y) \ , \\[4pt]
\delta(x_0 - y_0)[v_4^i(x), a_\mu^j(y)] = -\varepsilon_{ijk}a_\mu^k(x)\delta^4(x - y) \ , \\[4pt]
\delta(x_0 - y_0)[a_4^i(x), a_\mu^j(y)] = -\varepsilon_{ijk}v_\mu^k(x)\delta^4(x - y) \ .
\end{array}\right\} \tag{22}$$

By integration $\int d_3x$ one obtains commutators between total charges and current densities. Further integration $\int d_3y$, and setting $\mu = 4$ gives commutators between charges. These are the commutators that one obtains if one uses

$$v_\mu^i(x) = i\left(\bar\Psi(x)\gamma^\mu\frac{\tau^i}{2}\Psi(x)\right) \tag{23}$$

$$a_\mu^i(x) = i\left(\bar\Psi(x)\gamma^\mu\gamma^5\frac{\tau^i}{2}\Psi(x)\right) \tag{24}$$

and the canonical anticommutation rules for equal times

$$\{\Psi_{\alpha a}(x), \bar\Psi_{\beta b}(y)\}_{x_0=y_0} = \gamma_{\alpha\beta}^4\delta_{ab}\delta(\mathbf{x} - \mathbf{y}) \ . \tag{25}$$

In here a and b refer to isospin. Note that v_μ^i and a_μ^i are Hermitian for $\mu = 1, 2, 3$, anti-Hermitian for $\mu = 4$. In the quark model where one has only quark fields and currents like in (23) and (24) one derives indeed (22), with the structure constants f^{ijk} of SU_3 rather than the ε_{ijk} of SU_2 written here.

Unfortunately (22) cannot be correct. Consider

$$\sum_{l=1,2,3} [v_0^3(x), \partial_l v_l^3(y)]_{x_0=y_0}$$

which is zero according to (22). However, using $\partial_\mu v_\mu^3 = 0$ we obtain

$$= -[v_0^3(x), \partial_0 v_0^3(y)]_{x_0=y_0} \ .$$

Consider the vacuum expectation value

$$0 = \sum_c \langle 0|v_0^3(x)|c\rangle\langle c|\partial_0 v_0^3(y)|0\rangle - \langle 0|\partial_0 v_0^3(y)|c\rangle\langle c|v_0^3(x)|0\rangle$$

$$= i\sum_c E_c\{\langle 0|v_0^3(x)|c\rangle\langle c|v_0^3(y)|0\rangle + \langle 0|v_0^3(y)|c\rangle\langle c|v_0^3(x)|0\rangle\} \ ,$$

where we used $\partial_0 j_0(y) = i[H, j_0(y)]$, H is the total energy operator, and $E_c \geq 0$ is the energy of the state $|c\rangle$. In this equation we set $x = y$ and obtain

$$0 = 2i\sum_c E_c\langle 0|v_0^3(x)|c\rangle\langle c|v_0^3(x)|0\rangle$$

$$= 2i\sum_c E_c|\langle 0|v_0^3(x)|c\rangle|^2$$

since v_0^3 is Hermitian. Clearly the right-hand side is nonzero unless $\langle 0|v_0^3(x)|c\rangle = 0$ if $|c\rangle \neq |0\rangle$. Thus unless $v_0^3(x)$ is a c-number we have a contradiction.

Thus the right-hand side of (22) has to be supplemented by extra terms, whose precise nature is not known. They are commonly called Schwinger terms [although their existence was first demonstrated by Goto and Imamura.[b]] In one model, the algebra of fields model, the Schwinger terms can be computed explicitly.

As emphasized originally by Gell-Mann, the last equation of (22) fixes the scale between axial and vector currents and should therefore in principle be useful to determine $g_a(= G_A/G_v)$ in β-decay. Indeed, using this relation Adler and Weisberger have been able to determine g_a. This same scale is contained in the last term of (15). Note that the sign of g_a is not determined.

3. Low-energy Theorems

As a first application we will consider η-decay into three pions. This decay is forbidden for strong interactions because of G-parity (= combined invariance under C, charge conjugation and a rotation in isospin space), and can proceed only if either C or I-spin invariance are broken. We will assume that C is good and that I-spin breaking comes about by virtual e.m. interactions. Neglecting the weak terms we employ (15):

$$\partial_\mu a_\mu^i = \lambda m_\pi^3 \pi^i - \varepsilon_{ijk} e A_\mu^k a_\mu^k \ . \tag{26}$$

More specifically we need this equation only for $i = 3$. Then the second term is zero (as A^1 and A^2 are zero) and we have

$$\partial_\mu a_\mu^0 = \lambda m_\pi^3 \pi^0 \tag{27}$$

[b] Goto and Imamura, *Progr. Theor. Phys.* **14**, 396 (1955).

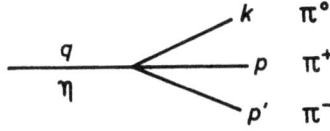

which is supposedly correct including e.m. interactions. The Fourier transform of (27) tells us in fact that any amplitude involving a π^0 of four momentum k should be proportional to k [from the derivative ∂_μ on the left-hand side of (27)]. To see this in detail we use the reduction formula:

$$\langle\ \rangle \equiv {}_{\text{out}}\langle \pi^+\pi^-\pi^0|\eta\rangle_{\text{in}}$$

$$= -i \int d_4 x \frac{1}{\sqrt{2k_0 V}} e^{-ikx}(\Box - m_0^2)_{\text{out}}\langle\pi^+\pi^-|\pi^0(x)|\eta\rangle_{\text{in}}$$

$$= i \int \frac{d_4 x}{\sqrt{2k_0 V}} e^{-ikx}(k^2 + m_0^2)_{\text{out}}\langle\pi^+\pi^-|\pi^0(x)|\eta\rangle_{\text{in}} \qquad (28)$$

k is the π_0 four momentum, and V is the volume in which the system is enclosed. Knowing this matrix element the decay rate can be computed; with M given by

$$_{\text{out}}\langle\pi^+\pi^-\pi^0|\eta\rangle_{\text{in}} = M\delta_4(q - k - p - p') \qquad (29)$$

one has

$$\frac{1}{\tau} = \Gamma = \int \frac{V}{(2\pi)^3}d_3 k \frac{V}{(2\pi)^3}d_3 p \frac{V}{(2\pi)^3}d_3 p' \frac{V}{(2\pi)^4}\delta_4(q - k - p - p')|M|^2\ .$$

In the η-rest system this can be worked out further with

$$|M|^2 = \frac{(2\pi)^8}{2^4 V^4 q_0 k_0 p_0 p_0'}|F[m_\eta^2, m_0^2, m_+^2, m_-^2, (qk), (qp)]|^2 \qquad (30)$$

where m_η, m_+ and m_0 are η, $\pi^\pm$ and π^0 mass respectively, one has:

$$\Gamma = \frac{1}{2^6\pi^3 q_0}\int_{m_0}^{a} dk_0 \int_{a_-}^{a_+} dp_0|F|^2\ , \qquad (31)$$

$$a = \frac{m_\eta^2 + m_0^2 - (m_+ + m_-)^2}{2m_\eta}\ ,$$

$$a_\pm = \frac{(m_\eta^2 + m_0^2 + m_+^2 - m_-^2 - 2m_\eta k_0)(m_\eta - k_0) \pm |\mathbf{k}|\sqrt{}}{2\{m_\eta^2 + m_0^2 - 2m_\eta k_0\}}\ ,$$

$$\sqrt{} = \sqrt{(m_\eta^2 + m_0^2 - m_+^2 - m_-^2 - 2m_\eta k_0)^2 - 4m_+^2 m_-^2}\ ,$$

$$|\mathbf{k}| = \sqrt{k_0^2 - m_0^2}\ .$$

If $F = $ constant, one finds from a width $\Gamma = 750$ eV the dimensionless value $F = 0.41$.

Using now (27) we get:

$$\langle\,\rangle = \frac{i}{\lambda m_\pi^3} \int \frac{d_4 x}{\sqrt{2k_0 V}} e^{-ikx}(k^2 + m_0^2)\partial_{\mu\,\text{out}}\langle\pi^+\pi^-|a_\mu^0|\eta\rangle_{\text{in}}$$

$$= \frac{i}{\lambda m_\pi^3} \int \frac{d_4 x}{\sqrt{2k_0 V}} e^{-ikx} ik_\mu(k^2 + m_0^2)_{\text{out}}\langle\pi^+\pi^-|a_\mu^0|\eta\rangle_{\text{in}}$$

$$= \frac{i}{\lambda m_\pi^3} \int \frac{d_4 x}{\sqrt{2k_0 V}} ik_\mu(k^2 + m_0^2)e^{ix(q-k-p-p')}{}_{\text{out}}\langle\pi^+\pi^-|a_\mu^0|\eta\rangle_{\text{in}} \qquad (32)$$

where q, p, and p' are the four momenta of η, π^+ and π^- respectively. The matrix-element

$$_{\text{out}}\langle\pi^+\pi^-|a_\mu^0|\eta\rangle_{\text{in}}$$

will display a singularity for $k^2 = -m_\pi^2$, but not for $k_\mu = 0$ since none of the in- or outgoing particles can, through an axial current, go over into a particle with the same mass. In the example in the figure the only particle X with the same mass as the pion is the pion itself, but $\langle\pi|a_\mu|\pi\rangle = 0$ because of parity.

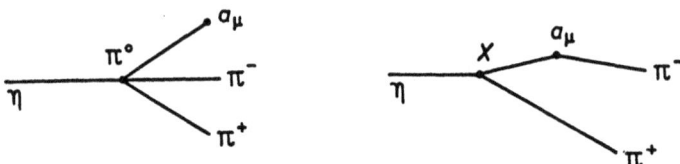

Singularity of a_μ for $k^2 = -m_\pi^2$ Example of possibly singular diagram for $k_\mu = 0$

Thus (27) tells us simply that the matrix element for the decay of $\eta \to \pi^0\pi^+\pi^-$ has a zero for zero four-momentum of the pion. As usual we must supplement this result with a smoothness assumption. The amplitude for η-decay can be written as a function of (kq), (pq) and k^2, keeping $q^2 = -m_\eta^2$, $p^2 = -m_+^2$ and $p'^2 = -m_-^2$. Our smoothness assumption is [see (29) and (30)]:

$$F(\eta \to \pi^+\pi^-\pi^0) = a + b(kq) , \qquad (33)$$

where a and b may be functions of (pq), but not of (kq) or k^2. Actually experimentally where k^2 is fixed to $-m_\pi^2$ one finds a and b to be constants, and

$$b \simeq \frac{a}{2m_\eta m_0} , \qquad (34)$$

that is the amplitude decreases with increasing π^0 energy (starting from $k_0 = m_0$) to zero for $k_0 \sim 2m_0$. But (27) now implies $a = 0$ (i.e. $F = 0$ if $k = 0$), and for any value of b that would imply a matrix-element *increasing* with increasing π^0-energy. Thus η-decay contradicts (27).

The argument can be carried further by considering also $\eta \to 3\pi^0$. We refer to Bell and Sutherland,[c] for a detailed analysis.

We now turn to a more complicated case, compton scattering on nucleons:

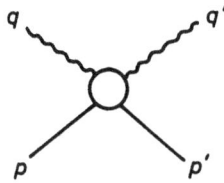

$$W + N \to W' + N' .$$

Here W and W' represent an isovector photon, or a lepton pair (only coupling to vector currents considered) depending on which isovector components of the currents are to be investigated. For the actual calculation we will take W_v and W'_v to be some vector boson isospin triplet. The coupling constant is g, and may be e or G as required. Let e^i_μ and e'^j_ν describe spatial polarization and I-spin for W and W' (thus for instance for isovector photon scattering we must set $e^1_\mu = e^2_\mu = e'^1_\nu = e'^2_\nu = 0$ for all μ and ν while e^3_μ and e'^3_ν are the polarization vectors for in- and outgoing photon).

A set of functions $M^{ij}_{\mu\nu}$ $(i, j = 1, 2, 3;\ \mu, \nu = 1, \ldots 4)$ are defined by

$$_{\text{out}}\langle q', p' | qp \rangle_{\text{in}} = _{\text{out}} \langle q', p' | qp \rangle_{\text{out}} + ig^2 \frac{e^i_\mu e'^j_\nu}{\sqrt{4V^2 q_0 q'_0}} M^{ij}_{\mu\nu}(q, q', p, p') . \qquad (35)$$

According to the reduction formula we have:

$$_{\text{out}}\langle q'p' | qp \rangle_{\text{in}} = _{\text{out}} \langle q'p' | qp \rangle_{\text{out}} - i \int d_4 x \frac{e^i_\mu}{\sqrt{2q_0 V}} e^{iqx}$$
$$\cdot (\Box - m^2_w)_{\text{out}}\langle q'p' | W^i_\mu(x) | p \rangle_{\text{in}}$$
$$= _{\text{out}} \langle q'p' | qp \rangle_{\text{out}} + ig \int d_4 x \frac{e^i_\mu}{\sqrt{2q_0 V}} e^{iqx}{}_{\text{out}}\langle q'p' | v^i_\mu(x) | p \rangle_{\text{in}} \qquad (36)$$

In the following we will neglect the first term on the right-hand side, which is the no-scattering part of the S-matrix.

[c] Bell and Sutherland, *Nucl. Phys.* **B4**, 315 (1968).

Comparing (35) and (36) we have

$$g\frac{e_\nu^{\prime j}}{\sqrt{2Vq_0'}}M_{\mu\nu}^{ij} = \int d_4x e^{iqx}{}_{\text{out}}\langle q'p'|v_\mu^i(x)|p\rangle_{\text{in}} . \tag{37}$$

Multiplying with q_μ and summing over q we have:

$$g\frac{e_\nu^{\prime j}}{\sqrt{2Vq_0'}}q_\mu M_{\mu\nu}^{ij} = \int d_4x e^{iqx}q_\mu{}_{\text{out}}\langle q'p'|v_\mu^i(x)|p\rangle_{\text{in}}$$

$$= -i\int d_4x(\partial_\mu e^{iqx})_{\text{out}}\langle q'p'|v_\mu^i(x)|p\rangle_{\text{in}} = (\text{partial integration})$$

$$= i\int d_4x e^{iqx}{}_{\text{out}}\langle q'p'|\partial_\mu v_\mu^i(x)|p\rangle_{\text{in}} . \tag{38}$$

We now use Eq. (14)

$$\partial_\mu v_\mu^i(x) = -\varepsilon_{ijk}gW_\mu^j(x)v_\mu^k(x) \tag{39}$$

to get for (38):

$$(38) = -ig\varepsilon_{ijk}\int d_4x e^{iqx}{}_{\text{out}}\langle q'p'|W_\mu^j(x)v_\mu^k(x)|p\rangle_{\text{in}} . \tag{40}$$

We will work to lowest order in g, and may consider therefore the $W_\mu^j(x)$ field occurring in this matrix-element as a free field. This field can then absorb the W with momentum q' from the out state and we get, as factor, the corresponding wave function

$$(41) = -ig\varepsilon_{ijk}\frac{e_\nu^{\prime j}}{\sqrt{2q_0'V}}\int d_4x e^{i(q-q')x}{}_{\text{out}}\langle p'|v_\nu^k(x)|p\rangle_{\text{in}} . \tag{41}$$

We now have the matrix element of the isovector current between two nucleon states:

$$\langle p'|v_\nu^k(x)|p\rangle = \frac{e^{i(p-p')x}}{V} \cdot [\bar{u}(p')\frac{\tau^k}{2}\{iF_1(Q^2)\gamma^\nu + \frac{\kappa}{2m}\sigma^{\nu\lambda}Q_\lambda\}u(p)]$$

$$\equiv \frac{1}{V}e^{i(p-p')x}\bar{u}(p')\Gamma_\nu^k(p',p)u(p) . \tag{42}$$

In here $m =$ nucleon mass, $Q = p - p' \cdot F_1$ and F_2 are the nucleon form factors, κ is the isovector anomalous moment and still depends on Q^2. In the following we will need κ only for $Q^2 = 0$. Further $\sigma^{\nu\lambda} = \frac{1}{2}(\gamma^\nu\gamma^\lambda - \gamma^\lambda\gamma^\nu)$.

Inserting (42) in (41) and integrating over x gives

$$(40) = -ig(2\pi)^4\varepsilon_{ijk}\frac{e^{\prime j}}{\sqrt{2q_0'V}} \cdot \frac{1}{V}\bar{u}(p')\Gamma_\nu^k u(p)\delta_4(p+q-p'-q') . \tag{43}$$

Thus, finally we arrive at the equation

$$q_\mu M_{\mu\nu}^{ij} = \frac{-i}{V}(2\pi)^4\delta_4(q+p-q'-p')\varepsilon_{ijk}\bar{u}(p')\Gamma_\nu^k(p',p)u(p) . \tag{44}$$

We dropped also $e_\nu'^j$ on both sides, which is o.k. because no restriction exists on $e_\nu'^j$ and the equation has to hold for every permissible $e_\nu'^j$.

A similar formula can be derived for $q_\nu' M_{\mu\nu}^{ij}$. There is no need to do so here as $M_{\mu\nu}^{ij}$ must satisfy crossing symmetry

$$M_{\mu\nu}^{ij}(q, q', p, p') = M_{\nu\mu}^{ji}(-q', -q, p, p') \,. \tag{45}$$

Equation (44) is the central result. Knowing Γ_ν^k (which is nothing but the isovector nucleon vertex measured in e.m. and weak interactions) it turns out to be possible to determine $M_{\mu\nu}^{ij}$ from (44) up to and including first order in q and q'. As an example, consider the following equation

$$q_\mu F_\mu(q, p_1, p_2, \dots, p_n) = G(q, p_1, \dots, p_n) \tag{46}$$

with known $G(q, p_1, \dots, p_n)$, nonsingular for $q \to 0$. Furthermore, suppose that F_μ is singular for $q \to 0$, and that the singular part of F_μ is known:

$$F_\mu = \frac{a_\mu}{(qa)} f(p_1, \dots, p_n) + F_\mu'(q, p_1, \dots, p_n) \,, \tag{47}$$

where a_μ is some linear combination of the $p_1, \dots, p_n$ and where F_μ' is nonsingular for $q \to 0$. Inserting (47) into (46) we obtain a so-called consistency condition for $q \to 0$:

$$f(p_1, \dots, p_n) = G(0, p_1, \dots, p_n) \tag{48}$$

Let us assume that (48) holds. Inserting (48) and (47) we obtain:

$$q_\mu F_\mu'(q, p_1, \dots, p_n) = G'(q, p_1, \dots, p_n) \tag{49}$$

with

$$G'(q, p_1, \dots, p_n) = G(q, p_1, \dots, p_n) - G(0, p_1, \dots, p_n) \,. \tag{50}$$

Let us now assume that F' can be expanded in q around $q = 0$. (49) permits the determination of F' up to vectors that are orthogonal to q_μ (for any q_μ). Such vectors must necessarily contain q_μ itself. The simplest such vector is of the type (we assume that F_μ' is a vector, not an axial vector):

$$p_{1\mu}(qp_2) - p_{2\mu}(qp_1) \,. \tag{51}$$

This vector is linear in q. Thus (49) permits the complete determination of F' for $q = 0$.

In the actual case at hand we have a tensor $M_{\mu\nu}$, and two equations as (46):

$$q_\mu M_{\mu\nu} = G_1 \,,$$
$$q_\mu' M_{\mu\nu} = G_2 \,. \tag{52}$$

The simplest tensor satisfying $q_\mu T_{\mu\nu} = q'_\nu T_{\mu\nu} = 0$ is bilinear in q and q'; for example

$$\delta_{\mu\nu}(qq') - q_\nu q'_\mu \ . \tag{53}$$

Thus in our case $M_{\mu\nu}$ can be determined up to tensors of this kind, that is, from $M_{\mu\nu}$ we can determine the terms independent of q and q' and the terms linear in q or q'.

The actual calculation is quite cumbersome. First of all we must find the singular terms of M for q or $q' \to 0$. These are given by the diagrams

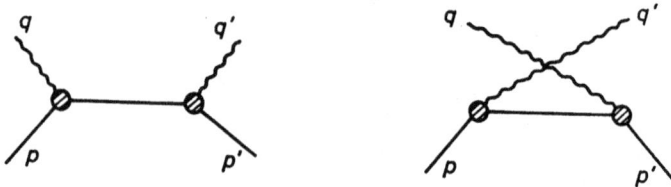

The contribution of these diagrams to the S-matrix is:

$$ig^2 \frac{(2\pi)^4}{V^2\sqrt{4q_0 q'_0}}\delta^4(p+q-p'-q')e^i_\mu e'^j_\nu [\bar{u}(p')\{\bar{\Gamma}^j_\nu(p',p+q)$$
$$\bar{S}_F(p+q)\bar{\Gamma}^i_\mu(p+q,p) + \bar{\Gamma}^i_\mu(p',p'-q)\bar{S}_F(p'-q)\bar{\Gamma}^j_\mu(p'-q,p)\}u(p)] \tag{54}$$

In here $\bar{\Gamma}$ and $\bar{S}$ are the fully dressed $W\bar{N}N$ vertex and N propagator. We write

$$\bar{S}_F(k) = \frac{1}{(i\gamma k + m) + (i\gamma k + m)^2 g(k^2, \gamma k)} = \frac{1}{i\gamma k + m} + \text{rest} \ ,$$
$$\equiv S_F(k) + \text{rest} \tag{55}$$

$$\bar{\Gamma}^j_\nu(k,k') = \Gamma^j_\nu(k,k') + (k^2 + m^2)h^j_\nu + (k'^2 + m^2)h'^j_\nu$$
$$\equiv \Gamma^j_\nu(k,k') + \text{rest}' \ . \tag{56}$$

Thus $\Gamma = \bar{\Gamma}$ if $k^2 = k'^2 = -m^2$. But the $W\bar{N}N$ vertex with the two nucleons on the mass shell is just the Γ defined in (42). We will use the definition (42) also for k^2 or $k'^2 \neq -m^2$, and assume the functions h and h' in (56) to be chosen such as to give the correct $\bar{\Gamma}$.

By changing in (54) $\bar{\Gamma}$ into Γ and $\bar{S}_F$ into S_F we obtain an expression whose pole term, location and residue, coincides with that of (54), since the singularity occurs if the momentum k of the intermediate nucleon satisfies $i\gamma k + m = 0$ and $k^2 + m^2 = (i\gamma k + m) \cdot (-i\gamma k + m) = 0$. This happens indeed for $q \to 0$ or $q' \to 0$ since in (54) $p'^2 = p^2 = -m^2$.

Thus as singularity term in $M^{ij}_{\mu\nu}$ we have

$$M^{ij}_{\mu\nu B} = \frac{(2\pi)^4}{V}\delta^4(p+q-p'-q')\{\bar{u}(p')[\Gamma^j_\nu(p',p+q)S_F(p+q)$$
$$\cdot \Gamma^i_\mu(p+q,p) + \Gamma^i_\mu(p',p'-q)S_F(p'-q)\Gamma^j_\nu(p'-q,p)]u(p)\} \ . \tag{56}$$

Next we write

$$M_{\mu\nu}^{ij} = M_{\mu\nu B}^{ij} + U_{\mu\nu}^{ij} \tag{57}$$

where U is nonsingular for q, $q' \to 0$. Equation (44) gives

$$q_\mu U_{\mu\nu}^{ij}$$
$$= -\frac{i}{V}(2\pi)^4 \varepsilon_{ijk} \bar{u}(p') \Gamma_\nu^k(p',p) u(p) \delta_4(q+p-q'-p') - q_\mu M_{\mu\nu B}^{ij} . \tag{58}$$

We compute the last term by using what is essentially the Ward identity

$$q_\mu \Gamma_\mu^i(p+q,p) = \frac{1}{2}\tau^i F_1(q^2) i\gamma^\mu q_\mu . \tag{59}$$

With $i\gamma q = -(i\gamma p + m) + [i\gamma(p+q) + m]$, $(i\gamma p + m)u(p) = 0$ and similar formulae for the second term of (56) we obtain

$$q_\mu M_{\mu\nu B}^{ij} = \frac{(2\pi)^4}{V}\delta_4(p+q-p'-q')$$
$$\left\{\bar{u}(p')\frac{1}{4}[\tau^j,\tau^i]F_1(q^2)\left[F_1(q'^2)i\gamma^\nu + \frac{\kappa}{2m}\sigma^{\nu\lambda}q'_\lambda\right]u(p)\right\} . \tag{60}$$

Substitution in (58) gives with $[\tau^j,\tau^i] = -2i\varepsilon_{ijk}\tau^k$:

$$q_\mu U_{\mu\nu}^{ij} = C_{ijk}\bar{u}(p')[\{F_1(q^2)F_1(q'^2) - f_1(Q^2)\}i\gamma^\nu$$
$$+ \frac{1}{2m}\sigma^{\nu\lambda}\{F_1(q^2)\kappa q'_\lambda - \kappa Q_\lambda\}]\frac{\tau^k}{2}u(p) \tag{61}$$

with

$$C_{ijk} = \frac{i(2\pi)^4}{V}\delta_4(p+q-p'-q')\cdot\varepsilon_{ijk}.$$

Another such relation can be derived by reducing the outgoing W from the out state instead of the ingoing W as in (36). This relation is

$$U_{\mu\nu}^{ij}q'_\nu = C_{ijk}\bar{u}(p')\{[F_1(q'^2)F_1(q^2) - F_1(Q^2)]i\gamma^\mu$$
$$- \frac{1}{2m}\sigma^{\mu\lambda}[F_1(q'^2)\kappa q_\lambda + \kappa Q_\lambda]\}\frac{\tau^k}{2}u(p) . \tag{62}$$

It can also be obtained from (61) with the help of crossing symmetry

$$U_{\mu\nu}^{ij}(q,q') = U_{\nu\mu}^{ij}(-q',-q) . \tag{63}$$

The situation is complicated by the fact that for fixed p the four-vectors q and q' are not independent, for we must have $p'^2 = -m^2$, or

$$(p+q-q') = -m^2$$

or

$$2(pq) - 2(pq') + q^2 + q'^2 - 2qq' = 0 . \tag{64}$$

Note also

$$\bar{u}(p)\{\gamma q' - \gamma q\}u(p) = \bar{u}(p)\{\gamma p - \gamma p'\}u(p) = 0 . \tag{65}$$

Let us first ask for the general solution of the equation

$$q_\mu T_{\mu\nu} = 0 , \tag{66}$$

where $T_{\mu\nu}$ may be constructed from q, q', p and γ-matrices. Notation: $X^1_\mu = q_\mu$, $X^2_\mu = q'_\mu$, $X^3_\mu = p_\mu$, $X^4_\mu = \gamma^\mu$. We have

$$T_{\mu\nu} = a_i(\delta_{\mu\nu}(X^i q) - X^i_\mu q_\nu) + b_{ij}(X^i_\mu(X^j q) - X^j_\mu(X^i q))Z^{ij}_\nu$$
$$+ c\{p_\mu[2(pq) - 2(pq') + q'^2] + q_\mu(pq) - 2q'_\mu(pq)\}Y_\nu + \ldots \tag{67}$$

Z_ν, Y_ν etc. are vectors constructed from the X. The a_i, b_{ij}, c etc. are arbitrary coefficients. The last term indicated shows how the identity (64) leads to further solutions of (66).

We now also require

$$T_{\mu\nu}q'_\nu = 0 . \tag{68}$$

Without any trouble one establishes that $T_{\mu\nu}$ consists of terms that contain explicitly at least two times q or q'. The simplest such term is

$$\delta_{\mu\nu}(q'q) - q'_\mu q_\nu . \tag{69}$$

Another one (remember $\gamma q = \gamma q'$):

$$[p_\mu(\gamma q') - \gamma^\mu(pq)][p_\nu(\gamma q') - \gamma^\nu(pq')] .$$

Thus from (61) and (62) it should be possible to determine $U^{ij}_{\mu\nu}$ up to terms explicitly containing twice q and q'. Without difficulty one establishes

$$U^{ij}_{\mu\nu} \simeq C_{ijk}\bar{u}(p)[2iF'_1(0)(q'_\mu\gamma^\nu + \gamma^\mu q_\nu)$$
$$- i\delta_{\mu\nu}F'_1(0)(\gamma q + \gamma q') + \frac{\kappa}{2M}\sigma^{\nu\mu}]\frac{\tau^k}{2}u(p) , \tag{70}$$

where $F'_1(0)$, the derivative of F_1, arises because of series expansion of F_1.

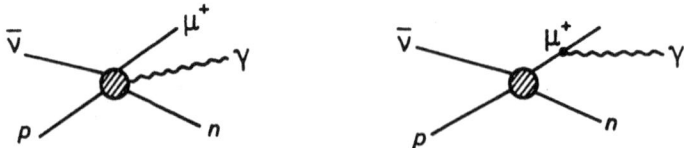

(70) is called a low energy theorem. For the time being it is pretty useless, as the type of reaction where (70) would manifest itself would be something like

$\nu + p \to \mu^+ + n + \gamma$, see figure. Note that a complete analysis of this reaction still involves the diagram where the photon is emitted by the muon. Also axial currents have to be considered. Besides the isoscalar photon emission has to be deduced also (only the Born term is important). By comparing (70) with experiment one tests C.V.C. and local conservation of change in a low energy limit.

In order to test (70) one may try to make an alternative calculation for $U_{\mu\nu}^{ij}$ and then to compare the results. First, one may compute $U_{\mu\nu}^{ij}$ by assuming that only the 3-3 resonance plays any role a s intermediate state:

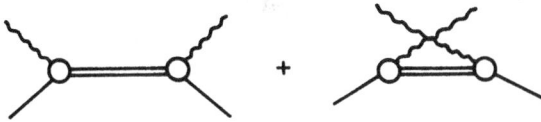

The necessary information concerning the $\gamma N N^*$ vertex may be obtained from the experimental data on N^* production by photons and nucleons.

Alternatively, using dispersion relations a more complete calculation can be done using, as input data, the total cross-sections of photons of all energies on nucleons. Setting equal the results (70) and the dispersion relation result one obtains a so-called sum rule.

Low energy theorems may be derived for a variety of processes. Except for the e.m. decays $\eta \to 3\pi$, $\pi^0 \to 2\gamma$ etc. they compare well with experiment. In particular the Adler–Weisberger relation, sensitive to the last term on the right-hand side of (15), agrees with experiment.

4. High Energy Results

Let us consider the vacuum exception-value of some current commutator

$$\langle 0|[v_\mu^i(x), v_\nu^j(y)]|0\rangle \ . \tag{71}$$

This commutator consists of two terms. The second term may be obtained from the first by exchanging i, μ, x with j, ν, y. For the first term we have

$$\langle 0|v_\mu^i(x)v_\nu^j(y)|0\rangle = \sum_n \langle 0|v_\mu^i(x)|n\rangle \langle n|v_\nu^j(y)|0\rangle$$

$$= \sum_n e^{ip_n(x_y)}\langle 0|v_\mu^i(0)|n\rangle \langle n|v_\nu^j(0)|0\rangle \ . \tag{72}$$

The summation extends over a complete set of states $|n\rangle$, p_n is the four-momentum of $|n\rangle$. First we sum over all states with given four-momentum p.
We have

$$(72) = \sum_{p_n} e^{ip_n(x-y)} F_{\mu\nu}(p_n) \cdot \delta_{ij}$$

with

$$\delta_{ij} F_{\mu\nu}(p_n) = \sum_s \langle 0|v_\mu^i(0)|s\rangle\langle s|v_\nu^j(0)|0\rangle , \tag{73}$$

where the $|s\rangle$ are all states with four-momentum p_n. We note the following facts: The right-hand side of (73) is zero unless $i = j$. This is simply because, in (71), the only operator that has nonzero vacuum expectation value must be an isoscalar, which implies δ_{ij}. Further, $F_{\mu\nu}(p)$ is a function of only one vector. Therefore

$$F_{\mu\nu}(p) = \frac{1}{V}[\rho_1(-p^2)\delta_{\mu\nu} + \rho_2(-p^2)p_\mu p_\nu] . \tag{74}$$

The summation in (73) goes over all states with positive energy ($p_0 > 0$). V is the volume in which the system is enclosed, and has been taken out for convenience. Thus we have (remember, $\sum_{\mathbf{p}} \to \frac{V}{(2\pi)^3} \int d_3p$)

$$\langle 0|v_\mu^i(x)v_\nu^j(y)|0\rangle = \frac{\delta_{ij}}{(2\pi)^3} \int d_4p\, e^{ip(x-y)}\Theta(p_j)$$
$$[\rho_1(-p^2)\sigma_{\mu\nu} + \rho_2(-p^2)p_\mu p_\nu] , \tag{75}$$

where ρ_1 and ρ_2 are unknown functions of $-p^2$. On account of the reality properties of (73) one may establish $F_{\mu\nu} \geq 0$ and real for $\mu = \nu \neq 4$; ≤ 0 for $\mu = \nu = 4$. This is because v_μ^i for $\mu = 1, 2, 3$ is Hermitian, and anti-Hermitian for $\mu = 4$. (75) is the Kaller–Lehman representation.

Furthermore to lowest order in e or G we have $\partial_\mu v_\mu = 0$, and consequently

$$p_\mu(\rho_1(-p^2) + p^2\rho_2(-p^2)) = 0 . \tag{76}$$

We then have

$$\langle 0|v_\mu^i(x)v_\nu^j(y)|0\rangle = \frac{\delta_{ij}}{(2\pi)^3} \int d_4p\, e^{ip(x-y)}\Theta(p_0)\rho(-p^2)\{\delta_{\mu\nu} - p_\mu p_\nu/p^2\} . \tag{77}$$

For the commutator we find (exchange $x \leftrightarrow y$, then $p \to -p$):

$$\langle 0|[v_\mu^i(x), v_\nu^j(y)]|0\rangle = \frac{\delta_{ij}}{(2\pi)^3} \int d_4p\, e^{ip(x-y)}\varepsilon(p_0)\rho(-p^2)\{\delta_{\mu\nu} - p_\mu p_\nu/p^2\} . \tag{78}$$

$\varepsilon(p_0) = 1$ for $p_0 > 0$, -1 for $p_0 < 0$.
For equal times, $x_0 = y_0$, the integration over p_0 may be performed. Setting moreover $\mu = 4$ we have

$$\int_{-\infty}^{\infty} dp_0\varepsilon(p_0)\rho(-p^2)\{\delta_{4\nu} - p_4p_\nu/p^2\} . \tag{79}$$

For $\nu = 4$ this is zero because of $\varepsilon(-p_0) = -\varepsilon(p_0)$. For $\nu \neq 4$ we find

$$(79) = -ip_\nu \int_{-\infty}^{\infty} dp_0 p_0\varepsilon(p_0)\rho(-p^2)/p^2$$
$$= \frac{ip_\nu}{2} \int_0^{\infty} d\mu^2 \rho(\mu^2)/\mu^2 \equiv ip_\nu \cdot \lambda , \tag{80}$$

where $\mu^2 = -p^2 = p_0^2 - \mathbf{p}^2$. We need that $\rho(-p^2) = 0$ for $-p^2 < 0$, because in (73) the physical states $|s\rangle$ have mass ≥ 0. Inserting (80) in (78) for $x_0 = y_0$, $\mu = 4$, we find

$$\delta(x_0 - y_0)\langle 0|[v_4^i(x), v_\nu^j(y)]|0\rangle$$
$$= \delta_{ij}\delta(x_0 - y_0)\frac{1}{(2\pi)^3}\int d_3p\, ip_\nu e^{ip(x-y)} \cdot \lambda$$
$$= \delta_{ij}\delta(x_0 - y_0)\frac{\partial}{\partial x_\nu}\delta(\mathbf{x} - \mathbf{y}) \cdot \lambda$$
$$= \lambda\delta_{ij}\partial_\nu\delta^4(x - y) \qquad \text{for } \nu \neq 4 \quad (= 0 \text{ for } \nu = 4) \tag{81}$$

with λ defined in (80). Since λ is obviously nonzero we have essentially another proof that (22) must be supplemented with Schwinger terms. (81) tells us what the vacuum expectation value of those Schwinger terms is.

Let us now make the assumption that the extra terms, the Schwinger terms in the current commutators (22) are c-numbers. That is, we assume that they equal their vacuum expectation values. Of course, λ may differ from commutator to commutator. However, Weinberg has proved that under this assumption $\lambda_v = \lambda_a$, where:

$$\delta(x_0 - y_0)[v_4^i(x), v_\nu^j(y)] = -\varepsilon_{ijk}v_\nu^k(x)\delta^4(x - y) + \lambda_v\delta_{ij}\partial_\nu\delta^4(x - y)$$
$$\delta(x_0 - y_0)[a_4^i(x), a_\nu^j(y)] = -\varepsilon_{ijk}v_\nu^k(x)\delta^4(x - y) + \lambda_a\delta_{ij}\partial_\nu\delta^4(x - y) . \tag{82}$$

Since for the axial current $\partial_\mu a_\mu^i \neq 0$ the equality (76) does not hold. One has, with the notation as in the right-hand side of (75):

$$\lambda_a = \frac{1}{2}\int_0^\infty d\mu^2(\mu^2)\rho_2(\mu^2)/\mu^2 . \tag{83}$$

To prove the assertion $\lambda_a = \lambda_v$ we write down the Jocobi identity

$$[v_4^i, [a_4^j, a_\nu^k]] + [a_\nu^k, [v_4^i, a_4^j]] + [a_4^j, [a_\nu^k, v_4^i]] = 0 \tag{84}$$

with $v_4^i = v_4^i(x)$, $a_4^j = a_4^j(y)$ and $a_\nu^k = a_\nu^k(z)$. Furthermore $x_0 = y_0 = z_0$.

For the innermost commutator we can forget about c-number Schwinger terms. We get, using (22):

$$- \varepsilon_{jkl}\delta(\mathbf{y} - \mathbf{z})[v_4^i(x), v_\nu^l(z)] - \varepsilon_{ijl}\delta(\mathbf{x} - \mathbf{y})[a_\nu^k(z), a_4^l(y)]$$
$$+ \varepsilon_{ikl}\delta(\mathbf{x} - \mathbf{z})[a_4^j(y), a_\nu^l(z)] = 0 .$$

Taking the vacuum expectation value of this expression we obtain

$$- \varepsilon_{jki}\delta(\mathbf{y} - \mathbf{z})\lambda_v \partial_\nu^x \delta(\mathbf{x} - \mathbf{z})$$
$$+ \varepsilon_{ijk}\delta(\mathbf{x} - \mathbf{y})\lambda_a \partial_\nu^y \delta(\mathbf{y} - \mathbf{z})$$
$$- \varepsilon_{ikj}\delta(\mathbf{x} - \mathbf{z})\lambda_a \partial_\nu^y \delta(\mathbf{y} - \mathbf{z}) = 0 .$$

Multiplying with some function $f(\mathbf{y})$ and integrating over all z and $\mathbf{y}$ we obtain

$$-\varepsilon_{ijk}\lambda_v \partial_\nu f(\mathbf{x}) + \varepsilon_{ijk}\lambda_a \partial_\nu f(\mathbf{x}) = 0$$

or

$$\lambda_v = \lambda_a \qquad (85)$$

or

$$\int_0^\infty \frac{\rho_v(\mu^2)}{\mu^2}d\mu^2 = \int_0^\infty \frac{\rho_{2a}(\mu^2)\cdot(\mu^2)}{\mu^2}d\mu^2 \ .$$

This is Weinberg's first sum rule. Let us now make some assumptions concerning the functions ρ defined by (73) and (74). We assume that in (73) for the states $|s\rangle$ only the ρ-meson contributes significantly. Now

$$\langle 0|v_\mu^i(0)|\rho\rangle = \frac{1}{\sqrt{2Vp_0}}g_\rho e_\mu^i \qquad p_0 = \sqrt{\mathbf{p}^2 + m_\rho^2} \ ,$$

where g_ρ is the ρ coupling constant and e_μ^i is the polarization vector of the ρ. In here g_ρ, apart from a factor e the strength with which the photon couples to the ρ-meson, is experimentally known to be somewhere between $m_\rho^2/5$ and $m_\rho^2/3.7$. The first number follows by assuming that photon-pion coupling is dominated by the ρ, see figure. At zero momentum transfer this coupling is e (the charge). Thus we require:

$$eg_\rho \cdot \frac{1}{q^2 + m_\rho^2} \cdot g_{\rho\pi\pi}(q^2) = e \quad \text{for } q^2 = 0 \ .$$

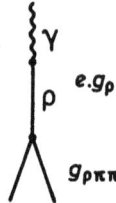

From the width of the $\rho \to 2\pi$ decay one derives $g_{\rho\pi\pi}(-m_\rho^2) = 5$. Assuming that $g_{\rho\pi\pi}(0) \sim g_{\rho\pi\pi}(-m_\rho^2)$ (which is quite an assumption) one finds $g_\rho = m_\rho^2/5$. Similar derivations on other processes give higher values, up to $g_\rho = m_\rho^2/3.7$ (from $\pi - N$ scattering).

Summing over ρ-polarization we have

$$\begin{aligned}
F_{\mu\nu}(p) &= \sum_{\text{pol.}}\langle 0|v_\mu^i(0)|\rho\rangle\langle\rho|v_\nu^j(0)|0\rangle \\
&= \frac{g_\rho^2}{V}\frac{1}{2p_0}(\delta_{\mu\nu} + p_\mu p_\nu/m_\rho^2) \\
&= \frac{g_\rho^2}{V}\Theta(p_0)\delta(p^2 + m_\rho^2)(\delta_{\mu\nu} + p_\mu p_\nu/m_\rho^2) \ , \qquad (88)
\end{aligned}$$

indeed of the form (74).

For the axial current matrix-element $\langle 0|a_\mu^i|s\rangle$ we assume that for s only the π-meson and the A_1 meson (mass 1070 ± 20 MeV, I-spin 1, spin-1, parity $+$, $C = +$, width 80 ± 35 MeV, decays almost entirely into 3π, probably $\rho\pi$. Existence still in doubt, some of these quantum numbers are not firmly established). The A_1 gives a contribution analogous to (88), with an unknown constant g_a instead of g_ρ. For the π-meson we have [see (11) till (12)]:

$$\langle 0|a_\mu^i(0)|\pi^i\rangle = i\lambda m_\pi p_\mu \cdot \frac{1}{\sqrt{2Vp_0}}, \qquad \lambda = \pm 0.924 \tag{89}$$

and

$$F_{\mu\nu}^A(p) = \frac{g_a^2}{V}\Theta(p_0)\delta(p^2 + m_A^2)(\delta_{\mu\nu} + p_\mu p_\nu/m_A^2)$$
$$+ \frac{\lambda^2 m_\pi^2}{V}\Theta(p_0)\delta(p^2 + m_\pi^2)p_\mu p_\nu \ .$$

Thus (remember $\mu^2 = -p^2$):

$$\rho_\nu(\mu^2) = g_\rho^2 \delta(\mu^2 - m_\rho^2) \ , \tag{90}$$

$$\rho_{2a}(\mu^2) = \frac{g_a^2}{m_A^2}\delta(\mu^2 - m_A^2) + \lambda^2 m_\pi^2 \delta(\mu^2 - m_\pi^2) \ . \tag{91}$$

We get for (86):

$$\frac{g_\rho^2}{m_\rho^2} = \frac{g_a^2}{m_A^2} + \lambda^2 m_\pi^2 \ . \tag{92}$$

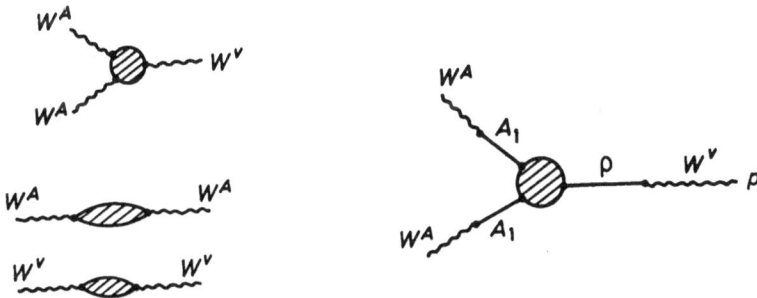

Next, one may consider a process involving three currents, one vector and two axial. Using divergence conditions (15) the amplitude may be related to the amplitudes of processes involving two axial or two vector currents. The latter involve factors of the form $\langle 0|v_\mu^i(x)v_\nu^j(y)|0\rangle$ and may be expressed in terms of the weight functions ρ_ν, ρ_{1a} and ρ_{2a}. Assuming now vector-meson dominance, i.e. a structure as indicated in the figure for the W^A, W^A, W^v vertex one finds behaviour as p^{-4}, if the four-momentum of one of the W^A is taken zero and the four-momentum of the other $= W^v$ four-momentum $= p$. This implies then

$$\int \frac{\rho(\mu^2)}{\mu^2 + p^2} d\mu^2 - \int \frac{\rho_{1a}(\mu^2)}{\mu^2 + p^2} d\mu^2 \sim 0\left(\frac{1}{p^4}\right) \quad \text{for } -p^2 \to \infty . \tag{93}$$

For large $-p^2$ we may expand

$$\frac{1}{\mu^2 + p^2} \simeq \frac{1}{p^2} - \frac{\mu^2}{p^4} \cdots$$

The coefficients of $1/p^2$ must be zero:

$$\int [\rho(\mu^2) - \rho_{1a}(\mu^2)] d\mu^2 = 0 . \tag{94}$$

Assuming again dominance by ρ and A_1 contributions one obtains

$$g_\rho = g_a$$

and together with (92) we get

$$\frac{m_\rho^2}{m_A^2} = 1 - \frac{\lambda^2 m_\pi^2 m_\rho^2}{g_\rho^2} ,$$

for $\frac{m_\rho^2}{5} < g_\rho < \frac{m_\rho^2}{3.7}$ we find

$$\frac{m_\rho^2}{m_A^2} = 1 - K^2 . \quad 3.7 < K < 5 .$$

One finds $980 < m_A < 1374$ MeV.
It may be noted that there is a low energy theorem[d] on the coupling constant $g_{\rho\pi\pi}$, and therefore on g_ρ.
This theorem states

$$g_\rho^2 = 2\lambda^2 m_\pi^2 m_\rho^2$$

giving

$$\frac{m_\rho^2}{m_A^2} = \frac{1}{2}, \quad m_A = \sqrt{2} m_\rho = 1103 \text{ MeV} .$$

The agreement with the experimental mass $m_A = 1070$ MeV is quite reasonable in view of the very crude approximations involved. Here we wish to point out that in itself the assumption that the matrix-element

$$\langle 0 | v_\mu^i | n \rangle$$

[d] K. Kawarabayashi and M. Suzuki, *Phys. Rev. Lett.* **16**, 255 (1966); Riazuddin and Fayyazuddin, *Phys. Rev.* **147**, 1071 (1966)

is nonzero only for $n = \rho^i$ is incompatible with the current commutators (22). For sandwiching the first commutator between $\langle 0|$ and $|\rho^k\rangle$ we find that $\langle \rho^i|v_\mu^j|\rho^k\rangle$ must be nonzero, which by crossing implies that $\langle 0|v_\mu^j|\rho^i\rho^k\rangle$ is nonzero. It is not possible to compute the contribution of the 2ρ intermediate state to the spectral function $\rho_v(\mu^2)$; first of all there is an unknown form factor involved, and secondly, assuming naively a constant form factor the 2ρ contribution is divergent.

What conclusion may be drawn from this? If we believe the sum rules (86) and (94), because of the numerical agreement given above then we must conclude that the vacuum-expectation values of $[v^i, v^i]$ and $[a^i, a^i]$ are equal, and that indeed the process involving three currents behaves asymptotically as indicated above. Thereby also divergence conditions (15) are used. [In fact, from (94) one can in first instance only conclude (93)]. Thus although the results obtained lent some support to the assumption of c-number Schwinger terms they are by no means a proof of that assumption.

We now want to discuss another type of sum rule, due to Björken.[e] These sum rules are in general highly model dependent, and are therefore of the greatest interest. The most important sum rules deal with lepton-nucleon collision processes in the high-energy limit, and backward scattering in the high-energy limit (that is, high energy and high momentum transfer). We will discuss the latter sum rules. They are based on the following expansion for large q_0:

$$\Theta(x_0)e^{iq_0x_0} = \frac{i}{q_0}\left[\delta(x_0) + \left(\frac{-i}{q_0}\right)\delta'(x_0) + \left(\frac{-i}{q_0}\right)^2\delta''(y_0)\dots\right]. \qquad (95)$$

For $\Theta(-x_0)\exp(iq_0x_0)$ the same expansion holds with $x_0 \to -x_0$ and $q_0 \to -q_0$ on the right-hand side. To see this we write:

$$\Theta(x_0) = \frac{1}{2\pi i}\int_{-\infty}^{\infty} d\tau e^{i\tau x_0}\frac{1}{\tau - i\varepsilon}$$

and

$$e^{iq_0x_0}\Theta(x_0) = \frac{-i}{2\pi}\int_{-\infty}^{\infty} d\tau \frac{e^{ix_0(\tau+q_0)}}{\tau - i\varepsilon} = \frac{-i}{2\pi}\int_{-\infty}^{\infty} d\tau \frac{e^{i\tau x_0}}{\tau - q_0 - i\varepsilon}.$$

For large q_0 we develop

$$\frac{-i}{\tau - q_0} \Longrightarrow \frac{i}{q_0}\left\{\frac{1}{1-\frac{\tau}{q_0}}\right\} = \frac{i}{q_0}\left\{1 + \frac{\tau}{q_0} + \frac{\tau^2}{q_0^2} + \dots\right\}. \qquad (96)$$

Of course, we must inspect that the region $\tau \sim q_0$ does not contribute too much to the integral, as otherwise this expansion is senseless. We have (P = principal value):

$$\frac{1}{\tau - q_0 - i\varepsilon} = P\frac{1}{\tau - q_0} + i\pi\delta(\tau - q_0)$$

[e] Varenna Lectures 1967, or *Phys. Rev. Lett.* **16**, 408 (1966), *Phys. Rev.* **148**, 1467 (1966), *Phys. Rev.* **163**, 1767 (1967)

and

$$-\frac{i}{2\pi}\int_{-\infty}^{\infty}d\tau e^{i\tau x_0}i\pi\delta(\tau - q_0) = \frac{1}{2}e^{iq_0 x_0} .$$

For large q_0 the factor $e^{iq_0 x_0}$ is a rapidly oscillating function. In applications we will have expressions of the type

$$\int dx_0 \Theta(x_0)e^{iq_0 x_0}f(x_0) . \tag{97}$$

In general one expects, or rather assumes that the functions $f(x_0)$ that occur in physics problems are smooth functions of their argument. Or, stated otherwise, that

$$\int dx_0 e^{iq_0 x_0}f(x_0) \tag{98}$$

goes rapidly to zero for $q_0 \to \infty$. We need behaviour at least like $1/q_0^2$ in order to be able to apply the expansion given above in any meaningful manner. In the following where necessary, we will assume that this condition holds, although that is neither trivial nor plausible. Also we may take the limit $q_0 \to i\infty$ instead; we wish to emphasize that for the usefulness of the expansion (95) the same condition on (98) must hold, because the expansion (96) requires small contributions of $f(x_0)$ for $\tau/q_0 \gtrsim 1$.

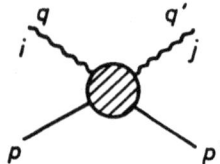

Let us now consider again compton scattering of isovector-vector bosons on nucleons (see figure). With the help of the reduction technique we obtain for the S-matrix element [see (35)]:

$$\langle q'p'|S|qp\rangle = 1 + ig^2 \frac{e_\mu^i e_\nu^{\prime j}}{\sqrt{4V^2 q_0 q_0'}}M_{\mu\nu}^{ij}(q,q',p,p')$$

with now

$$M_{\mu\nu}^{ij}(q,q',p,p') = i\int d_4 x d_4 y e^{iqx-iq'y}\{\langle p'|\Theta(x_0 - y_0)[v_\mu^i(x), v_\nu^j(y)]|p\rangle$$
$$+ \delta(x_0 - y_0)R_{\mu\nu}\} \tag{99}$$

In the following we will ignore the "contact-term" $R_{\mu\nu}$. We recall (41):

$$q_\mu M_{\mu\nu}^{ij} = -i\varepsilon_{ijk}\int d_4 x e^{i(q-q')x}\langle p'|v_\nu^k(x)|p\rangle . \tag{100}$$

(100) can also be derived from (99) with the help of the equal time commutators (22) and the relations:

$$\partial_\mu v^i_\mu(x) = 0 \;,$$

$$i\frac{\partial}{\partial x_4}\Theta(x_0 - y_0) = \delta(x_0 - y_0) \;.$$

Thereby one must ignore possible Schwinger terms in (22). In fact we see that if the divergence conditions (14), (15) hold than there must be a relation between Schwinger terms and the contact terms $R_{\mu\nu}$ in (99).

Consider now:

$$\langle p'|v^i_\mu(x)v^j_\nu(y)|p\rangle = \sum_n e^{ix(n-p')+iy(p-n)} \cdot \langle p'|v^i_\mu(0)|n\rangle\langle n|v^j_\nu(0)|p\rangle$$

$$= e^{iy(p-p')}\sum_n e^{i(x-y)(n-p')}\langle p'|v^i_\mu(0)|n\rangle\langle n|v^j_\nu(0)|p\rangle$$

$$= e^{iy(p-p')}\langle p'|v^i_\mu(x-y)v^j_\nu(0)|p\rangle \;. \tag{101}$$

Doing the same kind of trick for the other term of the commutator in (99), and introducing $z = x - y$ we obtain

$$M^{ij}_{\mu\nu}(q,q',p,p')$$
$$= i\int d_4y e^{iy(q+p-q'-p')} \cdot \int d_4z e^{iqz}\Theta(z_0)\langle p'|[v^i_\mu(z),v^j_\nu(0)]|p\rangle$$
$$= i(2\pi)^4\delta_4(q+p-q'-p')\int d_4z e^{iqz}\Theta(z_0)\langle p'|[v^i_\mu(z),v^j_\nu(0)]|p\rangle \;. \tag{102}$$

We specialize to the case of forward scattering, that is $q = q'$, $p = p'$. Furthermore we are interested in the limit of large q^2 which is what we obtain if we keep $\mathbf{q}$ fixed and let $q_0 \to i\infty$. Also then $(qp) \to -iq_0p_0$. By means of the expansion (95) we obtain

$$M^{ij}_{\mu\nu}(q,q,p,p) \underset{q_0\to i\infty}{\simeq} i(2\pi)^4\delta_4(q+p-q'-p')\frac{i}{q_0}$$
$$\int d_4z \delta(z_0)e^{i\mathbf{q}\mathbf{z}}\langle p|[v^i_\mu(z),v^j_\nu(0)]|p\rangle \;. \tag{103}$$

We now must deal with the equal time commutator. First of all we need this commutator for all values μ, ν and not only the case μ or $\nu = 4$, as in (22). The big question is: what are those commutators? They are very model dependent, and we will consider two models. First the quark-model. Consider a quark current

$$j(x) \equiv \bar\Psi^a_\alpha(x)0^b_{a\alpha\beta}\Psi_{\beta b}(x) \tag{104}$$

where a, b are quark indices, α, β spinor indices. $0^b_{a\alpha\beta}$ is some combination of γ-matrices and SU_3 matrices λ_i, $i = 1, \ldots, 8$ or the identity δ^b_a. The anticommutator rules for free fields are:

$$[\bar{\Psi}^a_\alpha, \Psi_{\beta b}(y)] = -i\delta^a_{\beta\alpha} S_{\beta\alpha}(x-y)$$

$$S_{\beta\alpha}(x-y) = \frac{-i}{(2\pi)^3}\int d_4 k e^{ik(x-y)}(i\gamma k - m)_{\beta\alpha}\delta(k^2 + m^2)\varepsilon(k_0) .$$

For equal times, $x_0 = y_0$:

$$\delta(x_0 - y_0)S_{\beta\alpha}(x-y) = i\gamma^4_{\beta\alpha}\delta(x_0 - y_0)\delta(\mathbf{x}-\mathbf{y}) .$$

Thus

$$\delta(x_0 - y_0)\{\bar{\Psi}^a_\alpha(x), \Psi_{\beta b}(y)\} = \delta^a_b \gamma^4_{\beta\alpha}\delta_4(x-y) . \tag{105}$$

With

$$j(y) \equiv \bar{\Psi}^c_\gamma(y)P^d_{c\gamma\delta}\Psi_{\delta d}(y)$$

we find

$$\delta(x_0 - y_0)[j(x), j(y)] = \delta_4(x-y)[\bar{\Psi}^a_\alpha(y)(0^b_{a\alpha\beta}\gamma^4_{\beta\gamma}P^d_{b\gamma\delta} - P^b_{a\alpha\beta}\gamma^4_{\beta\gamma}0^d_{b\gamma\delta})\Psi_{\delta d}(y)] . \tag{106}$$

In particular we may consider the cases $j(x) = v_\mu(x)$, $j(y) = v_\nu(y)$, that is:

$$0^b_{a\alpha\beta} = \frac{i}{2}b_i\gamma^\mu_{\alpha\beta}\lambda^b_{ia} ,$$

$$P^d_{b\gamma\delta} = \frac{i}{2}c_j\gamma^\nu_{\gamma\delta}\lambda^d_{jb} ,$$

where the b_i and c_j are certain numerical coefficients. We get

$$= -\frac{1}{4}b_i c_j \bar{\Psi}\{\gamma^\mu\gamma^4\gamma^\nu\lambda_i\lambda_j - \gamma^\nu\gamma^4\gamma^\mu\lambda_j\lambda_i\}\Psi .$$

Now:

$$[\lambda_i, \lambda_j] = 2if^k_{ij}\lambda_k, \qquad \{\lambda_i, \lambda_j\} = 2d^k_{ij}\lambda_k ,$$
$$\lambda_i\lambda_j = d^k_{ij}\lambda_k + if^k_{ij}\lambda_k, \qquad \lambda_j\lambda_i = d^k_{ij}\lambda_k - if^k_{ij}\lambda_k \tag{107}$$

so that we obtain for the commutator (106):

$$= \frac{1}{4}b_i c_j \bar{\Psi}[(\gamma^\mu\gamma^4\gamma^\nu - \gamma^\nu\gamma^4\gamma^\mu)d^k_{ij}\lambda_k + i(\gamma^\mu\gamma^4\gamma^\nu + \gamma^\nu\gamma^4\gamma^\mu)f^k_{ij}\lambda_k]\Psi$$

$$= -\frac{1}{2}b_i c_j \bar{\Psi}[\varepsilon_{4\mu\nu\alpha}\gamma^\alpha\gamma^5 d^k_{ij}\lambda_k + i(\delta_{\mu 4}\gamma^\nu + \delta_{\nu 4}\gamma^\mu - \delta_{\mu\nu}\gamma^4)f^k_{ij}\lambda_k]\Psi . \tag{108}$$

Notice that for μ and for $\nu = 4$ we obtain the result (22) with f^k_{ij} instead of ε_{ijk}, which is what one obtains by using $\tau_i/2$ instead of $\lambda_i/2$. For μ, $\nu \neq 4$ we have (only one of the b_i and c_j being nonzero):

$$\delta(x_0 - y_0)[v^i_\mu(x), v^j_\nu(y)] = id^k_{ij}\varepsilon_{4\mu\nu\alpha}a^k_\alpha(x)\delta_4(x-y) - f^k_{ij}\delta_{\mu\nu}v^k_4(x)\delta_4(x-y) .$$

In particular we are interested in the case where the currents are those encountered in weak interactions. Thus $\gamma_\mu(1+\gamma^5)$ instead of γ^μ and further

$$b_1 = \cos\Theta, \quad b_2 = i\cos\Theta, \quad b_4 = \sin\Theta, \quad b_5 = i\sin\Theta$$

[the SU_3 currents are

$$\frac{1}{2}\cos\Theta(\lambda_1 + i\lambda_2) + \frac{1}{2}\sin\Theta(\lambda_4 + i\lambda_5)] ,$$

where Θ is the Cabibbo angle. All other b's are zero as we consider only the currents coupled to a charged lepton pair. The c_i are given by

$$c_i = b_i^* .$$

We find with $j_\mu = b_i j_\mu^i$, $j_\mu^i = v_\mu^i + a_\mu^i = \bar{\Psi} i\gamma^\mu(1+\gamma^5)\lambda^i\Psi$:

$$\delta(x_0 - y_0)[j_\mu(x), j_\mu^+(y)]$$
$$= 2ib_i c_j d_{ij}^k \varepsilon_{4\mu\nu\alpha} j_\alpha^k(x)\delta_4(x-y) - 2b_i c_j f_{ij}^k \delta_{\mu\nu} j_4^k(x)\delta_4(x-y) . \quad (109)$$

Rather than looking up the table for d's and f's we consider the matrices A and $A^\dagger$:

$$A = \frac{1}{2}b_i\lambda_i = \begin{pmatrix} 0 & \cos\theta & \sin\Theta \\ 0 & 0 & 0 \\ 0 & 0 & 0 \end{pmatrix}$$

$$A^\dagger = \begin{pmatrix} 0 & 0 & 0 \\ \cos\Theta & 0 & 0 \\ \sin\Theta & 0 & 0 \end{pmatrix}$$

$$V \equiv \{A, A^\dagger\} = \begin{pmatrix} 1 & 0 & 0 \\ 0 & \cos^2\Theta & \cos\Theta\sin\Theta \\ 0 & \cos\Theta\sin\Theta & \sin^2\Theta \end{pmatrix} = \frac{b_i c_j d_{ij}^k \lambda_k}{2}$$

$$W \equiv [A, A^\dagger] = \begin{pmatrix} 1 & 0 & 0 \\ 0 & -\cos^2\Theta & -\cos\Theta\sin\Theta \\ 0 & -\cos\Theta\sin\Theta & -\sin^2\Theta \end{pmatrix} = \frac{b_i c_j f_{ij}^k \lambda_k}{2}$$

and, in obvious notation

$$\delta(x_0 - y_0)[j_\mu(x), j_\nu(y)] = 2i\varepsilon_{4\mu\nu\alpha} j_\alpha^v(x)\delta_4(x-y) - 2\delta_{\mu\nu} j_4^w(x)\delta_4(x-y) . \quad (111)$$

There are three quarks, p', n' and λ' (charges $\frac{2}{3}$, $-\frac{1}{3}$, $-\frac{1}{3}$) corresponding to the SU_3 indices 1, 2 1nd 3. A proton is composed out of two p' and one n'. For $\langle p|j^v|p\rangle$ we have a contribution 2 from the two p' quarks and $\cos^2\Theta$ from the n' quark:

$$2 + \cos^2\Theta . \quad (112)$$

For j^w:

$$2 - \cos^2 \Theta . \tag{113}$$

Going back to (103) with j_μ and $j_\nu^\dagger$ for v_μ^i and v_ν^j we have

$$M_{\mu\nu}(q, q, p, p) \xrightarrow[q_0 \to i\infty]{}$$

$$- 2i(2\pi)^4 \delta_{\mu\nu} \delta_4(q + p - q' - p') \frac{i}{q_0} \int d_4 z e^{iqz} \delta_4(z) \langle j_4^w(z) \rangle . \tag{114}$$

We dropped the j^v term because we are specializing to the case of identical configuration for in- and outgoing lepton combinations. Then only the part of $M_{\mu\nu}$ symmetrical in μ and ν contributes.

We must have the matrix-element of j_4^w between protons of the same four-momentum p, with $\mathbf{p} = 0$ in the laboratory system as understood here and we also stipulate identical polarizations. No axial current contributes then, and for zero momentum transfer the fourth component of a vector current gives just i (nonrenormalization of the charge). Thus

$$\int d_4 z e^{iqz} \delta_4(z) \langle j_{4(z)}^w \rangle = \frac{i}{V}(2 - \cos^2 \Theta) ,$$

where V is the volume in which the system is enclosed.

$$\lim_{q_0 = i\infty} M_{\mu\nu}(q, q, p, p) = \frac{2i}{V q_0}(2 - \cos^2 \Theta) \delta_{\mu\nu} (2\pi)^4 \delta_4(q + p - q' - p') . \tag{115}$$

An alternative expression for $M_{\mu\nu}$ for forward scattering may be obtained with the help of dispersion relations. We write

$$M_{\mu\nu}(q, q, p, p) = \frac{i(2\pi)^4}{V} \delta_4(q + p - q' - p')\{\delta_{\mu\nu} M + q_\mu q_\nu M_1 + \dots\} .$$

Only the $\delta_{\mu\nu}$ term is of interest to us. We assume an at most once subtracted dispersion relation in the center-of-mass energy $s = -(qp) = p_0 q_0$

$$M(s, q^2) = M(0, q^2) + \frac{s}{\pi} \int_{-\infty}^{\infty} \frac{\text{Im} M(s', q^2)}{s'(s' - s - i\varepsilon)} ds' . \tag{116}$$

As is well known the elastic cross-section in the forward direction may be related to the total cross-section (see figure). Experimentally these total cross-sections are measured by considering $\nu - p$ and $\bar{\nu} - p$ scattering. From these total cross-sections one must separate the part corresponding to M, that is the $\delta_{\mu\nu}$ part. We refer to Björken's work for the algebra involved.

In the limit $q_0 \to i\infty$, $s = q_0 p_0 \to i\infty$:

$$M(s, q^2) \to M(0, q^2) + \frac{q_0 p_0}{\pi} \int \frac{ds'}{s'^2} \text{Im} M(s', q^2) . \tag{117}$$

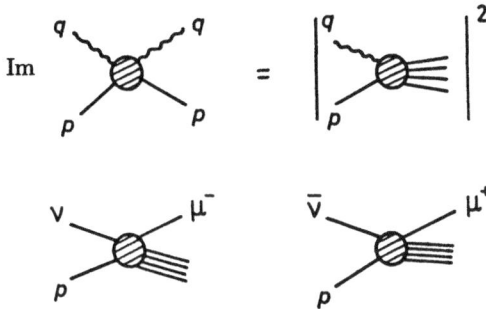

Here we used that Im $M(s', q^2)$ is nonzero only if there is sufficient energy to produce a final state with mass larger than the proton mass. Thus consider an incoming four-momentum Q and p,

$$-(Q+p)^2 > -p^2 = m^2 .$$

Of course we must have $Q^2 = q^2$, otherwise Q is in general not equal to q. We find

$$2s' = -2Qp > Q^2 = q^2 .$$

Since $s \sim q_0 p_0$, while $q^2 \sim \mathbf{q}^2 - q_0^2$ we have $|s'| \gg |s|$ for $q_0 \to i\infty$.

Equalizing the result from (115) with (117) we have

$$2(2 - \cos^2 \Theta) = -\frac{q^2 m}{\pi} \int \frac{ds'}{s'^2} \mathrm{Im} M(s', q^2) . \tag{118}$$

Note that $M(0, q^2)$ is even in q_0, and could thus correspond to the neglected terms $R_{\mu\nu}$ in (99). As stated before Im $M(s', q^2)$ is related to the $\delta_{\mu\nu}$ part of neutrino and antineutron cross-sections. The result (118) is valid in the limit $q_0 \to i\infty$, thus $q^2 \to +\infty$. This corresponds to neutrino's of high energy and backward muons, i.e. backward scattering. Then $q^2 = -2p_\nu p_\mu \simeq 2E_\nu(E_\mu - |\mathbf{p}_\mu|)$. We close this section by noting that the $\delta_{\mu\nu}$ terms in the commutator (111) are specific to currents constructed with the help of spin $\frac{1}{2}$ fields. In the so-called algebra of fields, where the vector and axial vector currents behave as spin 1 fields, one finds no $\delta_{\mu\nu}$ term. In this model the left-hand side of (118) would be zero.

The demonstrated technique is also useful to compute the leading part (from the point of view of divergencies) of a radiative contribution, where one is interested in elastic scattering of virtual photons in the limit of larger q^2. We refer the reader to the work of Abers et al.[f] and N. Cabibbo et al.[g]

5. Origin of the Current Equations

The major success of the current equations is the Adler–Weisberger relation, which can be considered as a test of the divergence condition (15) or the last commutator of (22) ignoring Schwinger term complications. We will take, in this section, the point of view that the divergence conditions (14), (15) are experimentally found to be correct with the possible exception of the existence of certain second order e.m. terms in (15) in view of the discrepancies in $\pi^0 \rightarrow 2\gamma$, $\eta \rightarrow 3\pi$ etc.,[h] while there is not yet sufficient evidence to conclude on one model or other on account of high energy results. If any, it seems that the quark-model is disfavored.

Divergence conditions (14), (15) may be derived on the basis of considerations of local gauge invariance as discussed by Yang and Mills[i] and Bell.[j] We therefore consider the study of Yang–Mills types of field theory of the highest importance. In this section we will first show how divergence conditions may be derived, next we will turn to a study of a field theoretical model. In the following lepton currents will not be considered, except as a tool to observe the intermediated boson (see figure)

Consider a Lagrangian L_s describing strong interactions and depending on a number of boson and fermion fields and their first derivatives

$$L_s = L_s(\varphi^i, \partial_\mu \varphi^i) \qquad i = 1, \ldots, n .$$

Let us suppose that this L_s is invariant under the infinitesimal gauge transformation

$$\varphi^i(x) \rightarrow \varphi^i(x) + \varepsilon^a T^i_{aj} \varphi^j(x) \qquad a = 1, \ldots, m \tag{119}$$

with the ε^a independent of x. Thus the $\partial_\mu \varphi^i$ transform identically

$$\partial_\mu \varphi^i(x) \rightarrow \partial_\mu \varphi^i(x) + \varepsilon^a T^i_{aj} \partial_\mu \varphi^j(x) , \tag{120}$$

where the ε^a are m small parameters, and the T^i_{aj} are a set of numbers. As an example we quote invariance under rotations in isospin space; if φ^i is an I-spin 1

[f] Abers et al., Phys. Rev. Lett. **18**, 676 (1967).

[g] N. Cabibbo et al., Phys. Lett. **25B**, 31 (1967); Phys. Lett. **25B**, 132 (1967).

[h] see M. Veltman, Proc. Roy. Soc. **A301**, 107 (1967) and D. Sutherland, Nucl. Phys. **B2**, 433 (1967).

[i] Phys. Rev. **96**, 191 (1954).

[j] Il. Nuovo Cimento **50**, 129 (1967).

field the T_{aj}^i are nothing but the ε_{aij}, while for I-spin fields we have the $i\tau_{aj}^i$. If one also considers axial currents in addition to vector currents we may have $i\tau_{aj}^i(1+\gamma^5)$ for a half integer spin field. If only invariance for rotations around the third axis is considered, one considers only ε^a with $\varepsilon^1 = \varepsilon^2 = 0$. For invariance under SU_3 with the φ^i for instance, representing quarks one has the matrices $i\lambda_a$ for the T_a. For a given field φ^i the precise form of the T_a will vary depending on which representation of the group one is dealing with; it will become clear that only the commutators of the T_a are important, and for any representation of a given group these commutators are described by the same structure constants. We now try to extend the invariance to a so-called local gauge invariance by requiring invariance under a transformation (119), however now with the ε^a depending on x:

$$\varepsilon^a = \varepsilon^a(x) .$$

Obviously $\partial_\mu \varphi^i$ will now transform differently from φ^i:

$$\partial_\mu\varphi^i(x) \rightarrow \partial_\mu\varphi^i(x) + \varepsilon^a(x)T_{aj}^i\partial_\mu\varphi^j(x) + \partial_\mu\varepsilon^a(x)T_{aj}^i\varphi^j(x) . \tag{121}$$

In general L_s will not be invariant under the transformations (119), (121) if L_s was invariant under (119), (120).

To compensate for the extra term proportional to $\partial_\mu\varepsilon^a(x)$ in (121) we introduce a set of m vector fields $W_\mu^a(x)$ and replace $\partial_\mu\varphi^i$ in L by $\nabla_\mu\varphi^i$ with

$$\nabla_\mu\varphi^i(x) = \partial_\mu\varphi^i(x) - gT_{aj}^i W_\mu^a(x)\varphi^j(x) . \tag{122}$$

In here, g is some constant. For the $W_\mu^a(x)$ we will describe transformation properties in such a way that $\nabla_\mu\varphi^i(x)$ transforms as $\varphi^i(x)$, i.e.

$$\nabla_\mu\varphi^i(x) \rightarrow \nabla_\mu\varphi^i(x) + \varepsilon^a(x)T_{aj}^i\nabla_\mu\varphi^j(x) . \tag{123}$$

This may be achieved by taking

$$W_\mu^a(x) \rightarrow W_\mu^a(x) + f_{cb}^a W_\mu^b(x)\varepsilon^c(z) + \partial_\mu\varepsilon^a(x)/g . \tag{124}$$

Indeed, inserting (119) and (124) into (122) one obtains (123) if one has (we drop the i, j indices on the matrices T_a)

$$[T_a, T_b] = f_{ab}^c T_c .$$

Notice that $f_{ab}^c = -f_{ba}^c$.

The f_{ab}^c are the representation independent structure constants of the group under which the original Lagrangian was invariant.

We may now add to the Lagrangian a piece L_W that depends only on the W and is invariant under the transformation (124). Consider:

$$G_{\mu\nu}^a \equiv \partial_\mu W_\nu^a - \partial_\nu W_\mu^a - gf_{de}^a W_\mu^d W_\nu^e . \tag{125}$$

We get under the transformation (124):

$$G^a_{\mu\nu}(x) \to G^a_{\mu\nu}(x) + f^a_{cb}(\partial_\mu W^b_\nu - \partial_\nu W^b_\mu)\varepsilon^c - f^a_{cb}[W^b_\mu(x)\partial_\nu\varepsilon^c(x) - W^b_\nu(x)\partial_\mu\varepsilon^c(x)]$$
$$- gf^a_{de}f^d_{bc}W^b_\mu W^e_\nu\varepsilon^c - gf^a_{bd}f^d_{ec}W^b_\mu W^e_\nu\varepsilon^c - f^a_{de}\partial_\mu\varepsilon^d W^e_\nu - f^a_{de}W^d_\mu\partial_\nu\varepsilon^e \ .$$

In view of the relation

$$f^a_{de}f^d_{bc} + f^a_{bd}f^d_{ce} + f^a_{dc}f^d_{eb} = 0 \tag{126}$$

we find

$$G^a_{\mu\nu}(x) \to G^a_{\mu\nu}(x) + f^a_{cb}G^b_{\mu\nu}(x)\varepsilon^c(x) \ . \tag{127}$$

Next we define the quantities g_{ab}:

$$g_{ab} = -f^c_{ad}f^d_{bc}, \qquad g_{ab} = g_{ba} \ . \tag{128}$$

With the help of the relation (126) it may be shown that $g_{ac}f^c_{db}$ is antisymmetric in a, b and d; in particular

$$g_{ac}f^c_{db} = -g_{bc}f^c_{da} \tag{129}$$

which then may be used to demonstrate that the combination

$$g_{ab}G^a_{\mu\nu}G^b_{\mu\nu} \tag{130}$$

is invariant under the transformation (127). It may be remarked that in practical applications one chooses things in such a way that g_{ab} is diagonal, i.e. $g_{ab} = 1$ if $a = b$ and zero otherwise. In particular this is true for the structure constants $f^a_{bc} = i\varepsilon_{abc}$ of SU_2 and the f^a_{bc} of SU_3 as used in Sec. 4 with the usual definition of the matrices λ_i (notice that $f^a_{bc} = if^a_{bc}$ of SU_3, Sec. 4).

Consider now the Lagrangian

$$L' = -\frac{1}{4}g_{ab}G^a_{\mu\nu}(x) \cdot G^b_{\mu\nu}(x) + L_s(\varphi^i, \nabla_\mu\varphi^i) \ , \tag{131}$$

where the second piece is obtained from the original Lagrangian by the replacement $\partial_\mu \to \nabla_\mu$. This L' is invariant under the simultaneous transformation (119), (124) with x dependent infinitesimal $\varepsilon^a(x)$. In the case that the structure constants f^a_{bc} have been chosen so that the $g_{ab} = \delta_{ab}$, and in the limit $g = 0$ the W^a_μ fields satisfy the equations of motion for a free massless vector field

$$\partial_\mu[\partial_\mu W^a_\nu(x) - \partial_\nu W^a_\mu(x)] = 0 \ . \tag{132}$$

To be more realistic we add a mass-term to the Lagrangian (131):

$$L'' \equiv L_w + L_s$$
$$= -\frac{1}{4}g_{ab}G^a_{\mu\nu}(x)G^b_{\mu\nu}(x) - \frac{1}{2}M^2 g_{ab}W^a_\mu(x)W^b_\nu(x) + L_s(\varphi^i, \nabla_\mu\varphi^i) \ . \tag{133}$$

This mass-term is invariant under the transformation (124) only for constant ε^a, so that $\partial_\mu \varepsilon^a = 0$, but it breaks the invariance of L under the general gauge transformation with nonzero $\partial_\mu \varepsilon^a$. That is, we are back to the original situation, apart from the fact that we have been able to introduce a vector-boson field coupled to the hadrons in such a way that the original invariance is not broken. This in itself is a doubtful achievement since we know that nonleptonic weak interactions (supposedly mediated by a vector-boson, thus from this point of view second order processes) do break isospin and SU_3 invariance.

What can be deduced from the invariance properties of the Lagrangian (133)? Let us first derive the W_μ^a equations of motion. We find, for nonzero g

$$\frac{\delta L''}{\delta W_\mu^a} - \partial_\nu \frac{\delta L''}{\delta \partial_\nu W_\mu^a} = 0$$

or

$$g_{ab}\partial_\nu G_{\nu\mu}^b + g f_{ac}^b g_{bd} G_{\mu\nu}^d W_\nu^c - M^2 g_{ab} W_\mu^b = -\frac{\delta L_s}{\delta W_\mu^a} \tag{134}$$

or

$$g_{ab}[\partial_\nu(\partial_\nu W_\mu^{\prime b} - \partial_\mu W_\nu^b) - M^2 W_\mu^b]$$
$$= g_{ab} f_{cd}^b g \{\partial_\nu W_\nu^c W_\mu^d + 2 W_\nu^c \partial_\nu W_\mu^d - W_\nu^c \partial_\mu W_\nu^d - g f_{ef}^c W_\mu^e W_\nu^f W_\nu^d\} - \frac{\delta L_s}{\delta W_\mu^a}$$

or

$$\partial_\nu(\partial_\nu W_\mu^b - \partial_\mu W_\nu^b) - M^2 W_\mu^B$$
$$= g f_{cd}^b [\partial_\nu(W_\nu^c W_\mu^d) + W_\nu^c(\partial_\nu W_\mu^d - \partial_\mu W_\nu^d) - g f_{ef}^c W_\mu^f W_\nu^f W_\nu^d] - \frac{\delta L_s}{\delta W_\mu^a} \cdot \tag{135}$$

We used that L_s does not depend on derivatives of the W_μ^a field. The expression (we keep the index a up contrary to the transformation properties of j_μ)

$$\frac{\delta L_s}{\delta W_\mu^a} \equiv g j_\mu^{;a} \tag{156}$$

is by definition just g times the hadron current that is coupled to the W_μ^a field; it may or may not depend on the W_μ^a field itself, but not on derivatives of the W_μ^a field. We will show now that this current satisfies divergence equations of the type (14). As has been noted before L_s is invariant under transformations with constant ε; that is

$$\left. \begin{array}{l} \varphi^i \to \varphi^i + \varepsilon^a T_{aj}^i \varphi^j \\ \partial_\mu \varphi^i \to \partial_\mu \varphi^i + \varepsilon^a T_{aj}^i \partial_\mu \varphi^j \\ W_\nu^a \to W_\nu^a + f_{cb}^a W_\nu^b \varepsilon^c \end{array} \right\} \cdot$$

Thus

$$\varepsilon^a \left\{ \frac{\partial L_s}{\delta \varphi^i} \cdot T_{aj}^i \varphi^j + \frac{\delta L_s}{\delta \partial_\mu \varphi^i} \cdot T_{aj}^i \partial_\mu \varphi^j + \frac{\delta L_s}{\delta W_\nu^c} f_{ab}^c W_\nu^b \right\} = 0 \, .$$

Using the φ equation of motion

$$\frac{\delta L''}{\delta \varphi^i} = \partial_\mu \left(\frac{\delta L''}{\delta \partial_\mu \varphi^i} \right)$$

and the definition (136) we get

$$\varepsilon^a \left\{ \partial_\mu \left(\frac{\delta L_s}{\delta \partial_\mu \varphi^i} T^i_{aj} \varphi^j \right) + g f^c_{ab} j^c_\nu W^b_\nu \right\} = 0 . \tag{138}$$

Because

$$\nabla_\mu \varphi^i = \partial_\mu \varphi^i - g T^i_{aj} W^a_\mu \varphi^j ,$$

we have

$$g j^a_\mu \equiv \frac{\delta L_s}{\delta W^a_\mu} = (-g T^i_{aj} \varphi^j) \frac{\delta L_s}{\delta \partial_\mu \varphi^i} .$$

Using this we find for (138)

$$\varepsilon^a \{ -\partial_\mu j^a_\mu + g f^c_{ab} j^c_\nu W^b_\nu \} = 0 .$$

This must hold for arbitrary ε^a, and we arrive at the divergence condition (14)

$$\partial_\mu j^a_\mu = g f^c_{ab} j^c_\nu W^b_\nu . \tag{139}$$

The following remarks may be made:

(a) (139) has been derived using only invariance of the original Lagrangian L under the transformation (119) with constant ε (that is isospin invariance or SU_3 invariance), and that the W field occurs exclusively in combination with derivatives as shown in (122). The properties of the part L_W of the total Lagrangian are of no importance.

(b) Thus there is a recipe whereby, given a Lagrangian invariant under certain transformation, currents may be defined and vector bosons introduced such that (139) holds. The big question is whether the vector and axial vector currents found in nature are indeed currents constructed in this way. The experimental verification of the divergence conditions seem to indicate so. We therefore take the point of view that the vector and axial vector currents found in nature are constructed according to the technique described above.

(c) The experimental verification of the divergence conditions does not imply the existence of vector bosons; more precisely, the experimental results do not depend on the mass of the intermediate bosons.

(d) If the original L_s is only partially invariant under the transformations employed there will be extra terms in the right-hand side of (139). We refer to the work of Gell-Mann and Levy[k] and J. S. Bell, see beginning of this section.

[k] *Il. Nuovo Cimento* **26**, 705 (1960)

The big question now is: what is so special about this way of constructing currents? Nobody knows. But, assuming that vector-bosons exist we may reasonably, safely assume that the properties of field theories of vector-bosons coupled to currents so constructed are relevant to the physics of weak interactions. For this reason we now study some aspects of such a theory. We are not interested in questions of selection rules, that is how isospin or SU_3 are broken; our main interest are the properties of perturbation theory. In particular we are interested in questions of renormalization.

There are many problems in a field theory of vector-mesons. There is the question of subsidiary condition

$$\partial_\mu W_\mu^a = 0$$

which becomes troublesome for zero mass mesons; furthermore the interaction Hamiltonian, defined by

$$H = i\pi^i \partial_4 \varphi^i - L ,$$

where π is the momentum associated with the field φ, is in general not explicitly covariant (i.e. H is not of the form $\delta_{44}X$, where X is some invariant operator). This apparent noncovariance is compensated by noncovariance in the evaluation of time ordered products as appearing in the S-matrix,

$$S = Te^{iH_I}$$

so that the S-matrix itself is covariant. Under these circumstances there is no simple relation between interaction Lagrangian and Feynman rules. In some cases it has been shown that despite these troubles the Feynman rules are such as one would guess from the interaction Lagrangian[1] but no general procedure exists. We will also ignore this kind of trouble, and assume vertices as obtained by simply taking the lowest order nonvanishing matrix-elements of the interaction Lagrangian. For the vector-boson propagators we take the usual expression

$$\frac{\delta_{\mu\nu} + k_\mu k_\nu / M^2}{k^2 + M^2 - i\varepsilon} . \tag{140}$$

Let us now consider a world without strange particles, and without strong and e.m. interactions, but with proton and neutron, and with a triplet of real vector-bosons coupled to a nucleon vector current and to each other:

$$L = -\frac{1}{4}G_{\mu\nu}^a G_{\mu\nu}^a - \frac{1}{2}M^2 W_\mu^a W_\mu^a - \bar\Psi\left\{\gamma^\mu\left(\partial_\mu - igW_\mu^a\frac{\tau_a}{2}\right) + m\right\}\Psi \quad a = 1,2,3, \tag{141}$$

where Ψ and $\bar\Psi$ are 8 component spinors describing the nucleon isodoublet. Recall:

$$\left[i\frac{\tau_a}{2}, i\frac{\tau_b}{2}\right] = -\varepsilon_{abc}i\frac{\tau_c}{2}; \quad f_{ab}^c = -\varepsilon_{abc}; \quad g_{ab} = +\delta_{ab} \tag{142}$$

[1] Lee and Yang, *Phys. Rev.* **128**, 885 (1962)

and

$$G_{\mu\nu}^a = \partial_\mu W_\nu^a - \partial_\nu W_\mu^a + g\varepsilon_{dea}W_\mu^d W_\nu^e \ . \tag{143}$$

In detail, the interaction Lagrangian is

$$L_I = -g\varepsilon_{dea}W_\mu^d W_\nu^e \partial_\mu W_\nu^a - \frac{1}{4}g^2\varepsilon_{dea}\varepsilon_{bca}W_\mu^d W_\nu^e W_\mu^b W_\nu^c$$
$$+ gW_\mu^a\left(\bar{\Psi}i\gamma^\mu\frac{\tau_a}{2}\Psi\right) \ . \tag{144}$$

The hadron current is

$$j_\mu^a = \left(\bar{\Psi}i\gamma^\mu\frac{\tau_a}{2}\Psi\right) \tag{145}$$

and, as may be verified by using the nucleon equation of motion

$$\partial_\mu j_\mu^a = -g\varepsilon_{abc}j_\nu^c W_\nu^b \ . \tag{146}$$

The vertices corresponding to this interaction are:
(1) a vertex with three vector-bosons, and proportional to the four-momentum of one of the bosons;
(2) a momentum independent vertex with four bosons;
(3) a momentum independent vertex with one boson and two nucleons.

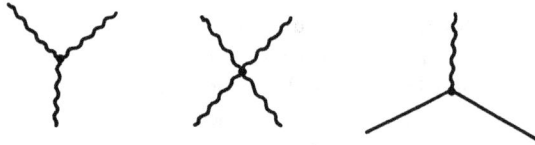

If the boson propagator behaved like $1/k^2$ for large four momentum k we would have a renormalizable theory. However, the boson propagator behaves as a constant for large k. Still it might be that the effect of the $k_\mu k_\nu/M^2$ term in the propagator is not what it seems because of the fact that the divergence of the current to which the boson is coupled is zero (remember that the boson is coupled to hadron *and* to itself.) Indeed, the total Lagrangian is invariant under the gauge transformation

$$\Psi \to \Psi + i\varepsilon^a\frac{\tau_a}{2}\Psi \ .$$
$$W_\mu^c \to W_\mu^c - \varepsilon_{abc}W_\mu^b\varepsilon^a \tag{147}$$

which leads to the result stated. Alternatively, the equations of motion for the W are [see also (134)]

$$\partial_\nu G_{\nu\mu}^a - M^2 W_\mu^a = g\varepsilon_{acb}G_{\mu\nu}^b W_\nu^c - gj_\mu^a \ . \tag{148}$$

Let us take the divergence of the right-hand side (the divergence of the part proportional to g in $\partial_\nu G_{\nu\mu}^a$ is trivially zero):

$$g\varepsilon_{acb}(\partial_\mu G_{\mu\nu}^b W_\nu^e + G_{\mu\nu}^b \partial_\mu W_\nu^c) + g^2\varepsilon_{abc}j_\nu^c W_\nu^b \ .$$

Using again the W-equation of motion (148) for $\partial_\mu G_{\mu\nu}^b$ we find another term with j_ν that cancels against the one present already, and one is left with

$$g\varepsilon_{acb}[g\varepsilon_{bde}G_{\nu\lambda}^e W_\lambda^d W_\nu^c + \frac{1}{2}G_{\mu\nu}^b(\partial_\mu W_\nu^c - \partial_\nu W_\mu^c)] \ , \tag{149}$$

where we employed also $G_{\mu\nu} = -G_{\nu\mu}$. Further, inserting

$$\partial_\mu W_\nu^c - \partial_\nu W_\mu^c = G_{\mu\nu}^c - g\varepsilon_{dec}W_\mu^d W_\nu^e$$

and using the identity

$$\varepsilon_{acb}\varepsilon_{dec} = -\varepsilon_{ace}\varepsilon_{bdc} - \varepsilon_{acd}\varepsilon_{ebc}$$

we see that the second term of (149) drops against the first.

On the basis of this fact we may suspect that the theory is not as divergent as one would think by using the full propagator (140) and doing straightforward power counting. Let us try to estimate the degree of divergence of any diagram. To this purpose we introduce a new technique which may be described roughly as follows. First we introduce an isotriplet of fields $\varphi^a(x)$. The particles associated with this field are made to interact with the nucleons and the hadrons by replacing everywhere (except in the W-mass term) W_μ^a by $\Omega_\mu^a = W_\mu^a - \frac{\kappa}{M}\partial_\mu\varphi^a$, with constant κ. The mass of the φ-particle is taken to be equal to that of the W-mesons. The Feynman rules change insofar that everywhere the W-propagator must be replaced by the Ω-propagator:

$$\frac{\delta_{\mu\nu} + (1 + \lambda)k_\mu k_\nu/M^2}{k^2 + M^2 - i\varepsilon} \tag{150}$$

with $\lambda = |\kappa|^2$. The S-matrix will depend on this λ, and if one takes λ negative or complex the corresponding S-matrix will be nonunitary. However, we subsequently remove the φ-field from physics by introducing additional terms in L such that the φ-field satisfies the free field equations of motion

$$(\Box - m^2)\varphi^a = 0 \ . \tag{151}$$

We can than be sure that states without such φ-particles will not be connected, by the S-matrix, to states containing φ-particles. The additional terms, that will be different from the ones already present in L, give rise to new vertices and the theory will become nonrenormalizable again (except when the W-mass is zero). The so-constructed S-matrix will depend on λ only, and one can show that (151) implies that the S-matrix is independent of λ. In that case we can make an analytical continuation in λ to the value $\lambda = -1$, without affecting the S-matrix. But for

$\lambda = -1$ the propagator (150) behaves like $1/k^2$, and although the newly introduced vertices still result in a nonrenormalizable field theory, the degree of divergence obtained from naive power counting for any given order of g, will be much less than before. In this way one implements the suspected cancellations due to the special properties of the W_μ source currents.

Thus, as a first step we construct a new Lagrangian from the Lagrangian (144)

$$L = -\frac{1}{4}(\partial_\mu W_\nu^a - \partial_\nu W_\mu^a)^2 - \frac{1}{2}(\partial_\mu \varphi^a)^2 - \frac{1}{2}M^2 W_\mu^a W_\mu^a - \frac{1}{2}M^2 \varphi^a \varphi^a$$
$$- \frac{1}{2}g\varepsilon_{dea}\Omega_\mu^d\Omega_\nu^e(\partial_\mu W_\nu^a - \partial_\nu W_\mu^a)$$
$$- \frac{1}{4}g^2\varepsilon_{dea}\varepsilon_{bca}\Omega_\mu^d\Omega_\nu^e\Omega_\mu^b\Omega_\nu^c + g\Omega_\mu^a(\bar{\Psi}i\gamma^\mu\frac{T_a}{2}\Psi) \,. \tag{152}$$

Note that trivially

$$\partial_\mu W_\nu^a - \partial_\nu W_\mu^a = \partial_\mu \Omega_\nu^a - \partial_\nu \Omega_\mu^a \tag{153}$$

with

$$\Omega_\mu^a = W_\mu^a - \frac{\lambda}{M}\partial_\mu\varphi^a \,. \tag{154}$$

We must write equations of motion. The W-equation of motion is as before, with everywhere Ω_μ^a instead of W_μ^a except in the mass term

$$\partial_\nu K_{\nu\mu}^a - M^2 W_\mu^a = g\varepsilon_{acb}K_{\mu\nu}^b\Omega_\nu^c - gj_\mu^a \tag{155}$$

with (see 143)

$$K_{\mu\nu}^a = \partial_\mu W_\nu^a - \partial_\nu W_\mu^a + g\varepsilon_{dea}\Omega_\mu^d\Omega_\nu^c \,. \tag{156}$$

The φ equation of motion is

$$(\Box - M^2)\varphi^a = \partial_\mu\frac{\delta L_I}{\delta\partial_\mu\varphi^a} = -\frac{\lambda}{M}\partial_\mu\frac{\delta L_I}{\delta\Omega_\mu^a}\left(=-\frac{\lambda}{M}\partial_\mu\frac{\delta L_I}{\delta W_\mu^a}\right) \,. \tag{157}$$

Before going on we wish to comment on this equation. The divergence of the W-source current is

$$\partial_\mu\left\{\frac{\delta L_I}{\delta W_\mu^a} - \partial_\nu\frac{\delta L_I}{\delta\partial_\nu W_\mu^a}\right\} \,. \tag{158}$$

Now the second term vanishes on account of the antisymmetry in μ and ν of $\delta L_I/\delta\partial_\nu W_\mu^a$. Thus we see that the source current of the φ field is just $-\lambda/M$ times the four-divergence of the W-source current. This divergence was zero before the introduction of the φ-field, see (148) and following, but unfortunately this is no longer true. Unfortunately, because if the right-hand side of (157) were zero our worries would be over and we would have a renormalizable theory.

Let us now evaluate the right-hand side of (157). To this purpose we take the four-divergence of the right-hand side of (155). We proceed as below (148) and obtain first

$$g\varepsilon_{acb}(\partial_\mu K_{\mu\nu}^b\Omega_\nu^e + K_{\mu\nu}^b\partial_\mu\Omega_\nu^c) + g^2\varepsilon_{abc}j_\nu^c\Omega_\nu^b \,. \tag{159}$$

Here we used the divergence condition

$$\partial_\mu j_\mu^a = -g\varepsilon_{abc}j_\nu^c\Omega_\nu^b \tag{160}$$

which is obviously true. Using (155) in (159) we get (149) with Ω instead of W, but with an extra term due to the fact that (155) equals (148) with Ω instead of W except in the mass term:

$$gM^2\varepsilon_{acb}W_\nu^b\Omega_\nu^c \tag{1161}$$

which is also the result since the rest gives zero.

Thus we have

$$\partial_\mu\frac{\delta L_I}{\delta W_\mu^a} = -gM^2\varepsilon_{acb}W_\nu^b\Omega_\nu^c \tag{162}$$

and (157) becomes

$$\begin{aligned}(\Box - M^2)\varphi^a &= \lambda M\varepsilon_{acb}W_\nu^b\Omega_\nu^c \\ &= -\lambda^2\varepsilon_{acb}W_\nu^b\partial_\mu\varphi^c = -\lambda^2\varepsilon_{acb}\Omega_\nu^b\partial_\nu\varphi^c \ .\end{aligned} \tag{163}$$

Here a few remarks are in order.

(a) For zero mass W's the right-hand side of (163) would have been zero, as desired. That the second line of (163) contains no mass is due to our choice of λ, see (154). Thus it appears that in that case the theory is renormalizable. Apart from the fact that zero-mass intermediate bosons are excluded experimentally there are also very difficult infrared problems for that case. Perhaps this is the right theory, the W acquiring mass by its interactions; we have no idea how this can be understood in a convincing way.

(b) Clearly, the fact that the W-mass term breaks local gauge invariance is intimately connected with the nonvanishing of the right-hand side of (163).

We will now try to add to L terms in such a way that the φ^a-field obeys the free field equations of motion. From (163) we suspect that such terms do not need to contain the W-field more than once. We make the ansatz:

$$L' = L + \lambda^2 g\Omega_\mu^a\partial_\mu\varphi^b f_{ab}(\varphi) \ , \tag{164}$$

where $f_{ab}(\varphi)$ is an unknown function of the φ-fields, to be determined. Perhaps also terms containing only φ-field have to be added; we will not worry about those for the moment.

The additional term affects both W and φ equations of motion. The W-equation of motion is now

$$\partial_\nu K_{\nu\mu}^a - M^2 W_\mu^a = g\varepsilon_{acb}K_{\mu\nu}^b\Omega_\nu^c - gj_\mu^a - \lambda^2 g\partial_\mu\varphi^b f_{ab}(\varphi) \ . \tag{165}$$

For future reference we derive from this by taking the four divergence of the whole equation

$$-M^2\partial_\mu W_\mu^a = g\varepsilon_{acb}[M^2 W_\nu^b - \lambda^2 g\partial_\nu\varphi^d f_{bd}(\varphi)]\Omega_\nu^c - \lambda^2 g\partial_\mu[\partial_\mu\varphi^b f_{ab}(\varphi)] \ . \tag{166}$$

The φ equation of motion contains a part proportional to the four-divergence of the right-hand side of (165), as given in the right-hand side of (166), and some extra terms due to that part of L where φ appears outside the Ω combination. We find:

$$(\square - M^2)\varphi^a = -\frac{\lambda}{M}(m^2\partial\mu W_\mu^a) + \lambda^2 g\left[\partial_\nu\left(\Omega_\mu^c\frac{\delta F_\mu^c}{\delta\partial_\nu\varphi^a}\right) - \frac{\delta F_\mu^c}{\delta\varphi^a}\Omega_\mu^c\right], \qquad (167)$$

where

$$F_\mu^c = \partial_\mu\varphi^d f_{cd}(\varphi) .$$

Let us assume that f_{cd} contain no derivatives of φ. We get

$$\frac{\delta F_\mu^c}{\delta\partial_\nu\varphi^a} = \delta_{\mu\nu}f_{ca}(\varphi) .$$

Further

$$\partial_\nu\frac{\delta F_\mu^c}{\delta\partial_\nu\varphi^a} = \partial_\nu(\delta_{\mu\nu}f_{ca}(\varphi)) = \frac{\delta f_{ca}}{\delta\varphi^d}\partial_\mu\varphi^d ,$$

$$\frac{\delta F_\mu^c}{\delta\varphi^a} = \partial_\mu\varphi^d\frac{\delta f_{cd}}{\delta\varphi^a} .$$

The requirement that the right-hand side of (167) is zero gives:

$$-\frac{\lambda}{M}M^2\partial_\mu W_\mu^a + \lambda^2 g(\partial_\mu\Omega_\mu^c)f_{ca} + \lambda^2 g\Omega_\mu^c\partial_\mu\varphi^d\frac{\delta f_{ca}}{\delta\varphi^d} - \lambda^2 g\partial_\mu\varphi^d\frac{\delta f_{cd}}{\delta\varphi^a}\Omega_\mu^c = 0 . \quad (168)$$

First we will consider only those terms that contain the W-field; eventual addition to L of term containing φ-fields only will later take care of the rest. Thus we require, using (166):

$$-g\lambda^2\varepsilon_{acb}W_\nu^b\partial_\nu\varphi^c - \frac{\lambda^3 g^2}{M}\varepsilon_{acb}W_\nu^c\partial_\nu\varphi^d f_{bd}$$

$$-\frac{\lambda^2 g}{M^2}[g\varepsilon_{cde}(-\lambda M W_\nu^e\partial_\nu\varphi^d - \lambda^2 g\partial_\nu\varphi^g f_{eg}W_\nu^d)f_{ca}]$$

$$+\lambda^2 g W_\mu^c\partial_\mu\varphi^d\frac{\delta f_{ca}}{\delta\varphi^d} - \lambda^2 g W_\mu^c\partial_\mu\varphi^d\frac{\delta f_{cd}}{\delta\varphi^a} = 0$$

or

$$g\lambda^2 g W_\nu^b\partial_\nu\varphi^c[-\varepsilon_{acb} - \frac{\lambda g}{M}\varepsilon_{abd}f_{dc} + \frac{\lambda g}{M}\varepsilon_{dcb}f_{da} + \frac{\lambda^2 g^2}{M^2}\varepsilon_{dbe}f_{ec}f_{da} + \frac{\delta f_{ba}}{\delta\varphi^c} - \frac{\delta f_{bc}}{\delta\varphi^a}] = 0$$

or

$$\varepsilon_{abc} + \frac{\delta f_{ba}}{\delta\varphi^c} - \frac{\delta f_{bc}}{\delta\varphi^a} = \frac{\lambda g}{M}(\varepsilon_{abd}f_{dc} - \varepsilon_{cbd}f_{da}) + \frac{\lambda^2 g^2}{M^2}\varepsilon_{deb}f_{ec}f_{da} . \qquad (169)$$

To zeroth order in g

$$f_{ba} = \frac{1}{2}\varepsilon_{bac}\varphi^c . \qquad (170)$$

To first order in g

$$f_{ba} = \frac{1}{2}\varepsilon_{bac}\varphi^c + \frac{g\lambda}{M}f_{ab}^1$$

and

$$\frac{\delta f_{ba}^1}{\delta\varphi^c} - \frac{\delta f_{bc}^1}{\delta\varphi^a} = \frac{1}{2}(\varepsilon_{abd}\varepsilon_{dce} - \varepsilon_{cbd}\varepsilon_{dae})\varphi^e$$

$$= -\frac{1}{2}\varepsilon_{ebd}\varepsilon_{dae}\varphi^e$$

which gives

$$f_{ba}^1 = -\frac{1}{6}\varepsilon_{ebd}\varepsilon_{dag}\varphi^e\varphi^g . \tag{171}$$

There is not much sense in going beyond this; thus we have now, up to corrections of order g^3 (and terms containing φ-fields alone):

$$L_2' = L + \lambda^2 g\Omega_\mu^a\partial_\mu\varphi^b\left\{\frac{1}{2}\varepsilon_{abc}\varphi^c + \frac{\lambda g}{6M}\varepsilon_{aed}\varepsilon_{bgd}\varphi^e\varphi^g\right\} . \tag{172}$$

We now consider possible terms without a W-field in L'. Thus we write:

$$L' = L_2' + f(\varphi) . \tag{173}$$

The right-hand side of (167) will acquire an extra term

$$-\frac{\delta f}{\delta\varphi^a} + \partial_\mu\frac{\delta f}{\delta\partial_\mu\varphi^a} .$$

Also (168) will get this term. The equation for the W-free part of (168) becomes:

$$\frac{\lambda^4 g^2}{M^2}\varepsilon_{acb}\partial_\nu\varphi^d\partial_\nu\varphi^c f_{bd} - \frac{\lambda^3 g}{M}\partial_\mu(\partial_\mu\varphi^b f_{ab}) - \frac{\lambda^3 g}{M}\partial_\mu(\partial_\mu\varphi^b f_{ba})$$

$$- \frac{\lambda^2 g}{M^2}\left\{+\frac{\lambda^3 g^2}{M}\varepsilon_{cde}\partial_\nu\varphi^g\partial_\nu\varphi^d f_{ge}f_{ca} - \lambda^2 g\partial_\mu(\partial_\mu\varphi^b f_{cb})f_{ca}\right\}$$

$$+ \frac{\lambda^3 g}{M}\partial_\mu\varphi^d\partial_\mu\varphi^c\frac{\delta f_{cd}}{\delta\varphi^a} - \frac{\delta f}{\delta\varphi^a} + \partial_\mu\frac{\delta f}{\delta\partial_\mu\varphi^a} = 0 . \tag{174}$$

Using (170) one notes that there is no term of order g in f, because of the antisymmetry of f_{ab} in lowest order. Here we will not consider this equation any further.

We conclude that with the Lagrangian (172) the field φ is a free field up to order g. Thus

$$(\Box - M^2)\varphi^a = 0(g^2) . \tag{175}$$

We must investigate the consequences of this fact. To this purpose we note that the φ-field in L, except in the propagator, is accompanied by a factor λ. In the S-matrix the φ-field will only enter in the combination

$$\lambda\varphi^a .$$

The source current of the φ-field may also be found as[m]

$$j^a(x) = iS^\dagger \frac{\delta S}{\delta \varphi_{\text{in}}^a(x)} \tag{176}$$

This current is exactly what one obtains for the right-hand side of (175), and, assuming that the treatment given above can be extended to all orders in g we thus have

$$iS^\dagger \frac{\delta S}{\delta \varphi^a} = 0 \ . \tag{177}$$

Multiplication with S on the left, and using unitarity $SS^\dagger = 1$, we find

$$\frac{\delta S}{\delta \varphi^a} = 0 \ . \tag{178}$$

With $\Lambda^a = \lambda \varphi^a$ we have

$$\frac{\delta S}{\delta \lambda} = \frac{\delta S}{\delta \Lambda^a} \frac{\delta \Lambda^a}{\delta \lambda} = \frac{1}{\lambda} \frac{\delta S}{\delta \varphi^a} \frac{\delta \Lambda^a}{\delta \Lambda} = 0 \ . \tag{179}$$

Thus indeed the S-matrix is independent of λ if φ satisfies a free field equation. But it must be emphasized that this is only true for the S-matrix, and not if one takes one or more W's from the mass-shell. Continuation to $\lambda = 0$ gives the original S-matrix, while continuation to $\lambda = i$ gives the propagator without $k_\mu k_\nu$ term for the Ω field. The price we pay for that, to order g, is a new vertex

$$\frac{\lambda^2 g}{2} \varepsilon_{abc} \Omega_\mu^a \partial_\mu \varphi^b \varphi^c \ . \tag{180}$$

There are many curious features to be observed. For instance, the theory with $\lambda \neq 0$ is certainly not identical to the original theory, as becomes clear by inspection of (166). We see that $\partial_\mu W_\mu^a \neq 0$, even in order g:

$$\partial_\mu W_\mu^a = -g\varepsilon_{abc}\Omega_\nu^b W_\nu^c = \frac{g\lambda}{M}\varepsilon_{abc}\partial_\nu \varphi^b W_\nu^c \ . \tag{181}$$

Further, one may ask the following question: what happens if we drop certain terms containing even powers of λ from L? That would certainly imply that φ is no longer a free particle, but that need not bother us, as long as the S-matrix remains unitary for $\lambda = i$. We have no answer to that question.

---

[m] see M. Veltman, *Physica* **29**, 186 (1963) for the meaning of this expression in perturbation theory, and the connection with the equations if motion.

UM − TH − 92 − 11
February 24, 1992

Landau Poles, Violations of Unitarity
and a Bound on the Top Quark Mass

F.J. Yndeuráin

Randall Laboratory of Physics
University of Michigan
Ann Arbor, Michigan 48109-1120
and
Departamento de Fisica Teórica
Universidad Autónoma de Madrid
*Canto Blanco, 28049- Madrid**

Contents:

---

* *Permanent address*

1. Introduction

A field theoretic Lagrangian, such as that of QED is formally self-adjoint, so suggesting that the S-matrix

$$S = T \exp i \int d^4x \mathcal{L}_{int}, \tag{1.1}$$

will be unitary. However, an expression like (1.1) is merely formal; local field theories require regularization and renormalization and it is not obvious that these will respect unitarity. Indeed, regularization and renormalization can, in the present state of the art, only be carried over in perturbation theory and there is no guarantee that formal proofs of unitarity of the series will imply unitarity of the sum.

In these notes I will discuss three types of situations where indications of violation of unitarity show up. The first, which occurs for scattering involving longitudinal W bosons in the Standard Model, is due to one particle, Higgs or chiral fermion becoming very heavy. If it is the Higgs mass, μ, that is large, then the tree level amplitudes for $W_L W_L$ (L for longitudinal) scattering grow linearly in s so the corresponding partial waves would appear to violate the unitary bound, $|f_\ell^2| \leq 1$. If the mass m of a chiral fermion, such as the top quark or an unlikely fourth generation, decreases too much, then the one loop corrections to $\overline{f}_1 f_2 \rightarrow W_L W_L$ with the f_i light fermions also grow like s/M^2.

In this first situation there are indications that summing the perturbative series may restore unitarity in spite of the fact that each of there terms violate the bound $|f_\ell| \leq 1$.

A second type of possible violations of unitarity is that due to Landau poles in $U(1)$ theories, say QED. These singularities appear when summing special kinds of graphs (essentially, leading logarithms); and it has been argued that perhaps adding the whole perturbative series will restore unitarity. We will show that this is very unlikely in the realistic case where one has not one single fermion, but a host of them: an unphysical singularity appears to be really generated by the existence of large numbers of charged fermions.

The corresponding poles are located at very high energies. For the $U(1)$ of the standard model one has

$$\Lambda \sim 10^{20} \text{ GeV to } 10^{42} \text{ GeV}, \tag{1.2}$$

depending on whether supersymmetry is excited or not. Thus the existence of these poles is mostly of academic interest in as much as (1.2) is above Planck's energy, $G_N^{-1/2} \sim 10^{19}$ GeV, where the Standard Model is certainly no more valid.

The third situation occurs again in the Standard Model. An unphysical pole appears in the Higgs propagator when the top quark (or any other chiral fermion) becomes very heavy, and there are many of them, as is the case in the limit of large number of colors, N_c. Specifically, in the limit

$$g_w^2/4\pi \to 0, \quad N_c \to \infty, \quad m_t^2/M^2 \to \infty;$$
$$(m_t^2/M^2)\, N_c\, g_w^2/4\pi \to \tilde{\alpha}_t = \text{finite} \neq 0, \tag{1.3}$$

the Higgs propagator can be calculated exactly, and it shows a Landau-type pole.

In the physical situation, i.e., with the known values for N_c, g_w and the expected ones for m_t, the location of the Landau pole is at such low energy that it allows us to reject t quarks heavier than some 170 GeV, and to conclude that a fourth generation of fermions is excluded, provided the Higgs is "light", i.e., $\mu \leq$ a few TeV.

In the second and third situations one can show that in some limits not too far removed from the real world Landau poles do really exist. One can then ask the question, what would the observable consequences be of the ensuing violations of unitarity. In the absence of a consistent formulation of probability-nonconserving quantum mechanics we cannot really say in detail. Nevertheless, one can calculate the potential generated by exchange of the particle whose propagator has a Landau pole between external, nonrelativistic bodies: not surprisingly, the potential turns out complex. We can interpret the expectation value of the imaginary past as indicating probability violation in the from of spontaneous creation or disappearance of matter. If the Landau pole is too light, the rates for these weird processes will be too large, thus incompatible with experiment. This is what yields the above quoted bound for m_t or the indication of absence of a fourth generation.

2. Heavy Higgs and Heavy Quarks In $W_L W_L$ Scattering and Production, and the Unitarity Limit

It is known since the beginning of gauge theories that the tree level amplitude for scattering of longitudinal $W_L W_L$'s grows linearly with the energy. Letting a, b, c, d be weak isospin indices,[*] and s, t, u the standard Mandelstam variables, we have

$$F^{(0)}(a+b \to c+d) \underset{s \gg M^2}{\simeq} \frac{g_w^2}{16\pi^2 M^2} \left(s\delta_{ab}\delta_{cd} + t\delta_{ad}\delta_{bc} + u\delta_{ac}\delta_{bd} \right), \tag{2.1}$$

and the growth continues until $s, t, u \simeq \mu^2$. The scattering amplitude is normalized so that, for $a = b = c = d$,

$$F = 2!\, \frac{2\sqrt{s}}{\pi\, |\vec{k}|} \sum_\ell (2\ell + 2)\, P_\ell(\cos\theta) f_\ell, \tag{2.2a}$$

---

[*] In this section we work in the approximation of neglecting $W^\circ - \gamma$ mixing. Thus in particular all three W^a have the same mass, M.

$$| f_\ell |^2 \leq 1. \qquad (2.2b)$$

The 2! in (2.2a) comes from the identity of the particles and, in the elastic region,

$$f_\ell = \sin \delta_\ell \, e^{i\delta_\ell}.$$

$\vec{k}$ is the c.m. momentum, $| \vec{k} | \simeq \frac{1}{2} s^{\frac{1}{2}}$.

The expression (2.1) violates the unitarity limit (2.2b) for $\mu \gg s^{\frac{1}{2}} > 4M/\alpha_w^{\frac{1}{2}} \simeq 2$ TeV. One would be tempted to conclude that this implies $\mu < 2$ TeV, but such need not be the case. Loop corrections[1] become comparable to the tree amplitude (2.1) before the unitarity limit is reached. Summing a ladder in the s-channel, (Fig. 1) gives an amplitude for e.g. $a = b \neq c = d$

$$F_{a=b\neq c=d}^{\text{ladder}} \simeq \frac{g_w^2}{16\pi^2 M^2} \frac{s}{1 - \left[(N_L^2 - 1) \, \alpha_w/32\pi M^2 \right] \log \mu^2/(-s)} \delta_{ab}\delta_{cd}. \qquad (2.3)$$

Here we have made the calculation for a $SU(N_L)$ group; of course in the real world $N_L = 2$. Now, (2.3) respects the unitary limit for all values of μ^2.

Certainly, this is not sufficient to conclude that the theory respects unitarity; one can worry about mixed terms (corresponding to one loop terms containing products $st \log s$) not taken into account in our summation (2.3). One cannot really tell, but there are indications that unitarity would indeed be restored. This can be proved in limiting situations and so Einhorn[2] has shown that such is the case in a $SU(N_L)$ theory for $N_L \to \infty$, where (2.3) and corresponding t and u channel ladders are the dominant terms.

A subtler violation of unitarity occurs at the one loop level[3] in processes

$$\overline{f}_1 f_2 \to W_L^a \, W_L^b.$$

What now happens is that the cancellations between the diagrams of Fig. 2A and B cease to occur due to the fermion loop corrections to the first (Fig. 2C). This effect is operative for energies

$$M^2 \ll s \ll m_t^2 \qquad (2.4)$$

if you had e.g. a t quark loop. Considering for example $u\overline{d} \to W^+W^\circ$ scattering we find that the amplitude behaves, in the energy region (2.4), as

$$F^{(2)} \left(u\overline{d} \to W^+W^\circ \right) \simeq \frac{\sqrt{2}N_c \alpha_w^2}{96\pi^2} \cdot \frac{s}{M^2} \cos\theta. \qquad (2.5)$$

θ is the scattering angle and $\alpha_w = g_w^2/4\pi$. Note that (2.5) is independent of the Higgs mass, but it depends implicitly on m_t: when $s \gg m_t^2$, F will decrease and eventually will drop below the unitary limit.

Because the process is now inelastic, the partial wave expansion reads

$$F\left(u\bar{d} \to W^+W^\circ\right) = \frac{2\sqrt{s}}{\pi\,|\,\vec{k}\,|} \sum (2\ell + 1)\, P_\ell(\cos\theta)\hat{f}_\ell, \qquad (2.6a)$$

and the $\hat{f}_\ell$ now satisfy the unitary bound

$$|\,\hat{f}_\ell\,| < 1/4. \qquad (2.6b)$$

This is violated by (2.5) if

$$s > \sqrt{2}\,\frac{288\pi}{N_c} \cdot \frac{M^2}{\alpha_w^2},$$

i.e., for $s^{1/2} > 50$ TeV. The bound on m_t that would follow from this, $m_t < 50$ TeV is not precisely what one would call exciting. Particularly since the ratio of (2.5) to the tree level amplitude,

$$|F^{(2)}/F^{(\text{tree})}| = \frac{N_c\,\alpha_w}{72\pi} \cdot \frac{s}{M^2}, \qquad (2.7)$$

exceeds unity at $s^{1/2} \simeq 3.6$ TeV, from which energy perturbation theory ceases to be valid.

Similar considerations hold for quark[*] loop contributions[4] to $W_L W_L \to W_L W_L$ scattering: if the quark is too heavy perturbative unitarity will be violated even if the Higgs particle is light so that the tree amplitude stays well behaved. The loop corrections will also overwhelm the tree one long before the unitarity limit is reached.

All the previous situations had something in common: although terms in the perturbative series violate the unitary limit, it appears likely that the sum of the series will arrange matters; such was in fact the case in the large N_L limit for WW scattering. For the situations we will consider in the coming sections the state of affairs will be very different. We will argue that the reality of a breakdown of unitarity is very likely as, indeed, in certain limits not too far removed from the real world one can prove rigorously that violations of unitarity, in the form of Landau-type poles appear in the full theory.

3. The Landau Pole in QED and in the $U(1)$ Piece of the Standard Model

That unphysical poles may appear in field theory when summing an infinite number of graphs was discovered as early as 1954 by Landau and collaborators.[5]

---

[*] We here refer to quarks but the result is also valid for any chiral fermion.

They noticed that the photon propagator, evaluated in the leading log approximation, grows a pole (Landau pole) at spacelike momentum. For QED with only one fermion (electron) one has

$$D_{\mu\nu}(q) \underset{|q^2|\gg m_e^2}{\simeq} -i \frac{g_{\mu\nu} - q_\mu q_\nu/q^2}{q^2 \left[1 - (\alpha/3\pi)\log(-q^2/m_e^2)\right]}. \tag{3.1}$$

The Landau pole is thus located at

$$\Lambda_L \simeq m_e \exp \frac{3\pi}{2\alpha} \simeq 10^{227} \text{ GeV}. \tag{3.2}$$

This is larger than the mass of our universe: the existence of the pole may threaten the consistency of QED from a mathematical point of view, but it is unlikely that any physical consequence could be extracted from this. What is more, one can argue[6] that at large momentum the effective coupling is not α but $\alpha \log q^2$, which is of order unity for $|q^2|$ near Λ_L^2, so one should perhaps take the presence of the Landau pole as just an indication of breakdown of the perturbation expansion rather than proving inconsistency of the theory.

Both objections against the relevance of the Landau pole can be overrun. First of all, there are in the real world more charged elementary fermions than just the electron. Defining the quantity R,

$$R = \sum_i Q_i^2,$$

is the sum of the squares of the charges of the elementary fermions, in units of the proton mass, we have* $R = 8$. One should accordingly modify (3.1) by replacing $\alpha \to R\alpha$: the pole then appears located at Λ_γ,

$$\Lambda_\gamma \simeq m \exp \frac{3\pi}{2R\alpha} \simeq 10^{34} \text{ GeV}, \tag{3.3}$$

(m is an average fermion mass). This begins to appear of possible phenomenological interest.

The second objection against the relevance of the Landau pole (that it would disappear when summing all the perturbative series) can also be countered. In fact, one can consider the theory obtained from QED by taking the limit

$$\alpha \to 0, \qquad R \to \infty, \qquad R\alpha \to \tilde{\alpha} \quad \text{finite} \neq 0. \tag{3.4}$$

This limit is not so far removed from the real world, where $\alpha \sim 10^{-2}$, $R \sim 10^1$. Now, in the strict limit (3.4) the theory can be solved exactly. All radiative corrections

---

* I assume that a t quark with mass 100-200 GeV exists. There are other elementary particles with charge, certainly the $W^\pm$ and perhaps more if supersymmetry or grand unification are realized in nature. We will comment later on the implications that this has for our analysis.

vanish except the one loop corrections to the photon propagator and its interactions (Fig. 3). Only for these the vanishing of α is compensated by the exploding of R to give a nonzero result. Summing the series of Fig. 3 we find

$$\widetilde{D}_{\mu\nu}(q) = -i \, \frac{g_{\mu\nu} - q_\mu q_\nu / q^2}{q^2 \left[1 - (2\tilde{\alpha}/\pi) V_1 \left(-q^2/m^2\right)\right]}, \tag{3.5}$$

where the tilde means that the limit (3.4) has been taken. In (3.5) we have renormalized at $q = 0$, and assumed that all fermions have the same mass, m. (In a realistic situation we could have renormalized at $|q_0^2| \gg$ all m_i^2, and m^2 in (3.5) is replaced by q_0^2.) The function V_1 is a particular case of V_n with

$$V_n(\xi) \equiv \int\limits_0^1 dx \, [x(1-x)]^n \log(1 + x(1-x)\xi).$$

Now, the expression (3.5) is <u>exact</u>, in the limit (3.4), and in this limit the theory really presents an unphysical pole for $-q^2 \sim \widetilde{\Lambda}^2$, given by

$$1 - \frac{2\tilde{\alpha}}{\pi} V_1 \left(\widetilde{\Lambda}^2/m^2\right) = 0.$$

If we use the physical value for $\tilde{\alpha}$, and the expression, valid for large ξ,

$$V_1(\xi) \simeq \frac{1}{6} \log \xi,$$

a formula like (3.2) is found for $\widetilde{\Lambda}$, thus justifying the existence of a Landau pole; but it should be stressed that (3.5) is actually valid, in the limit (3.4) of course, for all $q^2, \tilde{\alpha}$. In fact, it can be used to study the strong coupling limit of the model.[7]

The dominant corrections to (3.5) due to the fact that α does not vanish, nor is R infinite can be evaluated easily. Likewise, one can take into account the fact that, when the loop in Fig. 3 is a quark loop, it gets renormalized by strong interactions. All corrections are small; they merely shift the location of the pole, as obtained by using (3.5), by a 10% (in the exponent, however). Up to these corrections we then find that the photon propagator really exhibits a pole at $-q^2 = \widetilde{\Lambda}^2$ with

$$\widetilde{\Lambda}^2 = \bar{m} \exp 3\pi/2\tilde{\alpha} \simeq 4 \times 10^{34} \text{ GeV}. \tag{3.6}$$

More interesting than the corrections described above are corrections stemming from sources <u>outside</u> pure QED. First of all we have $\gamma - Z$ mixing corrections and $W^\pm$ loop corrections due to the fact that, above some 100 GeV, QED becomes embedded into the standard $SU_L(2) \times U(1)$ theory. This can be taken into account by replacing QED by the $U(1)$ piece of the Standard Model. This is effected by just substituting $\tilde{\alpha}$ by $\tilde{\alpha}_1$ where

$$\tilde{\alpha}_1 = \frac{5}{\cos^2 \theta_w} \alpha,$$

$\sin^2 \theta_w \simeq 0.23$ being the weak mixing angle. Thus the $U(1)$ pole is really located at Λ_1,

$$\Lambda_1 = \bar{m} \exp 3\pi/2\tilde{\alpha} \simeq 5 \times 10^{42} \text{ GeV}. \tag{3.7}$$

<u>Supersymmetry</u> goes in the opposite direction. If the Standard Model becomes supersymmetric at an energy much smaller than the Planck energy, say in the TeV range, then sleptons and squarks, Higgses and shiggses and winos also contribute. In a minimal SUSY-$U(1)$ extension, this implies a drastic reduction of Λ_1 to $\Lambda_1(\text{SUSY})$,

$$\Lambda_1(\text{SUSY}) \simeq 10^{20} \text{ GeV}. \tag{3.8}$$

This does not fare too badly compared with the Planck energy, 10^{19} GeV or even a grand unification energy, some 10^{16} GeV.

These last scales are scales at which a modification of the Standard Model is likely to occur. Lacking a theory of gravitation at the quantum level there is not much to say about it. As for grand unification, it is sufficient to remove the Landau pole: the vector bosons contribute with a large coefficient (55 for $SU(5)$) to e.g., the photon propagator, which is sufficient to stabilize it. One can then interpret our result (3.8) as implying grand unification, or gravitational modifications of the Standard Model below Λ_1, say 10^{20} GeV. This value of Λ_1, however, is very sensitive to even small alterations of the theory. For example, if we have SUSY with <u>four</u> Higgses (as has been proposed at times to solve the strong CP problem) then

$$\Lambda_1(\text{SUSY}, 4) \simeq 10^{17} \text{ GeV},$$

barely compatible with grand unification.

4. The Landau Pole in the Higgs Propagator and a Bound on the t Quark Mass

The propagator of the photon, or more generally $U(1)$ fields, are not the only ones to show Landau-type poles. These also appear in the Standard Model in the Higgs propagator if a quark (or any other fermion) is very heavy. This fact was used some time ago by Cabibbo et al.[8] to put an upper bound in the mass of the top quark. The argument runs as follows. The Higgs-top quark coupling gets renormalized by radiative corrections due to strong interactions and Higgs and top quark exchanges (weak and e.m. interactions also contribute, but in a small measure and will therefore be neglected). Defining a coupling

$$\tilde{\alpha}_t \equiv N_c \frac{m_t^2}{M^2} \cdot \frac{g_w^2}{4\pi}, \tag{4.1}$$

we find it to satisfy the renormalization group equation

$$\frac{d\tilde{\alpha}_t}{d\xi} = \frac{1}{2\pi} \left[\frac{2N_c + 3}{8N_c} \tilde{\alpha}_t - 4\alpha_s(Q^2) \right] \tilde{\alpha}_t, \tag{4.2}$$

where α_s is the strong coupling constant and $\xi = \log Q^2/\Lambda_{QCD}^2$, $\Lambda_{QCD} \simeq 100$ MeV the QCD mass scale. (4.2) can be solved to give a running coupling $\bar{\alpha}_t(Q^2)$,

$$\bar{\alpha}_t(Q^2) = \left(\frac{\xi_0}{\xi} \right)^{2d_m} \tilde{\alpha}_t(\nu_0^2)$$

$$\times \left\{ 1 - \frac{1}{2}\xi_0 \left[1 - \left(\frac{\xi_0}{\xi} \right)^{2d_m - 1} \right] \frac{(3 + 2N_c)\tilde{\alpha}_t(\nu_0^2)}{8\pi N_c (2d_m - 1)} \right\}^{-1}, \tag{4.3}$$

$d_m = \frac{-2}{\beta_1} = \frac{4}{7}$; $\xi_0 = \log \nu_0^2/\Lambda_{QCD}^2$.

We will choose $\nu_0 = 2m_t$, i.e., in $\tilde{\alpha}_t(\nu_0^2)$ the mass of the top quark is defined as the mass at $\bar{t}t$ threshold. This is the definition we take form the physical t quark mass. (4.3) is derived under the assumption that the Higgs particle is not too heavy; the results we will report ar indeed valid provided it does not much exceed the TeV level.

(4.3) presents a pole for $\tilde{\alpha}_t(\nu_0^2)$ large enough. Requiring that this pole occurs for Q^2 larger than a critical value, Λ_{crit}, with Λ_{crit} a grand unification scale (10^{14-16} GeV), or the Planck mass, gives a bound $m_t \leq 190$ Gev.

It is a priori nuclear what the bound means: the reasons for grand unification are not totally convincing and a pole in (4.3) does in principle only imply failure of perturbation theory. In fact, the situation is similar to that of the Landau pole in QED. We will show that there exists a limit, reasonably close to the real world, in which the Higgs propagator can be calculated exactly;[9] it develops an unphysical pole, related to the singularity in (4.3). The limit is

$$g_w^2/4\pi \to 0, \quad N_c \to \infty, \quad m_t^2/M^2 \to \infty \tag{4.4a}$$

in such a way that

$$N_c g_w^2/4\pi \to 0, \quad (m_t^2/M^2) g_w^2/4\pi \to 0, \tag{4.4b}$$

but

$$N_c \frac{m_t^2}{M^2} \cdot \frac{g_w^2}{4\pi} \to \tilde{\alpha}_t t \quad \text{finite} \neq 0. \tag{4.4c}$$

In this limit the Higgs propagator $\Delta(q^2)$ is exactly given by the sum of t quark loops and one finds

$$\Delta(q^2) = i \left\{ q^2 \left[1 - \frac{3\tilde{\alpha}_t}{4\pi} V_1 \left(\frac{-q^2}{m_t^2} \right) \right] \right.$$
$$\left. - \mu^2 \left[1 - \frac{m_t^2}{\mu^2} \cdot \frac{3\tilde{\alpha}_t}{4\pi} V_0 \left(\frac{-q^2}{m_t^2} \right) \right] \right\}^{-1}. \tag{4.5}$$

The corresponding Landau pole is located at Λ_H, defined by the vanishing of the denominator in (4.5). If, as occurs in reality $\tilde{\alpha}_t < 1$ and moreover μ is small compared to Λ_H one finds

$$\Lambda_H \simeq m_t \exp \frac{4\pi}{\tilde{\alpha}_t}. \tag{4.6}$$

The estimate (4.6) gets corrections when returning to the real world. Because N_c, m_t^2/M^2 are not very large, nor is $g_w^2/4\pi$ that small, the corrections are bigger than in the photon case. Two loop QCD and Higgs corrections (*i.e.*, beyond the contribution to renormalization of $\tilde{\alpha}_t$, taken into account by using (4.3) into (4.6)) are also important. The details may be found in ref. 7. The resulting values for Λ_H are

$$\Lambda_H = \begin{cases} 3 \times 10^{29} \text{ GeV}, \ m_t = 160 \text{ GeV} \\ 1.3 \times 10^{19} \text{ GeV}, \ m_t = 170 \text{ GeV} \\ 2 \times 10^{13} \text{ GeV}, \ m_t = 180 \text{ GeV} \end{cases}$$

We get a bound on m_t which actually improves that of Cabibbo et al., using (4.3). One has $m_t < 170$ GeV if we require that Λ_H be larger than the Planck mass, or $m_t < 175$ GeV, for Λ_H larger than a grand unification scale, 10^{14-16} GeV.

Because these bounds are so close to the expected mass[*] of m_t, $m_t \simeq 140^{+60}_{-40}$, GeV, it is of interest to try to get directly observable consequences from the existence of Landau poles, independent of whether or not grand unification exists and of the exact energy at which gravitational interactions become operative. This we do in the next section; for now, we finish with a comment on a possible fourth generation. The results on the ρ parameter imply that, if such a generation were to exist, it should be practically degenerate in mass. If m_4 is that mass, then one must have

---

[*] A lower bound on m_t follows from nonobservation of t at the Tevatron; an upper one from lack of renormalization of Veltman's ρ parameter. The estimate requires information on B decays and the shape of the Z peak. See refs. 10 and 11 for details

$m_4 > 100$ GeV. This displaces the Landau pole to a location $\Lambda_4 < 10^{13}$ GeV, and this at the lower edge, $m_4 = 100$ GeV; already at $m_4 \sim 140$ GeV one would have $\Lambda_4 \sim 10^7$ GeV: this contributes to make the existence of a fourth generation extremely unlikely.

5. Violation of Causality-Unitarity Due to Landau Poles

The presence of Landau poles entail violations of causality and unitarity. To see what this means, we consider the modifications in the corresponding potentials, *i.e.*, we fall back on nonrelativistic quantum mechanics. We will carry the derivation in detail for QED, and indicate the results for the Higgs case.

We couple the photon to heavy external particles, say nuclei, with charges e_1, e_2. The scattering amplitude between these is given by exchange of a photon with the full propagator, $\widetilde{D}_{\mu\nu}(q)$, given by (3.5), which corrects a "free" photon exchange,

$$D_{\mu\nu}^{(0)}(q) = ig_{\mu\nu}/q^2.$$

(see Fig. 4). The equivalent potential is obtained as the Fourier transform of the scattering amplitude in the nonrelativistic limit, T:

$$U(\vec{r}) = -\frac{1}{2\pi} \int d\vec{q}\, e^{i\vec{q}\vec{r}} T(\vec{q}),$$

$$T(\vec{q}) = -\frac{\alpha_{12}}{\pi} \cdot \frac{1}{\vec{q}^2 \left[1 - (2\tilde{\alpha}/\pi)\, V_1\left(\vec{q}^2/m^2\right)\right]}$$

Here $\alpha_{12} = e_1 e_2/4\pi$. m is an average fermion mass which, for part of the calculation will have to be replaced by a realistic set of masses, m_i. the integral (5.1) can be calculated by performing first the angular integrations and then relating the radial one to a contour integral. The details may be found in ref. 7. One gets,

$$U(\vec{r}) = U_C(\vec{r}) + U_{\text{cut}}(\vec{r}) + U_\Lambda(\vec{r}). \tag{5.2}$$

Here U_C is the Coulomb piece,

$$U_C = \frac{\alpha_{12}}{r}. \tag{5.3a}$$

The U_{cut} is dominated by the contribution of the lightest fermion, that is to say, for it we replace

$$\tilde{\alpha} V_1\left(\vec{q}^2/m^2\right) \to \alpha \sum_i Q_i^2 V_1\left(\vec{q}^2/m_i^2\right)$$

and then

$$U_{\text{cut}} \simeq \frac{2\alpha\alpha_{12}}{3\pi r} \int_{2m_e}^{\infty} dE \frac{e^{-rE}}{E} \left(1 + \frac{2m_e^2}{E^2}\right) \sqrt{1 - \frac{4m_e^2}{E^2}}. \qquad 5.3b$$

Cognoscenti will identify (5.3b) as a more accurate form of the familiar Serber-Uehling potential,

$$U_{SU}(\vec{r}) = \frac{4\alpha\alpha_{12}}{15m_e} \delta(\vec{r}),$$

to which it reduces for wavefunctions that only vary in the scale $r \sim 1/m_e\alpha$. Finally, the interesting piece is the contribution of the Landau pole; if $\tilde{\alpha} < 1$, one can write it as

$$U_\Lambda(\vec{r}) = -\frac{3\pi\alpha_{12}}{\tilde{\alpha}} \cdot \frac{e^{i\Lambda r}}{r}. \qquad (5.3c)$$

This is complex , as could have been expected. A possible way to interpret its imaginary part is to consider that the expectation value

$$\langle \Psi \mid ImU_\Lambda \mid \Psi \rangle \equiv -\Gamma_\Lambda,$$

is connected with the probability of spontaneous creation (if $\Gamma_\Lambda < 0$) or disappearance of matter (for $\Gamma_\Lambda > 0$). We can take the quantity $\tau_\Lambda = |1/\Gamma_\Lambda|$ to be the mean time for these processes to occur.

Because the effect increases at shorter distances, we will consider the Coulomb interaction, with modification due to U_Λ, of a proton in a nucleus. We then have $\alpha_{12} = (Z - 1)\alpha$. The wavefunction of the proton will be taken to be

$$\Psi(\vec{r}) = \frac{m_\pi^{3/2}}{\sqrt{\pi}} e^{-m_\pi r},$$

m_π being the pion mass. We find then,

$$\Gamma_\Lambda = \frac{24\pi(Z - 1)}{R} \cdot \left(\frac{m_\pi}{\Lambda}\right)^3 m_\pi. \qquad (5.4)$$

Putting in numbers with $Z \sim 10$ and $R = R_{\text{eff}} \simeq 6.5$ (to be realistic we are considering $U(1)$ pole effect), we find $\tau_\Lambda \sim 10^{98}$ yr. In the minimal SUSY case this decreases dramatically to $\tau_\Lambda \sim 10^{29}$ yr.

These numbers are beyond what can be measured experimentally. The activity of matter due to the appearance or disappearance of particles would have been noticed in experiments dedicated to look for proton or double β decay if the characteristic time had been less than an "experimental" limit,

$$\tau^{\exp} \sim 10^{29} \text{ yr.} \tag{5.5}$$

However, the dependence of τ_Λ on the fermion content of the theory is very strong. So, if we had SUSY with four Higgses, as has been suggested at times, we would have got $\tau_\Lambda \sim 10^{23}$ yr; and a fourth generation would also have implied violation of the bound (5.5), even in a minimal SUSY model.

The case of the pole in the Higgs propagator is still more interesting. We now consider the Higgs-mediated interaction between two quarks in a nucleon. The calculating runs along the same lines as before; m_π is replaced by R_p^{-1}, the inverse radius of the proton, and α_{12} by $\frac{1}{4}\alpha_w(m_q^2/M^2)$, where m_q is the mass of a quark. Because this is an ill-defined concept for quarks bound in a nucleon we have ambiguity here. We can take "constituent" quark masses, $m_q \sim 300$ MeV; or "current" quark masses defined at $Q^2 = 1$ GeV^2, so $m_d \sim 20$ MeV. With constituent masses we get

$$\tau_H = \begin{cases} 10^{33} \text{ yr, } m_t = 160 \text{ GeV} \\ 10^{22} \text{ yr, } m_t = 170 \text{ GeV} \\ 10^{15} \text{ yr, } m_t = 180 \text{ GeV.} \end{cases} \tag{5.6}$$

With current masses, multiply the r.h.s. above by a factor $10^3 - 10^4$. Taking also into account the errors in the evaluation of Λ_H, it follows that we can stretch things to make (5.6), with current masses, compatible with the experimental bound, (5.5), if $m_t < 170$ GeV; $m_t = 180$ GeV is certainly out of the question. Because where $m_t = 170$ GeV one has $\Lambda_H \sim 10^{19}$ GeV, grand unification (or gravity, perhaps) could prevent the formation of a Landau pole. Thus, the value $m_t = 170$ GeV could be allowed; but, when $m_t = 180$ GeV, we have $\Lambda_H \sim 10^{13}$ GeV and grand unification arrives too late.

6. Some Speculations

For minimal supersymmetric $U(1)$ as well as for the Higgs propagator, for a top quark with $m_t \sim 165$ GeV, the Landau pole falls around the Planck mass. The experimental consequences are compatible with p decay bounds. It is of course possible that grand unification occurs at some 10^{16} GeV, thus disposing of the Landau poles; but it could also happen that the poles are really there. Although I lack definite ideas, I would like to mention a few points suggesting lines of speculation.

First of all, it is perhaps indicative that unphysical poles appear in the less satisfactory pieces of the Standard Model, the abelian factor $U(1)$ and the Higgs field.

Secondly, it may be of cosmological relevance the fact that the presence of Landau poles allow for spontaneous appearance/disappearance of matter, at low rates at laboratory energies, but copiously at ultrahigh ones.

Lastly, one may consider that the Landau poles represent a real failure of the underlying theory. That they appear near the Planck mass could point to gravity (perhaps in the form of extra dimensions) as the cure.

REFERENCES

1. M. Veltman and F.J. Yndúráin, Nucl. Phys. B325, 1(1989);
   S. Dawson and S. Willenbrock, Phys. Rev. D40 2880 (1989).

2. M.B. Einhorn, Nucl. Phys. B246 75 (1984).

3. C. Ahn et al., Nucl. Phys. B309 221 (1988).

4. S. Dawson and G. Valencia, Phys. Lett. B246 156 (1990).

5. L.D. Landau, A.A. Abrikosov and I.M. Khalatnikov, Dokl. Akad. Nauk SSSR,
   95 1177 (1954);
   L.D. Landau and I. Pomeranchuk, Dokl. Akad. Nauk. SSSR, 95 1329) (1954).

6. N.N. Bogoliubov and D.V. Shirkov, "Introduction to the Theory of Quantized
   Fields", Interscience, 1959.

7. F.J. Yndúráin, preprint FTUAM 91/03 (1991).

8. N. Cabibbo et al., Nucl. Phys. B158 295 (1979).

9. F.J. Yndúráin, Phys. Lett. B255 574 (1991).

10. ALEPH Collaboration, CERN preprint PPE/90-104 (1990);
    OPAL Collaboration, CERN preprint EP/90-81 (1990);
    F. Abe et al., Phys. Rev. Lett. 65 2243 (1990);
    For a summary of experiments, see e.g. F. Dydak, Proc. 1990 Singapore Conf.
    High Energy Physics; B. Schwartzschild, Phys. Today 43 No. 12, 20 (1990).

11. P. Langacker, Phys. Rev. Lett. 63 1920 (1988);
    J. Ellis and G.L. Fogli, Phys. Lett. B232 139 (1989);
    J. Maalampi and M. Roos, Particle World, 1 148 (1990);
    A summary of theoretical bounds and estimates may be found in G. Altarelli;
    CERN preprint TH-5590/89 (1989).

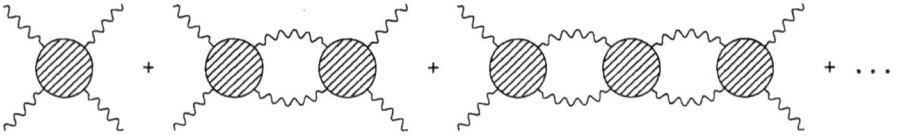

Fig. 1A

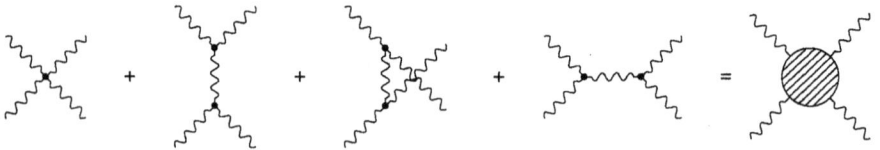

Fig. 1B

Fig. 1. (A) Sum of ladder in s-channel. The blob represents the effective vertex (B

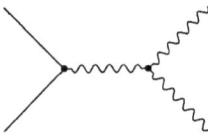

Fig. 2A

Fig. 2B

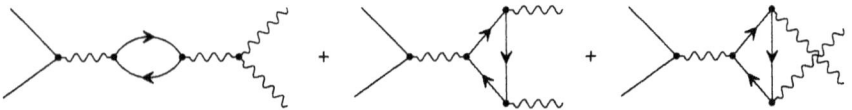

Fig. 2C

Fig. 2. Diagrams for $\bar{f}_1 f_2 \to W_L^a W_L^b$. (A,B) Tree level (C) Quark loop corrections.

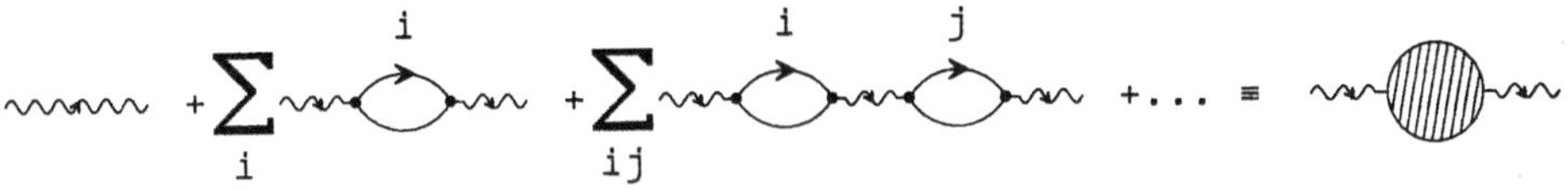

Fig. 3. Surviving contributions to the photon propagator in limit $\alpha \to \infty, R \to \infty, R\alpha \to \tilde{\alpha}$.

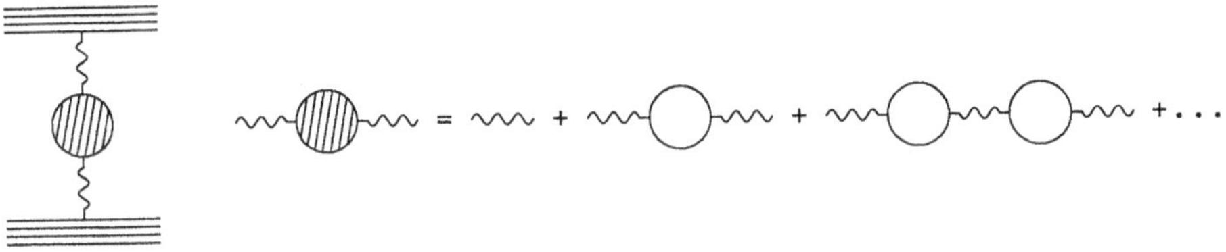

Fig. 4. Exchange of a dressed photon between external particles.

Conference on Gauge Theories – Past and Future

Schedule of Talks

Friday - May 17   Session 1: Chair, Y.-P. Yao

9:00 a.m. - 9:30 a.m.	H.A. Neal Univ. of Michigan	Martinus Veltman: A Colleague, Teacher and Scholar
9:30 a.m. - 10:10 a.m.	N. Cabibbo Univ. of Rome & INFN, Italy	Parallel Supercomputers for Lattice Gauge Theory
10:15 a.m. - 10:55 a.m.	B. Zumino Univ. of California, Berkeley	Non Cummutative Calculus and Quantum Field Theory
10:55 a.m. - 11:30 a.m.		**COFFEE BREAK**
11:30 a.m. - 12:10 p.m.	P. van Nieuwenhuizen SUNY at Stonybrook	Example of a Nonlinear Gauge Field Theory
12:10 p.m. - 12:50 p.m.	E. de Rafael CNRS, Marseilles	The Strong CP Problem in an Effective Chiral Lagrangian Approach
12:50 p.m. - 2:15 p.m.		**LUNCH**

Session 2: Chair, M. Consoli

2:15 p.m. - 2:55 p.m..	F.J. Yndurain Univ. Autonoma, Madrid	Landau Poles in the Standard Model, Instability of Matter and the Mass of the Top Quark
2:55 p.m. - 3:30 p.m.	L. Susskind Stanford University	The Cosmological Constant
3:30 p.m. - 4:00 p.m.		**COFFEE BREAK**
4:00 p.m. - 4:35 p.m.	J.J. van der Bij Univ. of Amsterdam	Avoiding the Veltman Theorem
4:35 p.m. - 5:15 p.m.	R. Thun Univ. of Michigan	The Veltman Vertex
5:15 p.m. - 5:50 p.m..	G. Passarino Univ. of Torino, Italy	The Last Ten Years of Radiative Corrections
7:30 p.m.		
Master of Ceremonies A. Krisch Univ. of Michigan	M. Schwartz Brookhaven National Laboratory	Banquet Speech

Saturday - May 18   Session 3: Chair, R. Gastmans

9:00 a.m. - 9:40 a.m..	B. de Wit Univ. of Utrecht	Gauge Independence in Quantum Gravity
9:40 a.m. - 10:20 a.m..	W.A. Bardeen FERMILAB	Mechanisms for Electroweak Symmetry Breaking
10:20 a.m. - 10:35 a.m.	A. Compagner Delft Univ.	The Best and Fastest Random Numbers Ever
10:35 a.m. - 11:00 a.m.		**COFFEE BREAK**
11:00 a.m. - 11:40 a.m.	M.K. Gaillard Univ. of California, Berkeley	Thresholds and Effective Theories in Particle Physics
11:40 a.m. - 12:20 p.m.	L. Lederman Univ. of Chicago	Higgs, Veltman, Waxachie & Beyond
12:20 p.m. - 1:00 p.m.	M. Veltman Univ. of Michigan	

Conference Participants

K. Adel	B. Haeri	L. Susskind
R. Akhoury	F. Halzen	R. Thun
D. Amidei	T. Hecht	S. Titard
V. Baluni	F. Hendel	E.T. Tomboulis
W.A. Bardeen	S. Hong	Y. Tomozawa
F. Berends	L. Jones	C. Uher
G. Bonvicini	G. Kane	P. van Baal
N. Cabibbo	A. Kataev	H. van Dam
M. Campbell	W. Kauffman	A. van de Ven
J. Chapman	C. Korthal Altes	J.J. van der Bij
A. Compagner	A. Krisch	J. van der Velde
M. Consoli	J. Krisch	P. van Nieuwenhuizen
H. Contopanagos	L. Lederman	H. Veltman
H.R. Crane	G.-L. Lin	M. Veltman
E. de Rafael	J. Liu	J. Vermaseren
B. de Wit	H. Neal	T. Weiler
L. de Wit	D. Nitz	J. Weyers
M. Duncan	G. Passarino	D. Williams
M. Einhorn	R. Phelps	A.C.T. Wu
K. Ellis	E. Remiddi	D. Wu
D. Errede	W. Repko	Y.-P. Yao
S. Errede	D. Robinson	F.J. Yndurain
P. Federbush	B. Roe	S. Zhang
K. Gaemers	S. Rudaz	B. Zumino
M.K. Gaillard	M. Schwartz	
R. Garisto	J. Smit	
R. Gastmans	H. Steger